# Joint Meeting of the 6th Conference of the Groupe De Recherche (GDR) 2501 and the 9th Anglo-French Physical Acoustics Conference (AFPAC)



Journal of Physics: Conference Series Volume 269

## Kendal, Cumbria, United Kingdom
## 18 – 22 January 2010

ISBN: 978-1-61782-371-8
ISSN: 1742-6588

# Joint Meeting of the 6th Conference of the Groupe De Recherche (GDR) 2501 and the 9th Anglo-French Physical Acoustics Conference (AFPAC)

Journal of Physics: Conference Series Volume 269

Kendal, Cumbria, United Kingdom
18 – 22 January 2010

ISBN: 978-1-61782-371-8
ISSN: 1742-6588

IOP Institute of Physics

**Printed from e-media with permission by:**

Curran Associates, Inc.
57 Morehouse Lane
Red Hook, NY 12571



Printed by Curran Associates, Inc. (2011)

For permission requests, please contact the Institute of Physics
at the address below.

Institute of Physics
Dirac House, Temple Back
Bristol BS1 6BE UK

Phone: 44 1 17 929 7481
Fax: 44 1 17 920 0979

techtracking@iop.org

**Additional copies of this publication are available from:**

Curran Associates, Inc.
57 Morehouse Lane
Red Hook, NY 12571 USA
Phone: 845-758-0400
Fax: 845-758-2634
Email: curran@proceedings.com
Web: www.proceedings.com

# TABLE OF CONTENTS


**ON THE MODELLING OF ULTRASONIC BULK WAVE PROPAGATION IN EPOXIES** ........ 1
*R E Challis, F Blarel, M E Unwin, X Guo*

**A DETAILED STUDY OF GUIDED WAVE PROPAGATION IN A VISCOELASTIC MULTILAYERED ANISOTROPIC PLATE** ........ 10
*L Taupin, A Lhémery, G Inquiété*

**MEASUREMENT OF GUIDED MODE WAVE VECTORS BY ANALYSIS OF THE TRANSFER MATRIX OBTAINED WITH MULTI-EMITTERS AND MULTI-RECEIVERS IN CONTACT** ........ 24
*Jean-Gabriel Minonzio, Maryline Talmant, Pascal Laugier*

**COHERENT ACOUSTIC RESPONSE OF A SCREEN CONTAINING A RANDOM DISTRIBUTION OF SCATTERERS: COMPARISON BETWEEN DIFFERENT APPROACHES** ........ 31
*J Dubois, C Aristégui, O Poncelet, A L Shuvalov*

**EXPERIMENTAL DETERMINATION OF PLATE PARAMETERS WITH AN AIR COUPLED INSTRUMENT** ........ 41
*Abdelhak El Mouhtadi, Jean Duclos, Hugues Duflo*

**PLATE MODELS AND HIDDEN INFORMATIONS** ........ 51
*Philippe Destuynder*

**THE EFFECTIVE WAVENUMBER OF A PRE-STRESSED NONLINEAR MICROVOIDED COMPOSITE** ........ 62
*W J Parnell, I D Abrahams*

**ULTRASONIC NDT OPTIMIZATION USING RANDOMIZED ADAPTIVE DIFFERENTIAL EVOLUTION** ........ 72
*B Puel, S Chatillon, D Lesselier*

**A 45° SILICON MIRROR FOR ACOUSTIC PROPAGATION PARALLEL TO THE PLANE OF THE SUBSTRATE** ........ 78
*Shengxiang Wang, Julien Carlier, Pierre Campistron, Weijiang Xu, Dorothée Callens-Debavelaere, Bertrand Nongaillard, Assane Ndieguene, Xingzhong Zhao*

**RELATION BETWEEN ULTRASONIC SCATTERING AND MICROSTRUCTURE OF POLYCRYSTALLINE MATERIALS** ........ 88
*O Dupond, N Feuilly, B Chassignole, T Fouquet, J Moysan, G Corneloup*

**TEMPERATURE DEPENDENCE OF BULK VISCOSITY IN WATER USING ACOUSTIC SPECTROSCOPY** ........ 95
*M J Holmes, N G Parker, M J W Povey*

**EVANESCENT CHOTS FOR THE OPTICAL GENERATION AND DETECTION OF ULTRAHIGH FREQUENCY SAWS** ........ 102
*A Arca, T Stratoudaki, R Smith, M Clark, M G Somekh*

**RECENT ADVANCES IN SEMI-ANALYTICAL SCATTERING MODELS FOR NDT SIMULATION** ........ 110
*M Darmon, S Chatillon, S Mahaut, P Calmon, L J Fradkin, V Zernov*

**ULTRASOUND CHARACTERIZATION OF RED BLOOD CELLS DISTRIBUTION: A WAVE SCATTERING SIMULATION STUDY** ........ 122
*Emilie Franceschini, Bruno Lombard, Joël Piraux*

**MULTIPLE SCATTERING BY CYLINDERS RANDOMLY LOCATED IN A FLUID: EFFECTIVE PROPERTIES** ........ 130
*Francine Luppé, Jean-Marc Conoir*

**MODELLING THE BACKSCATTER FROM SPHERICAL CAVITIES IN A SOLID MATRIX: CAN AN EFFECTIVE MEDIUM LAYER MODEL MIMIC THE SCATTERING RESPONSE?** ........ 137
*Valerie J Pinfield, Richard E Challis*

**THIN LAYER CHARACTERIZATION BY ZGV LAMB MODES** ........ 146
*Maximin Cès, Dominique Clorennec, Daniel Royer, Claire Prada*

**DIRECTIVITY OF GUIDED ULTRASONIC WAVE SCATTERING AT NOTCHES AND CRACKS** ........ 154
*P Fromme, C Rouge*

**A PRELIMINARY STUDY OF ACOUSTIC PROPAGATION IN THICK FOAM TISSUE SCAFFOLDS COMPOSED OF POLY(LACTIC-CO-GLYCOLIC ACID)** ........ 165
*N G Parker, M L Mather, S P Morgan, M J W Povey*

**MODELING AND EXPERIMENTAL INVESTIGATIONS OF LAMB WAVES FOCUSING IN ANISOTROPIC PLATES** ........ 174
*Bastien Chapuis, Nicolas Terrien, Daniel Royer*

**ELASTIC WAVE DIFFRACTION BY INFINITE WEDGES** ........ 182
*Larissa Fradkin, Arthur Gautesen, Victor Zernov, Michel Darmon*
**SIMULATION OF ULTRASONIC INSPECTION OF COMPOSITE USING BULK WAVES: APPLICATION TO CURVED COMPONENTS** ........ 196
*S Journiac, N Leymarie, N Dominguez, C Potel*
**PREDICTION OF THE POINT OF IMPACT IN AN ANISOTROPIC PLATE** ........ 205
*M Koabaz, T Hajzargarbashi, T Kundu, M Deschamps*
**PICOSECOND ULTRASONICS IN SINGLE CELLS: INTERFACE STEP MOTION FOR THIN ANIMAL CELLS AND BRILLOUIN SCATTERING FOR THICK VEGETAL CELLS** ........ 213
*M Ducousso, T Dehoux, B Audoin, O Zouani, C Chollet, M C Durrieu*
**Author Index**

# PREFACE

A joint meeting of two Anglo–French scientific research communities was held from 18 to 22 January 2010, in Kendal, Cumbria, UK. This was the sixth conference of the Groupe De Recherche (GDR) 2501, and the ninth conference of the Anglo–French Physical Acoustics Conference (AFPAC). This was the first time these two conferences have been run concurrently. The meeting was entitled GDR–AFPAC.

The interests of the GDR group focus on studying the propagation of ultrasound in inhomogeneous media, including the relevance to its use for Non Destructive Evaluation (NDE). The GDR was set up in France in 2002, as a network of university research groups and industrial organisations, with support from the French national funding agency CNRS, under the code 2501. The network was expanded to include membership of university research groups in the UK in 2008, and has support from the UK Research Centre in NDE (RCNDE). A significant extra benefit of the UK membership is that it brings together the UK researchers in applied mathematics with those in engineering. The GDR holds its conferences approximately every 18 months; all GDR conferences prior to this one have been held in France.

The interests of the AFPAC group pertain to studying physical acoustics. The AFPAC conference series is a collaboration between the Physical Acoustics Group (PAG) of the Institute of Physics and the Groupe d'Acoustique Physique, Sous-marine et UltraSonore (GAPSUS) of the Société Française d'Acoustique. First established in 2001, the aim of its annual conference is to provide a forum where the most recent research developments in the field of Physical Acoustics in the UK and France are reviewed. AFPAC alternates between venues in the UK and France, and the format has been designed to be 'small and friendly'. The conference attracts the main research leaders of both countries, and in particular aims to encourage research students to have their debut presentations at the event.

Thus both of these research communities have established collaborations between researchers in France and the UK, and both are concerned with the science and applications of ultrasound. Although there are some differences in scope, there is also very significant overlap, and much shared interest. Furthermore, the possibilities to extend contacts between the members in both countries, and between disciplines, was an additional benefit of a joint conference. The suggestion of a joint meeting was therefore met with enthusiasm from both sides.

The conference was arranged as a single-stream series of presentations and discussions in the style of a workshop. Four leading authorities in their fields were invited to give 50-minute keynote addresses; these were Joel Gilbert (Université du Maine, France), Peter Wells (Cardiff University, UK), John Willis (Cambridge

University, UK), and Philippe Destuynder (Conservatoire National des Arts et Métiers, France). Other participants gave 20-minute oral presentations or posters. There were 82 contributions in total. Participants were given the option to submit manuscripts to be peer-reviewed for publication in these proceedings. The cost of participation at the conference was part-supported by the GDR for all attendees.

The conference was a great success, and it was widely acknowledged that it achieved all the hopes for communication, collaboration, and new contacts and opportunities.

However, the attendees of the conference are also left with a very sad memory. This was the last working engagement of Andrew Temple, who died suddenly on 6 February 2010. Andrew was the Chairman of the PAG of the Institute of Physics and an enthusiast for both AFPAC and GDR activities. He was the central organiser of this conference, across all aspects, from his technical contributions as a member of the technical committee, to his provision of local information and plans for social activities and outings. He was a highly accomplished and respected scientist who was also a friend to many of us. These proceedings are dedicated to his memory.

Journal of Physics: Conference Series **269** (2011) 012001 doi:10.1088/1742-6596/269/1/012001

# On the Modelling of Ultrasonic Bulk Wave Propagation in Epoxies.

**R. E. Challis, F. Blarel, M. E. Unwin and X. Guo.**

Applied Ultrasonics Laboratory, Electrical Systems and Optics Division, Faculty of Engineering, University of Nottingham, University Park, Nottingham, NG7 2RD, UK.

E-mail: Richard.challis@nottingham.ac.uk

**Abstract**. Bulk wave attenuation in epoxies has been found experimentally to be a linear function of frequency. Anecdotal evidence suggests that the formulations used for bulk wave attenuation to be input to computational propagation models, including models for guided waves, can take many forms: Either the simple linear relationship between attenuation and frequency can be input directly, or any of a number of analytic models could, in principle, be used – examples being the Maxwell, Kelvin-Voigt and Zener anelastic solid models, or the hysteretic damping model sometimes employed for studies of structural vibrations. This paper considers the physical bases of these various models, as well as multiple degree of freedom models that have their origins in dielectric theory. It is shown that a simple two degree of freedom adaptation of the Zener model provides a good simulation of experimentally observed data for both attenuation and phase speed over a wide frequency band.

## 1. Introduction

In the context of non-destructive testing (NDT) and materials evaluation there is frequently a need to develop mathematical/computational models that simulate bulk and/or guided ultrasonic wave propagation in structures or their constituent materials. Epoxy polymers or composites formed of them represent an important class of materials used in safety critical applications such as aerospace, automotive, rail and marine, and it is therefore important to study wave propagation in them. Of particular interest is the fact that wave attenuation in epoxies is close to a linear function of frequency over a wide frequency band. This is true both for the fully cured material and throughout the cure cycle [1,2]. For simulation of this behaviour, the essential problem lies in the formulation of the complex elastic moduli which govern wave propagation, which lead to frequency dependent attenuation and, to a lesser extent, a frequency dependent phase speed through the Kramers-Kronig relationships [3,4]. This paper draws on results from a recent publication [5] to discuss a number of formulations for the complex elastic moduli, and their applicability to models of wave propagation.

## 2. Broad band behaviour of attenuation and phase speed

We have recently attempted to measure bulk shear wave attenuation in a cured epoxy [2]. Four techniques were used to measure attenuation either directly using conventional ultrasound transducers in pitch-catch, or indirectly using different methods across a wide frequency band – dynamic mechanical analysis (DMA) for frequencies below 200 Hz, a microwave excitation technique [6] for frequencies around 20 kHz, and a novel guided wave technique for frequencies around 500 kHz [2]. The results are shown on Fig.1 in which the dotted line represents a linear fit corresponding to an attenuation of $340\times10^{-6}$ Np $m^{-1}$ $Hz^{-1}$. The corresponding results for phase speed are shown on Fig. 2, on which the dotted line corresponds to a Kramers-Kronig calculation based on the attenuation data. Whilst these results are not 'perfect', they do indicate the close to linear

Journal of Physics: Conference Series **269** (2011) 012001 doi:10.1088/1742-6596/269/1/012001

relationship over five decades of frequency as well as reasonable compatibility with the Kramers-Kronig relationship.

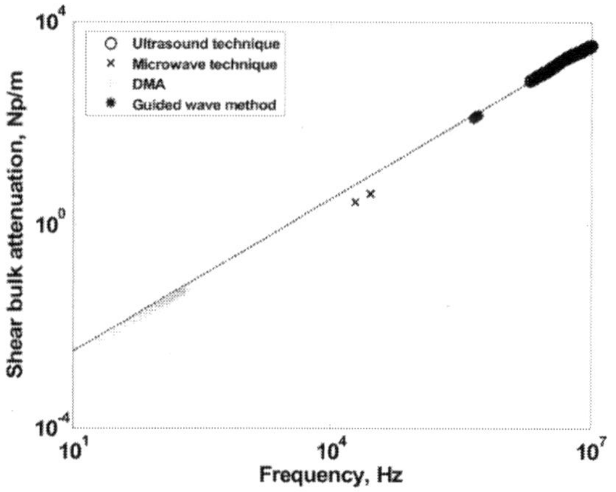


**Fig.1.** Bulk shear wave attenuation versus frequency for the 2011 epoxy, calculated from measurements using the four techniques. The dashed line on the attenuation curve is a straight line fit corresponding to $340 \times 10^{-6}$ Np Hz$^{-1}$.

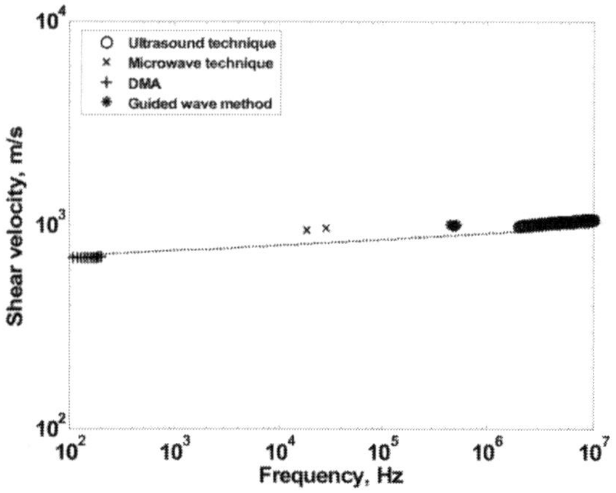


**Fig.2.** Bulk shear wave phase speed corresponding to the attenuation data on Fig.1. The dashed line was calculated on the basis of the Kramers-Kronig relationships applied to the attenuation data with the speed at 100 Hz taken as 705 ms$^{-1}$.

## 3. Basic wave propagation model

Our working follows the classical results of McSkimin [7] which define a complex elastic modulus governing bulk wave propagation

$$M(\omega) = M_R(\omega) + jM_I(\omega) \tag{1}$$

In terms of real quantities in epoxies the real part of the compression and shear moduli at radio frequencies (10 MHz, say) is found to be between one and three times the static (zero frequency) value, depending on the material. At the same high frequencies the values of the imaginary parts of the moduli are, typically, up to 5% of the real part for compression disturbances, and up to 12% for shear disturbances. It is therefore reasonable to make the assumption that $M_I(\omega) \langle\langle M_R(\omega)$, under which the phase speed is

$$c^2(\omega) = \frac{M_R(\omega)}{\rho} \tag{2}$$

Where $\rho$ is density. Under the same assumption the attenuation coefficient is

$$\alpha(\omega) = \frac{\omega M_I(\omega)}{2\rho c^3(\omega)} = \frac{\omega M_I(\omega)}{2c(\omega)M_R(\omega)} \quad (3a),$$

and the attenuation per wavelength is

$$\alpha\lambda = \pi\frac{M_I(\omega)}{M_R(\omega)} = \pi\tan\delta(\omega) \quad (3b),$$

where $\tan\delta(\omega)$ is the loss tangent. The complex modulus is sometimes expressed in an alternative form

$$M(\omega) = M_o\left[1 + r(1 - H(\omega))\right] \quad (4),$$

where $M_0$ is the low frequency value and the parameter $r$ represents the strength of the loss process. The frequency dependence incorporated in $H(\omega)$ represents the physics governing the loss process. The real and imaginary parts of $M(\omega)$ become

$$M_R(\omega) = M_o\left[1 + r(1 - H_R(\omega))\right] \quad (5) \text{ and}$$

$$M_I(\omega) = -rM_oH_I(\omega) \quad (6)$$

Given knowledge of $H(\omega)$ and $r$ the attenuation and phase speed are calculated using eqs. (2) and (3).

## 4. Basic loss models

We now consider five basic loss models that have been suggested by various workers in the context of wave modelling.

(i) *Maxwell series element* [8]

This arrangement is shown in Fig. 3a, and is commonly referred to as representing relaxation in a general sense; the corresponding complex modulus is

$$M(\omega) = M_o\frac{j\omega\tau}{1 + j\omega\tau} \quad (7)$$

where $\tau = \eta/M_0$ and $\eta$ represents viscous reaction proportional to frequency. Clearly, at low frequency such a modulus would not support bulk wave propagation, and so it cannot represent an epoxy.

(ii) *Kelvin-Voigt* [9,10]

This a parallel combination of a 'spring' and a 'dashpot', shown in Fig. 3b, and is sometimes referred to as representing the creep phenomenon; the complex modulus is

$$M(\omega) = M_o(1 + j\omega\tau) \quad (8)$$

This function results in a realistic phase speed but an attenuation that rises as the square of frequency – which does not match the linear function observed in epoxies.

(iii) *Zener anelastic solid* [11]

This model can be regarded as a combination of the foregoing series and parallel cases, Fig. 3c; the most convenient way to express the modulus is through the function $H(\omega)$ in eq. (4), thus

$$\frac{M_\infty - M(\omega)}{M_\infty - M_o} = H(\omega) = \frac{1}{1 + j\omega\tau} \quad (9)$$

where the relaxation time constant is

$$\tau = \frac{\eta}{M_\infty - M_o} \qquad (10),$$

and the loss parameter is

$$r = \frac{M_\infty}{M_o} - 1 \;=\; \frac{c_\infty^2}{c_o^2} - 1 \qquad (11)$$

where $c_\infty$ and $c_0$ are the high- and low-frequency asymptotic limits to the phase speed. If $r$ is much less than unity, the attenuation is

$$\alpha(\omega) = \frac{r\omega^2\tau}{2c_o(1+\omega^2\tau^2)} \qquad (12)$$

and the phase speed is

$$c(\omega) = c_o\,\frac{1+(r+1)^{\frac{1}{2}}\omega^2\tau^2}{1+\omega^2\tau^2} \qquad (13)$$

The attenuation initially rises as frequency squared and, as we will demonstrate, this gives a reasonable but by no means perfect match to measured results over a narrow frequency band. The phase speed changes by a small amount between low- and high-frequency asymptotes.

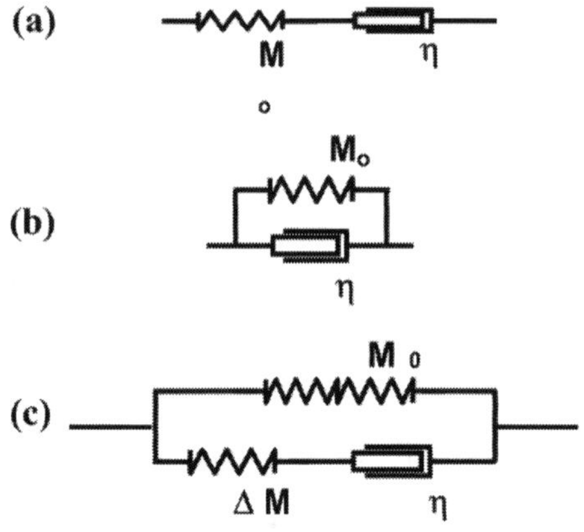


**Fig. 3.** Basic viscoelastic models: (a) Series - Maxwell [12], (b) Parallel - Kelvin Voigt [13, 14], and (c) Anelastic solid - Zener [19].

(iii) *Hysteretic damper* [12]

This model has found use in studies of vibration damping in civil structures. It consists of an elastic element ('spring') in parallel with a damper whose reaction does not depend on sliding speed; the modulus is

$$M = M_o + jm \qquad (14)$$

The attenuation from eq. (3) is

$$\alpha(\omega) = \frac{\omega m}{2c_o M_o} \qquad (15)$$

This is a linear function of frequency and could, in principle, represent propagation in an epoxy. However, the phase speed is

$$c(\omega) = c_o = \left(\frac{M_0}{\rho}\right)^{1/2} \tag{16}$$

This is independent of frequency, and does not comply with expectation on the basis of the Kramers-Kronig relationships, given the finite and frequency dependent attenuation. Thus, whilst such a model may be sufficient for practical purposes, it remains unsatisfactory in a more philosophical sense.

## 5. Multi-compartment models

There are many parallels between studies of mechanical loss and dielectric loss, and much past research has focussed on physical explorations and mathematical models which could explain the behaviour of complex dielectric permittivity where changes in the frequency domain extend over a much larger bandwidth than would be expected of a simple single time constant Debye formulation such as we have alluded to earlier in eq. (9). A more detailed treatment of these ideas is to be found in McCrum et al [13]. Three of the most commonly cited empirical formulas for the frequency function $H(\omega)$ are:

Davidson and Cole [14]
$$H(\omega) = \frac{1}{(1+j\omega\tau)^{\gamma}} \tag{17}$$

Cole and Cole [15]
$$H(\omega) = \frac{1}{1+(j\omega\tau)^{1-\gamma}} \tag{18}$$

Havriliak and Negami [16]
$$H(\omega) = \frac{1}{[1+(j\omega\tau)^{1-\gamma}]^{\beta}} \tag{19}$$

In a number of unpublished experiments we have found that the Cole-Cole formulation gives the best fit to measured attenuation and phase speed over the limited frequency range between 2 MHz and 20 MHz, see results, below. We now hypothesize that these models can be represented by groups of single time constant processes operating in parallel,

$$\text{thus } H(\omega) = \frac{1}{1+(j\omega\tau_o)^{1-\gamma}} = \sum_{n=1}^{N}\frac{g_n}{1+j\omega\tau_n} \tag{20}$$

where $g_n$ and $\tau_n$ are the weight and the time constant of the *n*th process; the weights $g_n$ sum to unity. We have used fitting between the right and left sides of eq. (20) to estimate a series of values for $g_n$ and $\tau_n$. Fig. 4 shows the imaginary part of $H(\omega)$ corresponding to a Cole-Cole distribution, with an equivalent set of Debye processes superimposed as vertical arrows where the heights correspond to the process weights $g_n$. Clearly, the seven processes shown would represent considerable complexity in the relaxation behaviour of an epoxy. It is therefore of interest to consider how few processes would be required to give an acceptable representation of attenuation over a limited frequency range that is typical of current laboratory measurements. Fig 5 shows attenuation as a function of frequency calculated on the basis of eq. (3) for three forms of $H(\omega)$: a direct implementation of the Cole-Cole model, the seven Debye processes shown on Fig. 4, and just two Debye processes, eq (20) with $N$=2. The parameters of these two processes are (*strength, frequency*) = (0.28, 2.51 MHz) and (0.72, 15.8 MHz). It is clear that the three calculated attenuation curves are practically coincident.

## 6. Experimental results

We now compare experimental measurements of attenuation and phase speed with expectation from a selection of the models described above. Compression wave attenuation over a limited frequency range was measured using techniques we have described previously [1,17,18]. The epoxy material used was *Araldite* 2015 (Huntsman Advanced Materials, Evenberg, Belgium); measurements were taken 20 minutes into cure at 40°C, at which point the elastic moduli would have developed to approximately 50% of their final (end of cure) values. Fig. 6 compares the experimental data with a best fit single time constant Debye process. The fitted parameters were $c_0$ = 1450 ms-1, $r = 0.31$ and $\tau$ = 40 ns. The attenuation prediction is acceptable at frequencies below 8 MHz whilst the phase speed fit is poor. Fig. 7 shows the equivalent data for a fitted full Cole-Cole function with the same parameter values and the coefficient $\gamma$ set to 0.51. The match to both attenuation and phase speed is much improved. Finally, referring to Fig. 5, we note that an equivalent fit to just two Debye processes with the parameter values given earlier would be equally effective.

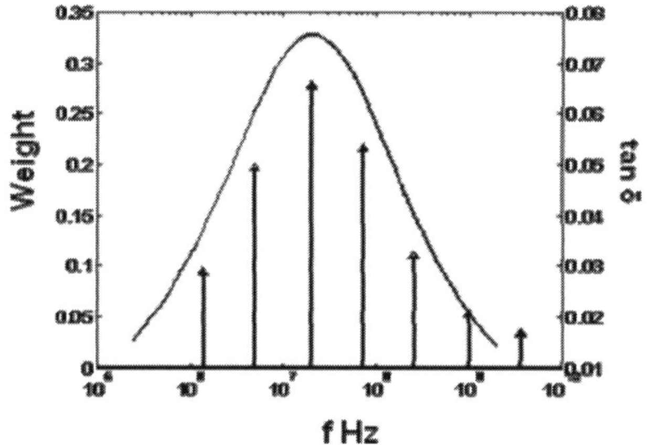


**Fig. 4.** Imaginary part of $H(\omega)$ corresponding to a Cole-Cole distribution, with an equivalent set of discrete relaxation process superimposed as vertical arrows where the heights represent the process weights $g_n$.

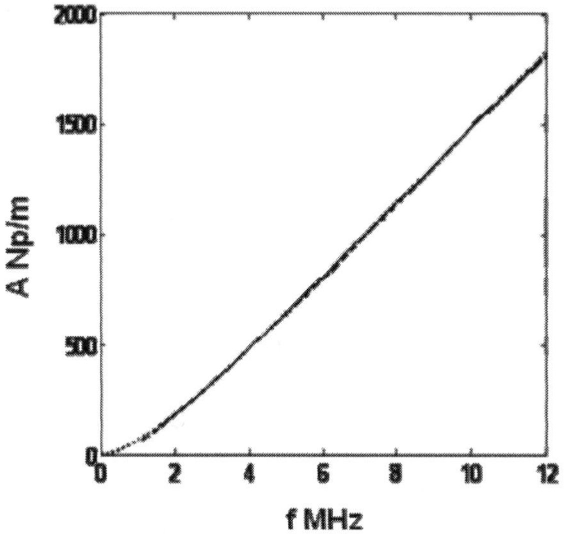


**Fig. 5.** Calculated attenuation on the basis of (solid curve) a direct implementation of the Cole-Cole model, (dots) the relaxation frequency distribution of figure 4, and (dashes) just two concurrent Debye single relaxation time processes of (strength, frequency) = (0.28, 2.51 MHz) and (0.72, 15.8 MHz). The curves are coincident.

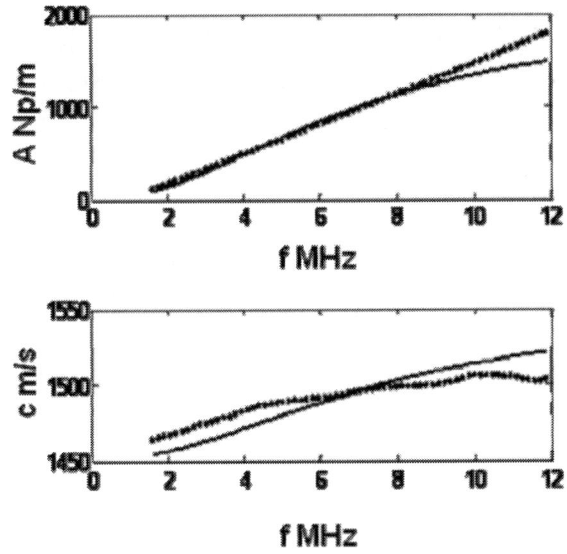


**Fig. 6.** 2015 epoxy, 20 minutes into cure at 40 C: Attenuation (top) and phase speed (bottom) versus frequency. The solid curves represent fits of a single Debye process to the experimental data which are represented as dots.

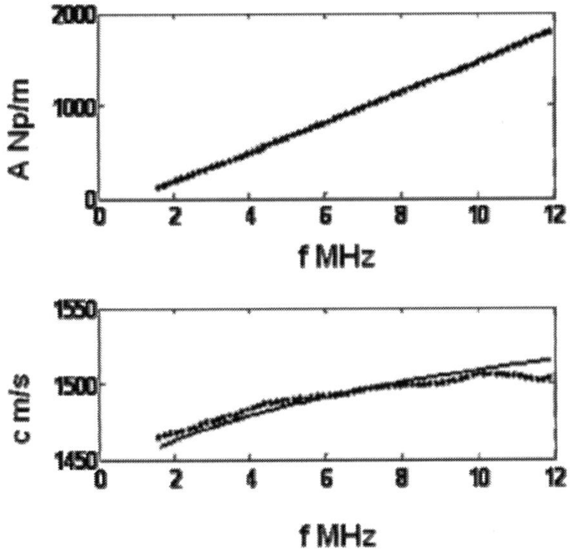


**Fig. 7.** The same experimental results as Fig. 6 (dots) together with the fitted Cole-Cole model (solid curves).

## 7. Discussion and concluding remarks

The motivation behind this work was to provide models of ultrasonic loss in epoxies which were both physically realistic and, at the same time, based on relatively uncomplicated physical ideas. We argued that the simple Maxwell and Kelvin-Voigt models were not appropriate. The hysteretic damper model, whilst giving attenuation as a linear function of frequency, would not simulate dispersion in phase speed. Whilst this would not lead to significant errors in most propagation modelling exercises, the conflict with expectation on the basis of the Kramers-Kronig relationships was philosophically unsatisfactory. Formulations for elastic loss based on ideas from dielectrics research, particularly that of Cole and Cole [15], gave accurate simulations of both attenuation and phase speed. They did, however, invoke underlying physical processes that were quite complicated in terms of their equivalence to a relatively large number of concomitant single time constant loss processes. However, a simple model consisting of just two Debye processes operating in parallel would give predictions of attenuation and phase speed over a narrow frequency band that are as accurate as those obtained with the more complex Cole-Cole model.

In this work we have not addressed the major scientific question as to what physical processes underpin the efficacy of the simple two-compartment model. To achieve this would require a major experimental programme involving many different epoxy materials, with different internal molecular structures. However, at this stage we are prepared to conjecture that the two degrees of freedom implied by the two compartment model are associated with two distinct aspects of the epoxy molecular arrangement. These *could* be the mean length of the epoxy polymer chains and the mean length of the cross-linking molecular segments.

The principal conclusion is that a simple two Debye process formulation will provide an effective and practical means to model bulk wave attenuation and phase speed in epoxy materials over the commonly used frequency band up to 20 MHz.

**References**

[1] R.E. Challis, R. J. Freemantle, R P. Cocker, D.L.Chadwick, D.J. Dare, C.M. Martin, A. Mendrasingham, and W. Fuller. Ultrasonic measurements related to the evolution of structure in curing epoxy resins. Plastics, Rubber and Composites, Vol. 29, pp. 109-118, 2000.

[2] Y. Wang, R.E. Challis, A.P-Y.Phang and M.E. Unwin. Bulk shear wave propagation in an epoxy: Attenuation and Phase Velocity over five decades of frequency. IEEE Trans. UFFC, Vol. 56, No. 11, pp 2504-2523, 2009.

[3] R. Kronig and H.A. Kramers. Absorption and dispersion in X-ray spectra. Zeits. F. Phys. Vol 48, pp. 174-179, 1928.

[4] M. O'Donnell, E.T. Haynes and J.G. miller. Kramers-Kronig relationship between ultrasonic attenuation and phase velocity. J. Acoust. Soc. Am. Vol. 69, pp. 696-701, 1981.

[5] R.E. Challis, F. Blarel, M.E. Unwin, J. Paul and X. Guo. Models of ultrasonic wave propagation in epoxy materials. IEEE Trans. UFFC, Vol. 56, No. 6, pp 1225-1237, 2009.

[6] A.R.H. Henni, C. Bacon and B. Hosten. In-plane vibration of thin circular structures submitted to pulse microwave. J. Acoust. Soc. Am, Vol. 119, No. 6, pp. 3782-3792, 2006.

[7] H.J. McSkimin, In: *Physical Acoustics*, Vol. 1A, ed W P Mason, Academic Press, New York, 1964.

[8] J.C. Maxwell. On the dynamical theory of gases. Philos. Trans. Roy. Soc. Vol. 157, pp. 49-86, 1867.

[9] W.P. Thompson (Lord Kelvin). Article on Elasticity, Encyclopedia Britannica, 1875.

[10] W. Voigt. Uber innere Reibung fester Korper, insbesondere der Metalle. Ann. D. Phys. Vol. 47, pp. 671-693, 1892.

[11] C. Zener. *Elasticity and Anelasticity of Metals*. University of Chicago Press, Chicago, 1948.

[12] D.J. Ewins. *Modal Testing: Theory and Practice*. Research Studies Press, John Wiley and Sons Inc. New York, 1986.

[13] N.G. McCrum, B.E. Read and G. Williams, *Anelastic and Dielectric Effects in Solids*, Wiley, New York, 1967.

[14] D.W. Davidson and R.H. Cole, Dielectric relaxation in glycol, propylene glycol, and n-propanol. J. Chem. Phys., Vol. 19, pp. 1484-1490, 1951.

[15] K.S. Cole and R.H. Cole, Dispersion and absorption in dielectrics I. Alternating current characteristics. J. Chem. Phys., Vol. 9, pp. 341-351, 1941.

[16] S. Havriliak and S. Negami, A complex plane representation of dielectric and mechanical relaxation processes in some polymers. Polymer, Vol. 8, pp. 161-211, 1967.
[17] R.E. Challis, T. Alper, A.K. Holmes and R.P. Cocker, Near-plane-wave acoustic propagation measurements in thin layers of adhesive polymer. Meas. Sci. Technol., Vol. 2, pp. 59-68, 1991.
[18] R.J. Freemantle and R.E. Challis. Combined compression and shear wave ultrasonic measurements on curing adhesive. Meas. Sci. Technol. Vol. 9, pp. 1291-1302, 1998.

Journal of Physics: Conference Series **269** (2011) 012002 doi:10.1088/1742-6596/269/1/012002

# A detailed study of guided wave propagation in a viscoelastic multilayered anisotropic plate

**L Taupin[1], A Lhémery[1] and G Inquiété[2]**

[1] CEA, institut LIST,
centre de Saclay, bât. 611, point courrier 120, F-91191 Gif-sur-Yvette *cedex*, France.
[2] EADS - Innovation Works,
12, rue Pasteur, BP76, F-92152 Suresnes *cedex*, France.

E-mail: alain.lhemery@cea.fr

**Abstract.** Guided waves (GW) are very attractive in nondestructive technique applications (eg. Structural Health Monitoring) because of their ability to propagate at long range. In a structure made of composite materials, their propagation is complex due to material anisotropy and to their dispersive and multi-modal nature. Interpreting measurements of GW in such a structure requires a sound grasp of their behaviour. Here, the Semi-Analytical Finite Element (SAFE) method is used for studying GW propagation in viscoelastic multilayered anisotropic plates. Beside classical post-processing techniques used to compute the displacement, dispersion and slowness curves, the Poynting vector is also obtained, allowing us to study energy propagation in complex plate structures. Then, GW propagation in multilayered viscoelastic composite (C-epoxy) plates is studied; different stacking sequences typical of those used to build aeronautical parts are considered. Phase, energy velocities and attenuation are studied for different propagation directions and frequencies. It appears that symmetries of GW behaviour are complex: the axes of symmetry depicting this behaviour do not coincide with those of stacking sequences and depend on frequency. Modes appearing above the first cut-off frequency have such a complex behaviour that they cannot be used in practical applications.

## 1. Introduction

Nowadays, composite materials are used to build many parts of aircrafts (structural components, wing and fuselage elements). Some can be subjected to impacts that must be both detected and characterized for operating the aircraft safely; nondestructive techniques are thus extensively used in aircraft maintenance. Structural Health Monitoring (SHM) is an alternative technique to classical non-destructive testing methods. It may be used either actively (for periodic examinations) or passively if operated while the component under examination is in-service. In the aeronautic industry, SHM could constitute a good choice of method to make the maintenance both easier and more efficient [1,2].

The present work is part of a more general study on the feasibility of SHM for composite parts using elastic guided waves led by EADS [2]. Guided waves (GW) can propagate at long range in plate-like geometries. Therefore, the number of transducers to generate or to measure such waves propagating in a plate could be limited in principle while an overall sensitivity to events (impact) or to the presence of a defect positioned arbitrarily in the structure could be ensured. Optimizing the number, type and position of permanently attached transducers to ensure full sensitivity requires simulation tools to deal with the variety of constituting materials, thicknesses involved etc. and to deal with the complex behaviour of GW in general (multimodal and dispersive natures) and specifically as they propagate in a multilayered viscoelastic anisotropic plate.

The present study aims at better understanding the behaviour of GW in multilayered composite plates typical of those used at EADS. At first, a simulation tool is developed; a model is first derived then validated. Then, this tool is used for a detailed study of GW behaviour in composite plates of different stacking sequences and at various frequencies. Special attention is paid for determining their degree of anisotropy as far as GW propagation is concerned.

## 2. Theoretical model for GW in composite plates and its implementation

There is a very large body of literature offering various theoretical or numerical methods to predict GW propagation in multilayered anisotropic plate-like structures, see [3] for a recent review.. Among the methods available, the SAFE method (Semi-Analytical Finite Element, see [4] for the reference used in this paper for its specific relevance to our concerns) is rather easy to develop and to extend for addressing complex cases (propagation in guides of complex geometry and complex material structure). In plate geometry, the discretization by finite elements is limited solely to the thickness of the plate; the method is therefore very efficient computationally. A general tool is under development at CEA LIST for the simulation of NDT techniques based on the use of guided elastic waves [5], to be included in the CIVA software platform [6]. This tool uses the SAFE method for computing modes in.

### *2.1. Viscoelasticity of a multi-layered plate*

Before deriving a SAFE model for the problem in hand, we have to decide how to model the viscoelastic behaviour of a multilayered composite plate. The viscoelastic model chosen in this study consists in writing elastic stiffness constants as complex numbers. Constants are defined for a uni-directional layer where Carbon fibres are aligned and randomly positioned in a viscous Epoxy matrix. Plates are modelled by stacking sequences of such layers (figure 1) defined by their orientation and thickness. Solid-solid boundary conditions at interfaces between layers are further assumed.

For one given layer, the real part of stiffness constants is assumed to be independent of the frequency. The imaginary part results from the use of a hysteretic model to account for losses in the layer. Other models together with this standard model are described in [7]; they could be equally used at no additional cost. Under this assumption, the imaginary part of stiffness coefficients is independent of frequency. According to the chosen sign convention, the complex-valued stiffness tensor is written

$$\tilde{\mathbf{C}} = \mathbf{C} + i\boldsymbol{\eta} \tag{1}$$

where $\mathbf{C}$ stands for the elastic tensor in a purely elastic medium and $\boldsymbol{\eta}$, for the viscosity tensor. The stiffness tensor of the layer is rotated by an angle $\theta$ corresponding to the layer orientation in the stacking sequence, using a standard matrix calculation, $\mathbf{R}_1$ and $\mathbf{R}_2$ being defined in the Appendix.

$$\tilde{\mathbf{C}}_\theta = \mathbf{R}_1(\theta)\,\tilde{\mathbf{C}}\,\mathbf{R}_2^{-1}(\theta)\,. \tag{2}$$

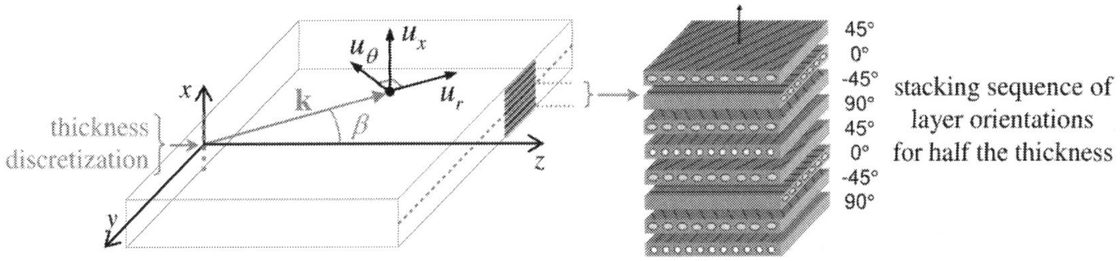


**Figure 1.** Plate geometry, constituting material and definition of notations. The stacking sequence shown for half the thickness is generally symmetrised relatively to the mid-plane, this being denoted by a subscript S. The sequence pictured here would be denoted by $[45°,0°,-45°,90°,45°,0°,-45°,90°]_S$.

### *2.2. SAFE method for GW propagation along an arbitrary direction in multi-layered plate*

The SAFE method consists in the decomposition of elastodynamic fields which relies on the discretization of fields by finite elements in the guide cross section while the propagation in guiding directions is calculated analytically (see [4] for the treatment of cases similar to present ones).

Specific attention must be paid to deal with a multilayered structure and with anisotropic properties of plates considered here. Contrary to GW propagation in isotropic materials, the behaviour of GW in anisotropic materials depends on the direction of propagation. The angle $\beta$ shown on figure 1 stands for that direction relatively to the $z$-axis of the reference frame; it must be introduced in the SAFE system. For this, the wave number is written as

$$\mathbf{k} = k\cos\beta\,\mathbf{z} - k\sin\beta\,\mathbf{y}\ , \tag{3}$$

so that the displacement $\mathbf{u}$ at a point $(x,y,z)$ at time $t$ can be decomposed in the form

$$\mathbf{u}(x,y,z,t) = \mathbf{U}(x)\exp\, i(\omega t + k\sin\beta\, y - k\cos\beta\, z)\ . \tag{4}$$

Standard formulae of the SAFE method must be slightly modified to account for the $\beta$-dependency. It is out of the scope of the present paper to give the detailed derivation of the SAFE method adapted with these modifications. This derivation is straightforward by following the various steps described in the literature (see [4] for example). Here, only specific formulae for the problem in hand are given.

The discretization is limited to the thickness of the plate ($x$-dependency) – this is the main reason for its high numerical performance –, while the propagation kernel depends on the two other space variables $y$ and $z$. The displacement at an arbitrary point in the plate is re-written as a function of the interpolation matrix $\mathbf{N}$ (see Appendix), nodal displacements, and the propagation term. One has

$$\mathbf{u}^{(e)}(x,y,z,t) = \mathbf{N}\mathbf{d}^{(e)}\exp\, i(\omega t + k\sin\beta\, y - k\cos\beta\, z)\ , \tag{5}$$

where
$$\mathbf{d}^{(e)} = \begin{bmatrix} d_x^{(e),1} & d_y^{(e),1} & d_z^{(e),1} & . & . & . & d_x^{(e),n} & d_y^{(e),n} & d_z^{(e),n} \end{bmatrix}^T \tag{6}$$

and $d_i^{(e),j}$ denotes the $i$-th component of the nodal displacement at the $j$-th node of element $(e)$, the element $(e)$ involved in this formula being the element corresponding to the same $x$-value. In our implementation, finite elements considered are segments of three nodes; the interpolation matrix $\mathbf{N}$ for this type of elements is given explicitly in the Appendix.

The strain vector $\boldsymbol{\varepsilon}$ for one element is then given by

$$\boldsymbol{\varepsilon} = \left[\mathbf{B}_1 - ik_y\mathbf{B}_2 - ik_z\mathbf{B}_3\right]\mathbf{d}^{(e)}\exp\, i(\omega t + k\sin\beta\, y - k\cos\beta\, z)\ , \tag{7}$$

where,
$$\mathbf{B}_1 = \mathbf{L}_\mathbf{x}\mathbf{N}_{,\mathbf{x}}\ ,\ \mathbf{B}_2 = \mathbf{L}_\mathbf{y}\mathbf{N}\ ,\ \mathbf{B}_3 = \mathbf{L}_\mathbf{z}\mathbf{N}, \tag{8}$$

the matrices $\mathbf{L}$ being given in the Appendix. $\mathbf{N}_{,x}$ denotes the derivative of $\mathbf{N}$ relatively to $x$.

Then, classical steps for obtaining the variational formulation and its discretized form are followed, as in [4]. The dependency on the direction of propagation $\beta$ makes it necessary to introduce the following stiffness matrices $\mathbf{K}_{mn}$ (in the FE sense of the word) and the mass matrix $\mathbf{M}$. These matrices are given here for each element along the thickness, assuming that one element does not belong to two different layers of the stacking sequence. For the element $(e)$, one has

$$\mathbf{k}_{mn}^{(e)} = \int_{(e)} \mathbf{B}_m^T\tilde{\mathbf{C}}_\theta^{(e)}\mathbf{B}_n dx \quad\text{and}\quad \mathbf{m}^{(e)} = \int_{(e)} \rho\mathbf{N}^T\mathbf{N}dx\,. \tag{9}$$

The overall matrices are obtained by standard assembling techniques

$$\mathbf{K}_{mn} = \bigcup_{e=1}^{n_e}\mathbf{k}_{mn}^{(e)} \quad\text{and}\quad \mathbf{M} = \bigcup_{e=1}^{n_e}\mathbf{m}^{(e)}\,. \tag{10}$$

Finally, a quadratic system of equations is obtained that is written

$$-\omega^2\mathbf{M}\mathbf{d}+[k^2(\cos^2\beta\,\mathbf{K}_{33}+\sin^2\beta\,\mathbf{K}_{22}-\cos\beta\sin\beta\,\mathbf{K}_{23}) \\ -i\,k(\cos\beta\,\mathbf{K}_{13}-\sin\beta\,\mathbf{K}_{12})+\mathbf{K}_{11}]\mathbf{d}=\mathbf{0}. \tag{11}$$

To simplify the resolution, the quadratic system is re-written as a linear system. Its eigenvalues are the mode wave numbers and its eigenvectors give corresponding modal displacement. This system is thus:

$$(\mathbf{A}-k\mathbf{D})\mathbf{Q}=0, \tag{12}$$

where

$$\mathbf{Q}=\begin{bmatrix}\mathbf{d}\\ k\mathbf{d}\end{bmatrix}, \tag{13}$$

and

$$\mathbf{A}=\begin{bmatrix}0 & \mathbf{K}_{11}-\omega^2\mathbf{M}\\ \mathbf{K}_{11}-\omega^2\mathbf{M} & -i(\cos\beta\,\mathbf{K}_{13}-\sin\beta\,\mathbf{K}_{12})\end{bmatrix}, \tag{14}$$

$$\mathbf{D}=\begin{bmatrix}\mathbf{K}_{11}-\omega^2\mathbf{M} & 0\\ 0 & -(\sin^2\beta\,\mathbf{K}_{22}+\cos^2\beta\,\mathbf{K}_{33}-\cos\beta\sin\beta\,\mathbf{K}_{23})\end{bmatrix}. \tag{15}$$

This system explicitly depends on the angle $\beta$. Predicting the anisotropic behaviour of GW properties makes it necessary to loop on $\beta$, equation (12) being solved for each direction considered.

### *2.3. Energy velocity*

For NDT applications operated for locating defects, the knowledge of the energy direction is crucial. In damping materials, it is well known that, due to viscosity, the standard derivative for defining the group velocity is no longer valid [8]; rather, the energy velocity must be considered which is defined as

$$\mathbf{v}_e=\frac{\mathbf{P}}{E_T},\ \text{with}\ E_T=E_C+E_P, \tag{16}$$

where $E_T$ stands for the total energy (see Appendix for its expression). Computing the energy velocity allows us to represent the direction of energy propagation in the composite plate and to predict wave front curves. Each component of the Poynting vector is computed for one given phase velocity direction. The Poynting vector is obtained by post-processing the SAFE results; it is written [9],

$$P_k=\frac{1}{2}\mathrm{Re}\ i\omega\sigma_{jk}u_j^*\ \ ,j\ \text{and}\ k\in\{1,2,3\}, \tag{17}$$

where $\boldsymbol{\sigma}$ is obtained through Hooke's relations from the displacement computed by the SAFE method.

### *2.4. Convergence and validation studies*

#### *2.4.1. Convergence study*

To study the convergence of the present SAFE model, wave numbers have been computed for various discretizations of the same multilayered viscoelastic plate. For example, results obtained with one element per layer were compared to results obtained with 10 elements per layer. Relative errors were computed and are in a range of $10^{-4}$ and $10^{-5}$. This accuracy was tested in a configuration typical of those of interest for the present study: overall plate thickness of a few mm and excitation frequencies less than 2 MHz were considered. More quantitative convergence studies have been conducted but these results are out of the scope of the present paper.

#### *2.4.2. Validation studies*

To validate the model, we have reproduced results of the literature obtained by means of methods which fundamentally differ from the SAFE method. Among many other authors, Wang and Yuan [10] developed a transfer matrix method for predicting GW; they considered multilayered plates of various stacking sequences, each layer being assumed to be purely elastic. They show dispersion curves (phase and group velocities as functions of the frequency) and polar plots of slowness and group velocities as functions of the direction of propagation. Figure 2 successfully compares our predictions to published results for the slowness of propagative modes in a plate with stacking sequence of $[45°_6/-45°_6]_S$ [10].

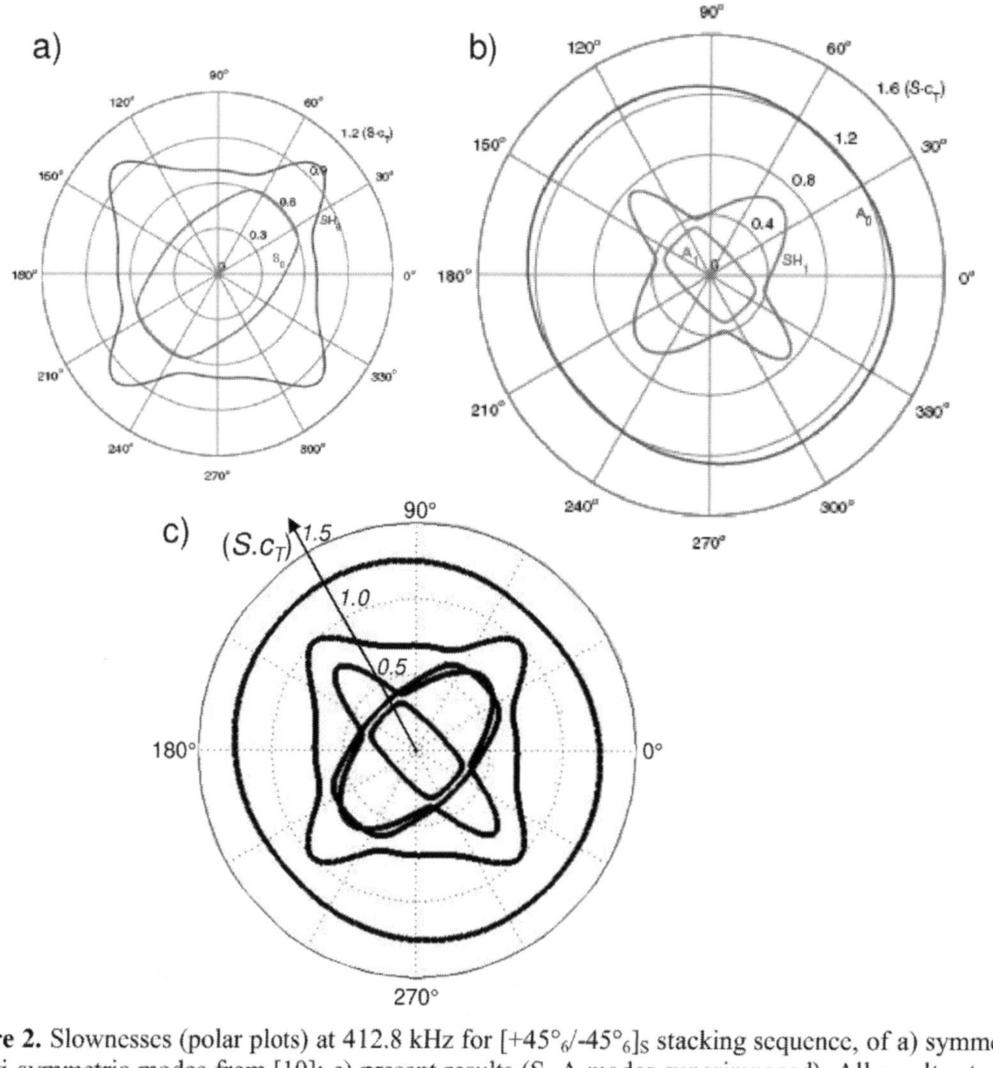


**Figure 2.** Slownesses (polar plots) at 412.8 kHz for $[+45°_6/-45°_6]_S$ stacking sequence, of a) symmetric b) anti-symmetric modes from [10]; c) present results (S, A modes superimposed). All results at same scale.

Since our model allows us to consider viscoelastic plates, further comparisons were made with original results obtained by Neau [11] who derived an analytical formulation for GW propagation in a single viscoelastic layer. The comparisons concern now both the attenuation and the energy velocity in a unidirectional layer of Carbon fibres in Epoxy matrix. They are shown in figures 3 and 4.

Again, our predictions are in perfect agreement with Neau's, while the two methods are radically different. Many similar comparisons have been made leading to the same conclusion of validity of our model and implementation. Future experiments will complete our validation work.

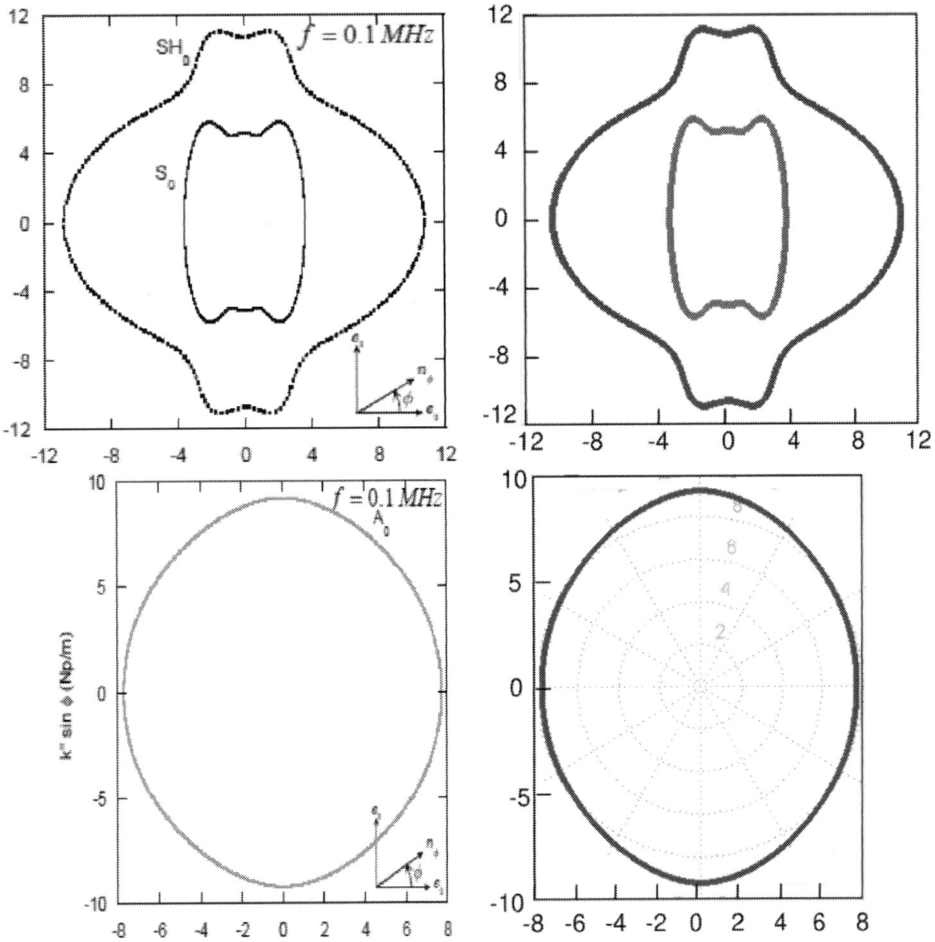


**Figure 3.** Attenuation polar plots (in Np/m) at 100 kHz in a unidirectional viscoelastic layer. Left: column: results in [11]. Right column: present model. Top: $S_0$ (green), $SH_0$ (blue): Bottom: $A_0$ mode.

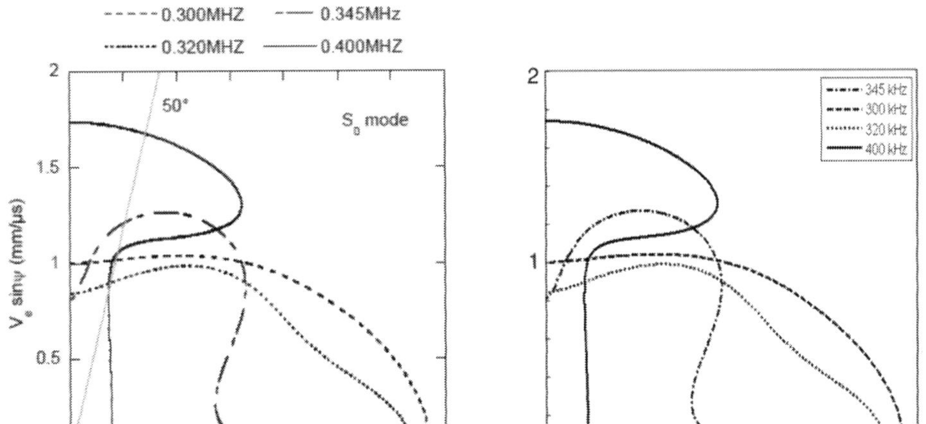


**Figure 4.** $S_0$ energy velocity (polar plots) at four frequencies. Left: from [11], right: present model. Curves are displayed in coordinates $V_e \cos\psi$ and $V_e \sin\psi$ (in mm/µs), $\psi$ denoting the energy direction.

## 3. Detailed study in multilayered viscoelastic anisotropic plate

The present paragraph is now dedicated to studies of GW propagation in plates typical of those used in practice. Aeronautic engineers use such plates in the design of many parts of an aircraft and may specify which stacking sequence is to be used according to the required structural performance at each location.

As far as SHM applications are concerned, it is important to have an excellent grasp of GW behaviour in typical plates to be used and to be tested. Our validated model allows us to study in detail this behaviour.

A 16-layer plate is first considered (unidirectional Carbon fibre-Epoxy layer 0.133 mm-thick each). Viscoelastic properties used herein are given by (manufacturer for **C**, postulated from [4] for $\eta$)

**Table 1**. Viscoelastic properties of a Carbon-Epoxy layer (stiffness in GPa, $\rho$ in kg.m$^{-3}$).

| $C_{11}$ | $C_{12}$ | $C_{13}$ | $C_{22}$ | $C_{23}$ | $C_{33}$ | $C_{44}$ | $C_{55}$ | $C_{66}$ | $\rho$ |
|---|---|---|---|---|---|---|---|---|---|
| 11.5 | 5.258 | 5.0 | 11.5 | 5.0 | 145.5 | 5.2 | 5.2 | 3.5 | 1494 |

| $\eta_{11}$ | $\eta_{12}$ | $\eta_{13}$ | $\eta_{22}$ | $\eta_{23}$ | $\eta_{33}$ | $\eta_{44}$ | $\eta_{55}$ | $\eta_{66}$ |
|---|---|---|---|---|---|---|---|---|
| 0.8 | 0.4 | 0.7 | 0.8 | 0.7 | 10 | 0.4 | 0.4 | 0.2 |

The stacking sequence of this plate is $[45°/0°/135°/90°/45°/0°/135°/90°]_S$ as shown in figure 1. Structural engineers consider such sequences lead to static isotropic properties in the plate plane. Thus, one question is to know whether this property is shared by GW propagating in it. Before answering this question, it is important to study the GW behaviour for one given direction of propagation in the plate. Therefore, the first results concern the prediction of phase and energy velocities and attenuation in the reference direction 0° as functions of the frequency.

### *3.1. Dispersion curves for the anisotropic plate*

Figure 5 shows the dispersive and multimodal behaviour of GW propagating in the 0° direction.

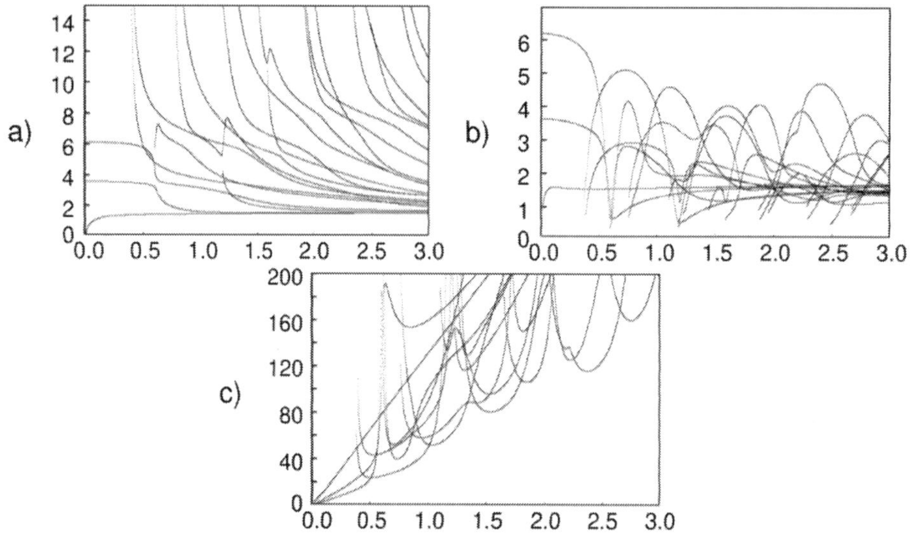


**Figure 5.** a) Phase velocities (mm/µs), b) energy velocities (mm/µs) and c) attenuations (Np/m) of the various modes as function of frequency (MHz).

Several comments can be made at this stage. Below the first cut-off frequency (the frequency above which more than three modes exist), phase velocities vary much less than above this frequency. This is an interesting property since time-dependent waveforms of frequency content below the first cut-off will be less affected by dispersion. Another point is that in this frequency range, attenuation is at its lowest. Moreover, it varies linearly with the frequency. Above the first cut-off frequency, attenuation is strong and varies in a complicated manner with frequency. Using GW for the examination of such plates seems therefore far easier in the frequency range below the first cut-off frequency.

A new question arises from these comments: does the first cut-off frequency vary with the direction of propagation? This has been studied and the following table gives its value for four different directions. It appears that the variation is very small (less than 3 kHz, or 0.7%).

**Table 2 .** 1st cut-off frequency for various directions of propagation.

| direction of propagation (°) | 0 | 45 | 90 | 135 |
|---|---|---|---|---|
| 1st cut-off frequency (kHz) | 372.09 | 372.24 | 373.74 | 374.77 |

*3.2. Study of wave propagation in the plate plane*

It has been observed that dispersion curves depend on the direction of propagation. This effect is quantified now for every possible directions of propagation in the plate. Figure 6 displays at four frequencies (three below the first cut-off, one above) the slowness curves of the various modes.

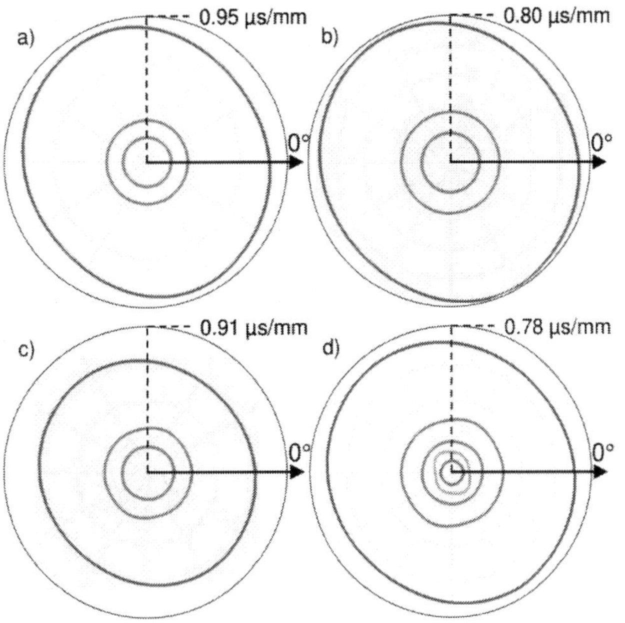


**Figure 6.** Slowness curves (polar plots) for the 16-layer plate at a) 100 kHz, b) 200 kHz, c) 350 kHz and d) 410 kHz. $A_0$ in blue, $SH_0$ in green and $S_0$ in red. Cyan and pink curves correspond to higher modes appearing above the 1st cut-off frequency.

For one of the three modes (in red) existing below the first cut-off frequency, the curve is almost circular. This almost isotropic mode is the symmetric $S_0$ which propagates with the smallest slowness (highest speed). The blue plot corresponds to the antisymmetric $A_0$ mode; it has the highest slowness but this slowness varies slightly with the direction of propagation. At 410 kHz, above the first cut-off frequency, new modes appear of higher speed, which are much more anisotropic.

In the presence of anisotropy, it is known that phase and energy velocities are no longer collinear. Figure 7 shows how the energy velocities vary with the direction of propagation (of energy).

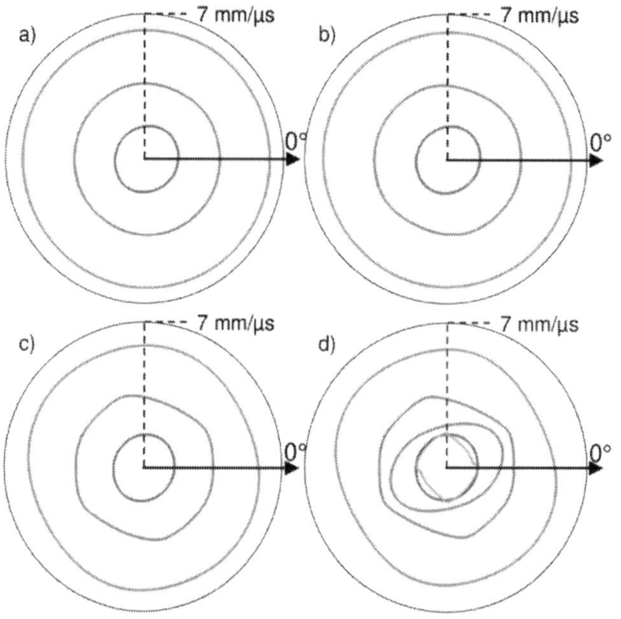


**Figure 7.** Same as figure 6, but energy velocity curves.

At the lowest frequency considered, anisotropy is already visible for the slowest $A_0$ mode. Then, as the frequency increases, the anisotropy becomes more and more marked. Above the first cut-off frequency, higher modes are strongly anisotropic. It is noticeable that symmetries exist: curves are centro-symmetric due to the stacking sequence which makes the plate invariant by a $\pi$-rotation.

It is also interesting to measure how the symmetry varies. For this, we have determined the directions of highest slowness, of highest energy velocity and of lowest slowness, for the three modes below the first cut-off. Values are given in the following table.

**Table 3.** Directions (angle relative to $z$) of maximum slowness, maximum energy velocity, minimum slowness for $A_0$, $SH_0$ and $S_0$ modes below the first cut-off frequency, for various frequencies.

| f (kHz) | angle (°) of maximum slowness | | | angle (°) of maximum energy velocity | | | angle (°) of minimum slowness | | |
|---|---|---|---|---|---|---|---|---|---|
| | $A_0$ | $SH_0$ | $S_0$ | $A_0$ | $SH_0$ | $S_0$ | $A_0$ | $SH_0$ | $S_0$ |
| 100 | 110.8 | 67.3 | 29.3 | 47.5 | 107.4 | 151.5 | 36.9 | 107.2 | 151.6 |
| 200 | 116.2 | 67.3 | 29.2 | 55.1 | 107.6 | 150.4 | 40.7 | 107.4 | 150.9 |
| 350 | 123.0 | 67.2 | 29.0 | 58.9 | 108.1 | 145.9 | 44.9 | 107.7 | 148.6 |

Angles of maximal values of phase and energy velocity differ, due to the fact that phase and energy velocity are not aligned. Moreover, variations with frequency differ for angles of maximal and minimal values of slowness: for the $A_0$ mode, the angles of maximal slowness values vary from 110.8° to 123.0° ($\Delta_{max}$= 12.2°), whereas, for the minimal value, angles vary from 36.9° to 44.9° ($\Delta_{min}$= 8.0°).

Interestingly, the various angles considered depend on frequency and they are not aligned with fibre orientations. This latter point must be understood. For this, the particle displacement of the various modes propagating in the plate is studied in detail in the next paragraph.

### *3.3. Particle displacement in the thickness*

This study aims at explaining why predicted anisotropy depends on frequency as observed in previous results. For this, the various components of the particle displacement of the three modes existing below the first cut-off frequency are computed and shown in figure 8 at two different frequencies (100 and 350 kHz). In both cases, the direction of phase velocity is taken in the 0° reference direction.

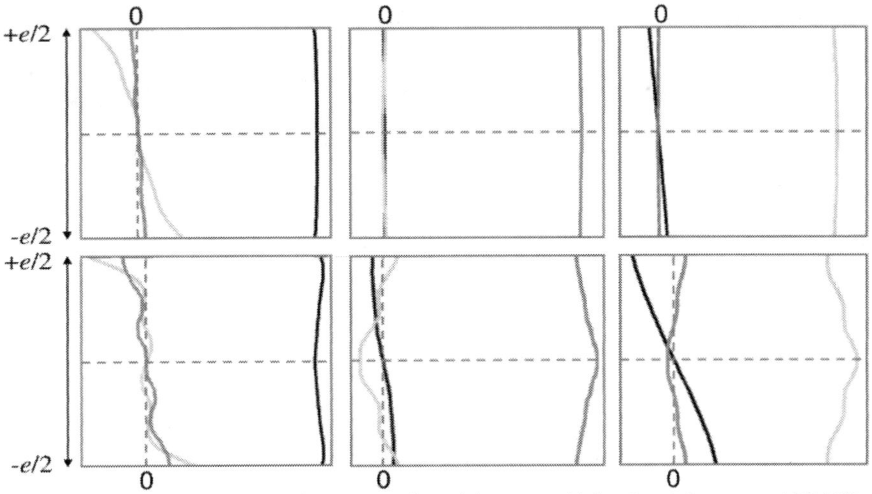


**Figure 8.** Relative amplitudes of displacement (in arbitrary unit) in the tickness at 100 kHz (top) or 350 kHz (bottom) for $A_0$ (left column), $SH_0$ (middle column) and $S_0$ (right column). $u_r$ (green), $u_\theta$ (pink) and $u_x$ (black).

As commented above, the $A_0$ mode is the most anisotropic even at low frequency. Its $u_\theta$ and $u_x$ components at 100 kHz are almost constant in the thickness of the plate but its $u_r$ component takes its highest values (anti-symmetrically) in the layers close to the outer surfaces. Interestingly, the inner multilayered structure is not perceptible in these results where the amplitude of all components varies smoothly in the thickness. At 350 kHz, this last remark is no longer true: the inner structure is clearly "seen" by the guided modes, especially for the $A_0$ one. The highest amplitude of $u_r$ component is even more concentrated in the external layers. This influences directly the global anisotropy of the mode: each layer being unidirectional, it is strongly anisotropic and favours the propagation along fibres.

The lower anisotropy of the two other modes $SH_0$ and $S_0$ can be explained according to the same detailed study of amplitude variation of the various components of the particle displacement. Note again that the multilayered structure of the plate is not at all perceptible at the lowest frequency and is only slightly visible for a few components of the displacement at the highest frequency considered.

Symmetries observed in the various polar curves (slowness, energy velocity, and as it will be seen farther, attenuation) and their dependence upon the frequency clearly show that a plate made up of typical stacking sequences involving layers of 0°, ±45°, 90° fibre orientations with equal partition and thickness cannot be modelled as a transversely isotropic plate in the plate plane, as far as guided waves are concerned. Moreover, even at relatively low frequencies (lower than the first cut-off frequency), the multilayered structure cannot be ignored and is actually responsible for both the global anisotropy observed and its dependency upon the frequency.

### *3.4. Comparison between two different stacking sequences*

To demonstrate the importance of the inner structure of composite plates, we now compare guided wave propagation in two plates with two different stacking sequences: the same as that in previous results $[45°/0°/135°/90°/45°/0°/135°/90°]_S$ (16 layers) and the sequence $[45°/0°/135°/90°]_S$ (8 layers).

A first comparison concerns the angle of maximal slowness; the same product thickness x frequency is considered for easy comparisons. Results are shown in Table 4.

**Table 4.** Angles (°) of maximal slowness of $A_0$ for the 8-(left) and the 16-layer plate (right).

| 8-layer plate ($e$ = 1.07 mm) | | 16-layer plate ($e$ =2.14 mm) | |
|---|---|---|---|
| frequency (kHz) | angle (°) | frequency (kHz) | angle (°) |
| 200 | 118.1 | 100 | 110.8 |
| 700 | 132.5 | 350 | 123.0 |

The maximal slowness values are significantly different for the 8-layer and 16-layer plates for the $A_0$ mode. Similar results were obtained for the two other modes, even if the difference is smaller. Likewise, the orientations of maximal energy velocity and minimal slowness differ in the two plates.

Now, energy velocity curves are plotted for the 8-layer sequence on Figure 9. Singularities in the form of loops appear below the 1st cut-off frequency at 700 kHz. Such singularities were not observed for the 16-layer plate at 350 kHz. The presence of loops in the energy velocity curves is a phenomenon which would make very difficult the use of such a guided mode in SHM application: it would generate complex time-dependent signals strongly varying in a limited range of propagation angles.

Above the first cut-off frequency (figure 9), new modes appearing have a particularly complex behaviour that prohibits their use in SHM applications.

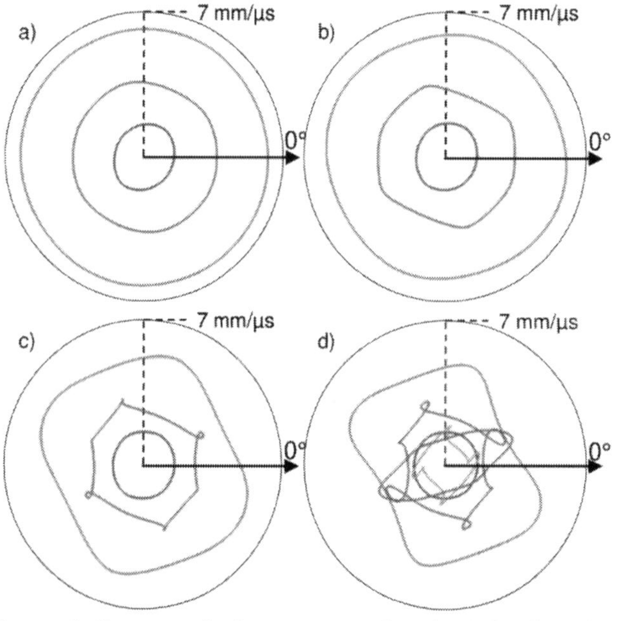


**Figure 9.** Energy velocity curves (polar plots) for the 8-layer plate at a) 200 kHz, b) 400 kHz, c) 700 kHz and d) 820 kHz.

Other results for plates made of a larger number of layers have also been computed which are not shown here for conciseness. As the analysis of previous results suggested, it has been observed that for an increasing number of layers, energy velocity curves for modes below the first cut-off frequency eventually tend to become more and more isotropic (in the plate plane). At the lowest values of the frequency x thickness product, whatever the sequence considered, the energy velocities are roughly the same for a given mode in a given direction (same property for phase velocities). This is an interesting fact for practitioners.

*3.5. Attenuation*

For completeness, figures 10-11 show the variation of GW attenuation in the propagation plane. Curves are displayed in Np/m. As far as attenuation is concerned, the absolute values obtained depend on the frequency: this explains why keeping the same frequency x thickness product for computing results in the two plates leads to higher attenuations for the 8-layer thinner plate (about twice that of the 16-layer plate at half the frequencies).

Interestingly, attenuation below the first cut-off frequency is all the more anisotropic since the frequency increases; in the case of the 8-layer plate, some curves show strong concavities. At frequencies above the first cut-off frequency, attenuation curves of new modes become very complex. The report of such behaviour reinforces the conclusion already reached that the use of higher (than the fundamental ones) guided modes cannot be recommended for SHM applications.

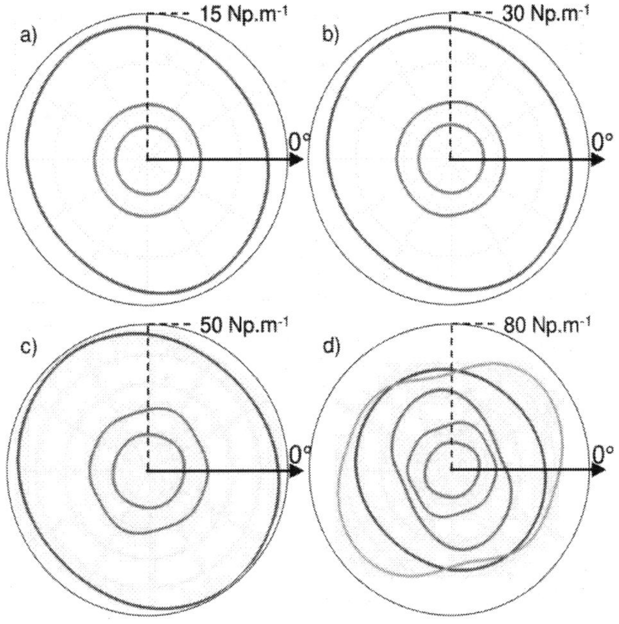


**Figure 10.** Attenuation curves (polar plots) for the 16-layer plate at a) 100 kHz, b) 200 kHz, c) 350 kHz and d) 410 kHz.

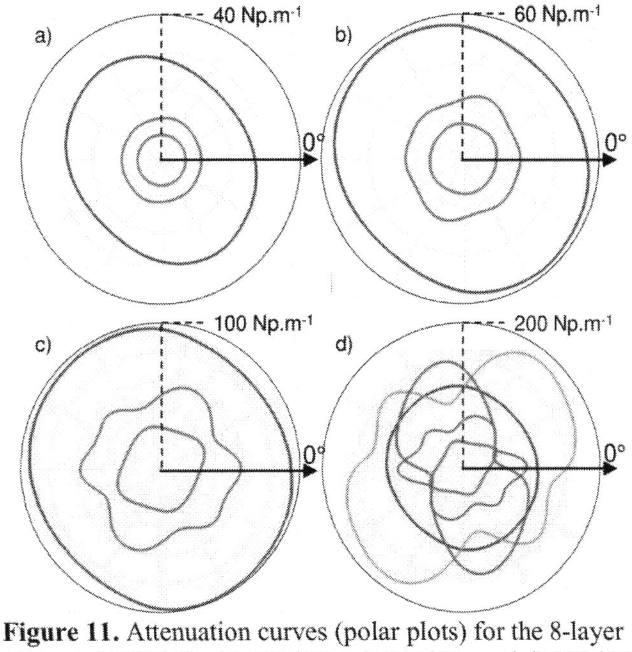


**Figure 11.** Attenuation curves (polar plots) for the 8-layer plate at a) 200 kHz, b) 400 kHz, c) 700 kHz and d) 820 kHz.

## 4. Conclusions

A detailed study of elastic guided wave propagation in multilayered viscoelastic plates has been conducted. For this, simulation tools have been developed based on the well-established SAFE method which allows fast and accurate predictions; the underlying theory of the SAFE method has been given in some details specific to our implementation dealing with multilayered viscoelastic anisotropic plates and computing quantities of interest by post-processing eigenmode raw results. These tools have been used to depict the complex behaviour of GW in plate structures typical of some found in aircrafts.

It may be concluded that in the usual frequency range for using elastic GW in non-destructive testing, i) stacking sequences of anisotropic layers assumed to have isotropic (in the plane) static properties lead to anisotropic behaviours of GW, ii) anisotropy increases with increasing frequency, iii) the generally strong attenuation is at its lowest for modes below the first cut-off frequency and varies linearly with the frequency in this range, iv) modes appearing at frequencies above the first cut-off have such a complex behaviour that they cannot be used in practice, v) designing NDT methods based upon GW propagation for multilayered plates makes it necessary to study the detailed behaviour of these waves; vi) eventually, only appropriate simulation tools can handle the intrinsic complexity of GW in these structures and vii) the SAFE method is particularly well suited for such simulations, combining accuracy, numerical performance with easy adaptability to deal with complex cases.

In a future work, the present model will be hybridised with a FE model to compute GW scattering in multilayered plates by various discontinuities, this generalising the approach described in [5, 12].

**References**

[1] Giurgiutiu V 2008 *Structural Health Monitoring with piezoelectric wafer active sensors* (Academic Press, Burlington)

[2] Petitjean B, Barut S and Simonet D 2007 Impact damage detection on aeronautical composite parts: issues and challenges, *in Proc. 6th Int. Workshop on SHM (Stanford, 2007)* 1127

[3] Su Z, Ye L and Lu Y 2006 Guided Lamb waves for identification of damage in composite structures: A review *J. Sound Vib.* **295** 753

[4] Bartoli I, Marzani A, Lanza di Scalea F and Viola E 2006 Modeling wave propagation in damped waveguides of arbitrary cross-section *J. Sound Vib.* **295** 685

[5] Lhémery A, Baronian V, Jezzine K and Bonnet-BenDhia A-S 2010 Simulation of inspections of elastic waveguides of arbitrary section containing arbitrary local discontinuities or defects *Review of Progress in QNDE* **29**, ed D O Thompson and D E Chimenti (Melville: AIP Conf. Proc. **1211**) 145

[6] CIVA software platform for simulating NDT techniques (UT, EC, RT) http://www-civa.cea.fr

[7] Challis R E, Blarel F, Unwin M E, Paul J and Guo X 2009 Models of ultrasonic wave propagation in epoxy materials *IEEE Trans. UFFC* **56** 1225

[8] Auld B A 1990 *Acoustic fields and waves in solids - Vol. II* (Malabar: Krieger Publishing Company) p. 155

[9] Wang L and Yuan F G 2007 Group velocity and characteristic wave curve of Lamb waves in composites: Modeling and experiments *Composites Science Tech.* **67** 1370

[10] Bernard A, Lowe M J S and Deschamps M 2001 Guided waves energy velocity in absorbing and non-absorbing plates J. Acoust Soc. Am. **110** 186

[11] Neau G 2003 *Lamb waves in anisotropic viscoelastic plates. Study of the wave fronts and attenuation*, PhD thesis Bordeaux I University, n° d'ordre 2634 (in English)

[12] Baronian V, Lhémery A and Bonnet-BenDhia A-S 2009 Simulation of non-destructive inspections and acoustic emission measurements involving guided waves *J. Phys.: Conf. Ser.* **195** 012001

**Appendix. Definition of terms appearing in the SAFE formulation**

The matrices of rotations of an angel $\theta$ (one angle per layer), $\mathbf{R}_1$ and $\mathbf{R}_2$, are given by (with $c = \cos\theta$ and $s = \sin\theta$):

$$\mathbf{R}_1 = \begin{bmatrix} 1 & 0 & 0 & 0 & 0 & 0 \\ 0 & c^2 & s^2 & -2cs & 0 & 0 \\ 0 & s^2 & c^2 & 2cs & 0 & 0 \\ 0 & cs & -cs & c^2-s^2 & 0 & 0 \\ 0 & 0 & 0 & 0 & c & s \\ 0 & 0 & 0 & 0 & -s & c \end{bmatrix}, \mathbf{R}_2 = \begin{bmatrix} 1 & 0 & 0 & 0 & 0 & 0 \\ 0 & c^2 & s^2 & -cs & 0 & 0 \\ 0 & s^2 & c^2 & cs & 0 & 0 \\ 0 & 2cs & -2cs & c^2-s^2 & 0 & 0 \\ 0 & 0 & 0 & 0 & c & s \\ 0 & 0 & 0 & 0 & -s & c \end{bmatrix}. \quad \text{(A1)}$$

The matrices $\mathbf{L}_x$, $\mathbf{L}_y$ and $\mathbf{L}_z$ are defined as:

$$\mathbf{L}_x = \begin{bmatrix} 1 & 0 & 0 \\ 0 & 0 & 0 \\ 0 & 0 & 0 \\ 0 & 0 & 0 \\ 0 & 0 & 1 \\ 0 & 1 & 0 \end{bmatrix}, \mathbf{L}_y = \begin{bmatrix} 0 & 0 & 0 \\ 0 & 1 & 0 \\ 0 & 0 & 0 \\ 0 & 0 & 1 \\ 0 & 0 & 0 \\ 1 & 0 & 0 \end{bmatrix}, \mathbf{L}_z = \begin{bmatrix} 0 & 0 & 0 \\ 0 & 0 & 0 \\ 0 & 0 & 1 \\ 0 & 1 & 0 \\ 1 & 0 & 0 \\ 0 & 0 & 0 \end{bmatrix} \quad \text{(A2)}$$

and $\mathbf{N}$ is the interpolation matrix which is written:

$$\mathbf{N}(x) = \begin{pmatrix} \varphi_1 & 0 & 0 & \varphi_2 & 0 & 0 & \varphi_3 & 0 & 0 \\ 0 & \varphi_1 & 0 & 0 & \varphi_2 & 0 & 0 & \varphi_3 & 0 \\ 0 & 0 & \varphi_1 & 0 & 0 & \varphi_2 & 0 & 0 & \varphi_3 \end{pmatrix} \quad \text{(A3)}$$

with the following interpolation functions,

$$\varphi_1 = 2(\xi(x)-1)(\xi(x)-0.5),\ \varphi_2 = 4\xi(x)(1-\xi(x)),\ \varphi_3 = 2\xi(x)(\xi(x)-0.5), \quad \text{(A4)}$$

where

$$\xi(x) = (x - x_1^{(e)})/\Delta \quad \text{(A5)}$$

and $x$ is the integration variable and $x_1^{(e)}$ denotes the coordinate of the first node of element $e$. $\Delta$ denotes the length of an element assuming a constant discretization step (not a necessary condition).

Finally, the total energy appearing in the expression of energy velocity (equation 16) is the sum of the two following terms (the detailed derivation of these terms may be found in [9], p. 52) where $h$ denotes the plate thickness:

$$E_C = \frac{\rho\omega^2}{4h} \int_{-h/2}^{h/2} \mathbf{u}^T \mathbf{u}\, dx, \quad \text{(A6)}$$

$$E_P = \frac{1}{4h} \int_{-h/2}^{h/2} \boldsymbol{\varepsilon}^{\mathbf{T}} \mathbf{C} \boldsymbol{\varepsilon}\, dx. \quad \text{(A7)}$$

Journal of Physics: Conference Series **269** (2011) 012003 doi:10.1088/1742-6596/269/1/012003

# Measurement of guided mode wave vectors by analysis of the transfer matrix obtained with multi-emitters and multi-receivers in contact.

**Jean-Gabriel Minonzio, Maryline Talmant, Pascal Laugier**
UPMC Univ Paris 06, UMR 7623, LIP,
15 rue de l'école de médecine
F-75005, Paris, France

E-mail: jean-gabriel.minonzio@upmc.fr

**Abstract.** Different quantitative ultrasound techniques are currently developed for clinical assessment of human bone status. This paper is dedicated to axial transmission: emitters and receivers are linearly arranged on the same side of the skeletal site, preferentially the forearm. In several clinical studies, the signal velocity of the earliest temporal event has been shown to discriminate osteoporotic patients from healthy subjects. However, a multi parameter approach might be relevant to improve bone diagnosis and this be could be achieved by accurate measurement of guided waves wave vectors. For clinical purposes and easy access to the measurement site, the length probe is limited to about 10 mm. The limited number of acquisition scan points on such a short distance reduces the efficiency of conventional signal processing techniques, such as spatio-temporal Fourier transform. The performance of time-frequency techniques was shown to be moderate in other studies. Thus, optimised signal processing is a critical point for a reliable estimate of guided mode wave vectors. Toward this end, a technique, taking benefit of using both multiple emitters and multiple receivers, is proposed. The guided mode wave vectors are obtained using a projection in the singular vectors basis. Those are determined by the singular values decomposition of the transmission matrix between the two arrays at different frequencies. This technique enables us to recover accurately guided waves wave vectors for moderately large array.

## 1. Introduction

Elastic waves are widely used in material characterization. For example, guided waves, such as Lamb waves, are used to characterize plates or tubes [1]. Structural and material properties can be characterized by fitting measured to theoretical guided mode wave vectors. The two dimensional spatio–temporal Fourier transform is one of the usual methods used to determine the guided mode wave vectors [2], [3]. Nevertheless this approach requires a large distance probed by the receivers. Practical constraints, as in clinical inspection of cortical bones, reduce the inspected spatial length and therefore the efficiency of this technique. The approach, presented in this paper, is based on the Singular Values Decomposition (or SVD) applied to the configuration of multi-emitter and multi-receiver arrays in contact with the object to be characterized. SVD, a widely used filtering tool [4], is adapted to the specific case of axial transmission. The aim of this paper is to introduce a method of measurement of the guided mode wave vectors in the context of ultrasonic cortical bone testing [5]-[8].

## 2. Theory

### *2.1. Response matrix*

A single probe consisting of two arrays is used in contact with a metallic plate, using coupling gel (figure 1). Superscripts E and R refer respectively to the emission and reception arrays. Thus the emission array contains $N^E$ emitters of position $x_i^E$, with $1 \leq i \leq N^E$. Likewise, the reception array contains $N^R$ receivers of position $x_j^R$, with $1 \leq j \leq N^R$. Define the $N^R \times 1$ vector **R** containing the $N^R$ Fourier transforms $R_j(f)$ of $r_j(t)$, with $f$ the temporal frequency. Likewise, the $N^E \times 1$ vector **E** contains the $N^E$ Fourier transforms $E_i(f)$. The relation between the two vectors **R** and **E** writes, at a given frequency

$$\mathbf{R} = \mathbf{H}.\,\mathbf{E}\,, \tag{1}$$

with **H** the $N^R \times N^E$ matrix containing the elements $H_{ji}(f)$ equal to the temporal Fourier transform of the impulse responses $h_{ji}(t)$. The **R** matrix, called afterward "response matrix", is characteristic of the "emission – scattering medium – reception" system.

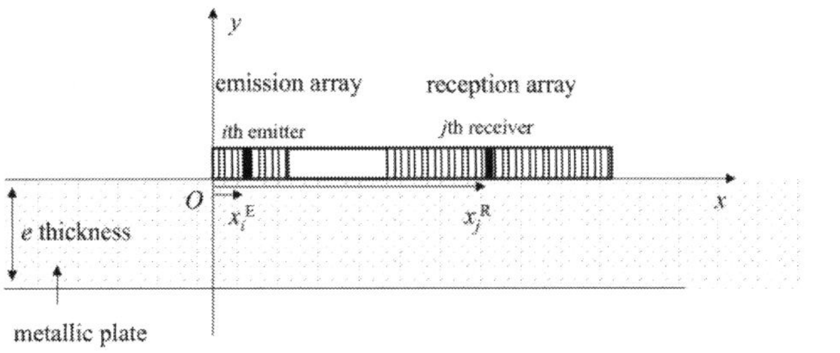


**Figure 1.** Geometry of the problem: the emission and reception arrays are in contact with the metallic plate.

### *2.2. Singular vectors basis*

The Singular Value Decomposition (SVD) of the transmission matrix **R** writes

$$\mathbf{R} = \sum_{n=1}^{N^E} \mathbf{U}_n \sigma_n {}^t\mathbf{V}_n^* \,, \tag{2}$$

where the notations $^t$ and $*$ denote the transpose and conjugation operations. The number of experimental singular values $\sigma_n$ is equal to minimum size of the arrays, i.e. the minimum between $N^E$ and $N^R$, equal in the following to $N^E$. The notation $\mathbf{V}_n$ refers to an emission singular vector, of dimension $N^E \times 1$, and $\mathbf{U}_n$ to a reception singular vector, of dimension $N^R \times 1$. These two vectors are associated with the singular value $\sigma_n$.

The singular vectors $\mathbf{U}_n$ (resp. $\mathbf{V}_n$) are normalized and define an orthogonal basis of the received (resp. emitted) signals [12]. In particular any spatial plane wave can be expressed in the reception basis. Consider a spatial plane wave $\mathbf{e}^{pw}(k)$ defined on the $j$th receiver as

$$e_j^{pw} = \frac{1}{\sqrt{N^R}} \exp\left(ikx_j^R\right) \,, \tag{3}$$

with $k$ the wave vector equal to $2\pi f / c$ and $c$ the phase velocity. The previous vector is defined normalized. The expression on the basis defined by the reception singular vectors $\mathbf{U}_n$, is

$$\mathbf{e}^{pw} = \sum_{n=1}^{N^E} \left\langle \mathbf{e}^{pw} \middle| \mathbf{U}_n \right\rangle \mathbf{U}_n \,, \tag{4}$$

with the notation $< \mathbf{e}^{pw} \mid \mathbf{U}_n >$ corresponding to the Hermitian scalar product, equal to ${}^t[\mathbf{e}^{pw}]^* .\mathbf{U}_n$ .

### *2.3. Noise and signal separation*

One of the advantages of the SVD approach is its ability to separate noise and signal subspaces [10], [11]. Indeed, using an intermediate order $m$ corresponding to the limit between the two subspaces, equation (4) can be rewritten with a truncated sum, with $n \leq m$. The order $m$ is defined at each frequency using a threshold $t_1$ applied to the singular values $\sigma_n$. In the following, the signal singular vectors are retained (for $\sigma_n \geq t_1$) whereas the noise singular vectors (for $\sigma_n < t_1$) are eliminated. Thus the norm of the spatial plane wave on the basis of the signal subspace becomes

$$\left\|\mathbf{e}^{pw}\right\|_{\{\mathbf{U}_{n\leq m}\}} = \sqrt{\sum_{n=1}^{m}\left|\left\langle \mathbf{e}^{pw} \middle| \mathbf{U}_n \right\rangle\right|^2}, \tag{5}$$

with the notation | | used for the modulus of complex numbers. The value of the threshold $t_1$ is heuristically chosen, based on an estimation of the signal to noise ratio and on a trade off between the estimated number of modes and the noise. That threshold may also vary with frequency and attenuation. An example is shown in section 3 and figure 2.

### *2.4. Evaluation of the guided mode wave vectors*

The norm of the spatial plane wave expressed on the signal (or retained) singular vectors in (5) is used to form an image in the $(k, f)$ plane. To enhance the contrast between the low and high values, the square of the norm is used in the following. As $m$, the dimension of the signal subspace, is less than the number of receivers $N^R$, i.e. the dimension of the vector $\mathbf{e}^{pw}$, the basis defined by the signal reception singular vectors $\mathbf{U}_{n\leq m}$ is incomplete. It implies that the norm defined by (5) is less than 1. Consequently, each pixel $(k, f)$ of the image ranges from 0 to 1.

The value of the pixel reflects how the spatial plane wave is represented in the basis of the signal subspace. On the one hand, if the value is small compared to 1, the spatial plane wave is absent of the received signals. On the other hand, if the value is close to 1, the spatial plane wave is present in the received signal and probably due to a guided wave. A second threshold $t_2$ is then applied to the image in order to reduce the range from $t_2$ to 1. Thus a function, denoted *Norm*, is defined in the $(k, f)$ plane as

$$\begin{cases} Norm(k,f) = \left\|\mathbf{e}^{pw}\right\|^2_{\{\mathbf{U}_{n\leq m}\}} & if \left\|\mathbf{e}^{pw}\right\|^2_{\{\mathbf{U}_{n\leq m}\}} \geq t_2 \\ Norm(k,f) = 0 & if \left\|\mathbf{e}^{pw}\right\|^2_{\{\mathbf{U}_{n\leq m}\}} < t_2 \end{cases}, \tag{6}$$

The maxima of the function *Norm*$(k, f)$ provide the wave vectors of the guided mode present in the signal subspace. In the examples shown in section 3, the value of that second threshold $t_2$ is heuristically chosen equal to 0.4 as illustrated in figure 4.

## 3. Experimental results and discussion

### *3.1. Experimental set up*

The probe developed at the LIP (Laboratoire d'Imagerie Paramétrique) for the bone characterization, consists of $N^E$ of emitters is equal to 5 and the number of receivers $N^R$ is equal to 32. Thus the number of emitters, which is less than the number of receivers, corresponds to the number of experimental singular values. A coupling gel is used in order to improve the acoustical impedance matching. Emitted signals are pulses with bandwidth about 0.5–2 MHz. The reception array pitch is equal to 0.80 mm. The experimental scattering medium is a 2 mm thickness copper plate. The free plate guided modes are given by the Rayleigh-Lamb dispersion equation [1]. No attenuation is considered in this model. The modes are labelled considering their symmetry or anti-symmetry and cut off frequencies, i.e. $S_n$ and $A_n$.

### *3.2. Resolution and evaluation domain of the system*

The resolution can be estimated considering the size of the main lobe of the sinc function, given by the scalar product between a reception singular vector $\mathbf{U}_n$ corresponding to a guided mode $k_n$ and a plane wave $\mathbf{e}^{pw}(k)$. With $L^R$ the size of the reception array, the resolution condition is

$$\Delta k \geq \frac{2\pi}{L^R} \cdot \qquad (7)$$

Furthermore, the Shannon criterion is verified when k is less than $\pi/p$, with $p$ the array pitch. Thus the range $2\pi/L^R \leq k \leq \pi/p$ corresponds to the evaluation domain, where the guided mode wave vectors $k_n$ can be evaluated.

### *3.3. Experimental results, comparison with the 2D Fourier transform*

Figure 2 shows the experimental singular values $\sigma_n$ as function of the frequency $f$. The number of experimental singular values, i.e. 5, corresponds to the number of emitters. At low frequency, only the two first singular values are significant, whereas for $f > 1$ MHz, the number of significant singular values ranges from 3 to 5. It corresponds to the number $m$ of significant singular vectors used in (6). It depends on the signal to noise threshold $t_1$, shown with a curved line, heuristically fixed at a level as shown in Figure 2. The different peaks correspond to thickness resonance of guided modes. For example, the thin peak for $f = 1.1$ MHz corresponds to the ZGV (zero group velocity) resonance [2].

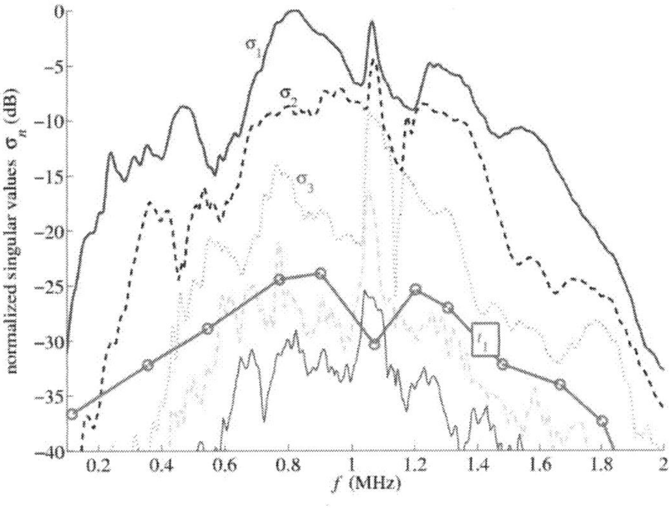


**Figure 2.** Singular values versus frequency (MHz) for a 2 mm thick copper plate. The signal to noise threshold $t_1$ is shown with a curved dashed line.

Figure 3($a$) shows the image $Norm(k, f)$ defined by (6) using the signal singular vectors, i.e. the singular vectors associated with singular values bigger than the threshold $t_1$ shown in figure 2. The value of the pixels ranges from $10^{-8}$ to 0.96 is in agreement with the expected range, *i.e.* 0 to 1. The guided mode wave vectors are shown as trajectories in the $(k, f)$ plane. The width of each trajectory corresponds to the resolution given by (7). Figure 3($b$) shows the two-dimensional spatio-temporal Fourier transform for comparison. Each line (or frequency) is normalized by its maximum.

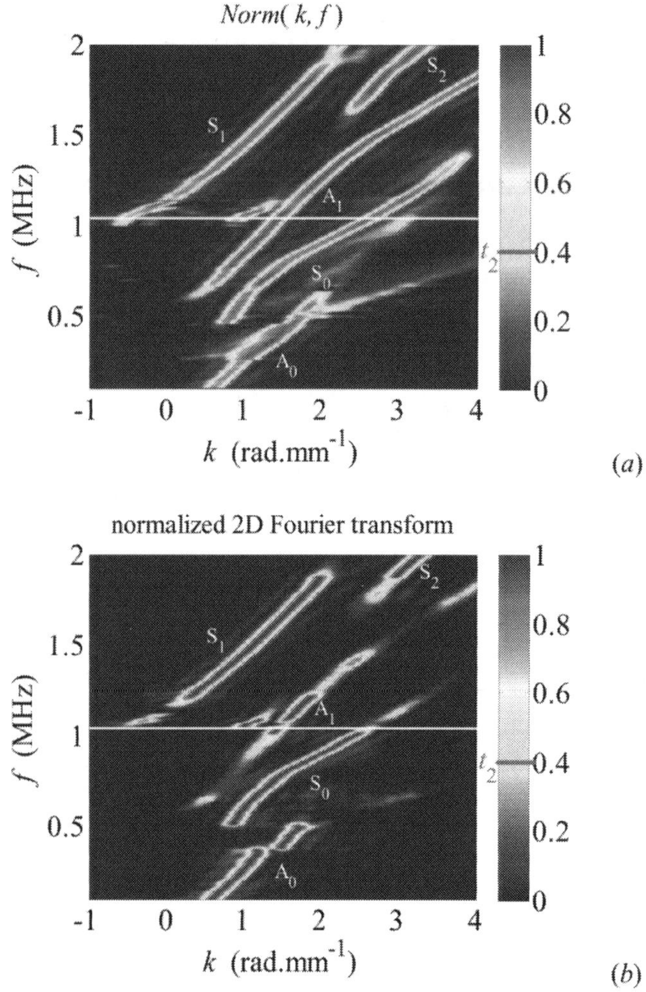


**Figure 3.** *Norm(k, f)* (*a*) and normalized spatio-temporal (2D) Fourier Transform (*b*) for a 2 mm thick copper plate. The maxima of the *Norm* function are more clearly defined. The second threshold $t_2$ is indicated on the colorbar. The frequency indicated with an horizontal line is shown on figure 4(a).

The results given by the two methods are similar. However, the *Norm* function allows a better ability to detect the guided mode wave vectors as illustrated in figure 4(*a*) for a fixed frequency. This frequency is shown as a dashed line in figure 3(*a*). The maxima of the *Norm* function ranging from 0.4 to 0.9 are clearly defined, in contrast with the maxima of the normalized spatio-temporal Fourier transform depending on the relative amplitude of each guided modes. As discussed in paragraph 2.4, the values of the maxima of the *Norm* function do not depend on the relative amplitude of the guide mode. The peak are more clearly defined compared to the sinc secondary maxima. Thus this original technique can be applied to noisy signals such as those acquired in cortical bone. The two methods are compared in the whole (*k*, *f*) domain in figure 4(*b*). The results are in good agreement with the theoretical Lamb (computed with $e = 2$ mm, $c_L = 4700$ m.s$^{-1}$ and $c_T = 2300$ m.s$^{-1}$) modes for the two methods. The norm function give more experimental values than the spatio-temporal Fourier transform using the same second threshold equal to 0.4.

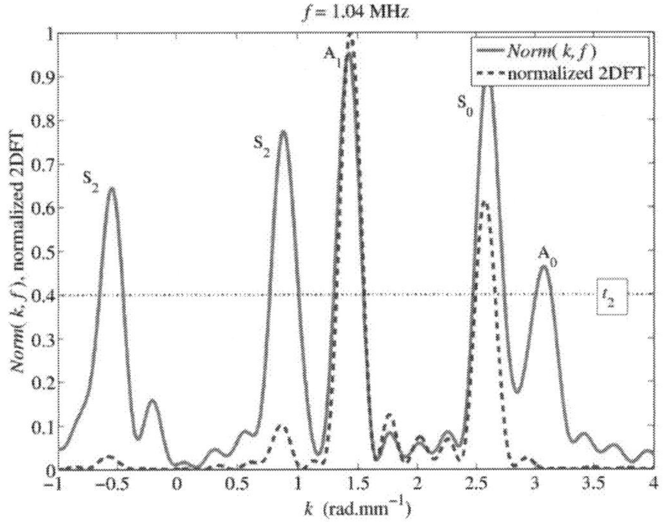


(*a*)

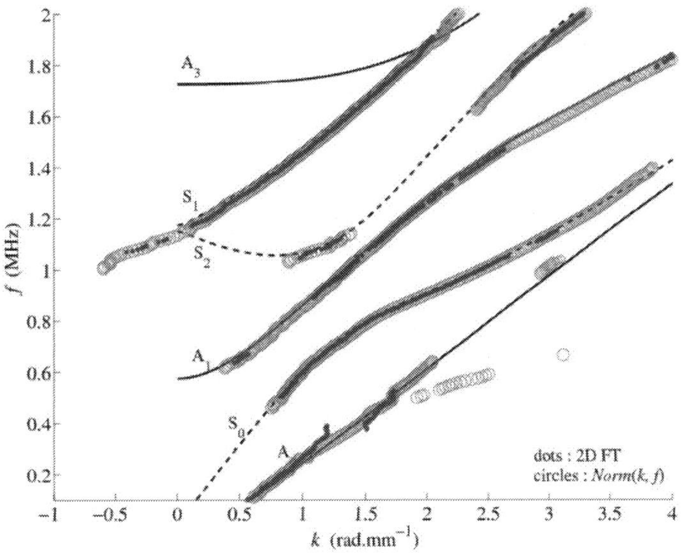


(*b*)

**Figure 4.** Comparison of both methods: *Norm*(*k*, *f*) (continuous line and circles) and normalized 2D Fourier transform (dashed line and dots), for *f* = 1.04 MHz (*a*) and in the whole (*k*, *f*) domain (*b*). The frequency 1.04 MHz is shown as a dashed line in figure 3(*a*). The second threshold $t_2$ is shown with a horizontal dashed line. The ability to detect phase velocity is improved using the *Norm* function as the maxima are better defined, compared to the sinc secondary maxima.

## 4. Conclusion

An original method for evaluating the guided mode wave vectors using arrays in contact (an emission array and a reception array) is exposed in this paper. The ability to detect the guided mode wave vectors is improved compared to the results obtained with the spatio-temporal Fourier transform as the values of the maxima of the *Norm* function do not depend on the relative amplitude of the guide mode. Further applications will concern in particular the evaluation of elastic properties of cortical bone with a reduced inspected spatial length due to practical constraints. These results open a perspective towards a good evaluation of the thickness and/or the transverse and longitudinal bulk wave velocities. Attenuation and non-isotropic medium will be taken into account.

**Acknowledgment**
This work has been funded by the ANR TecSan under EVA project.

## 5. References

[1] D. Royer, and E. Dieulesaint, "Elastics waves in solids 1 : Free and guided propagation," Springer, Berlin, p. 317 (1999).

[2] D. Alleyne and P. Cawley, "A two-dimensional Fourier transform method for the measurement of propagating multimode signals," J. Acoust. Soc. Am. **89**, 1159–1168 (1991).

[3] C. Prada, D. Clorennec, and D. Royer, "Local vibration of an elastic plate and zero-group velocity Lamb modes," J. Acoust. Soc. Am. **124**, 203–212 (2008).

[4] T.G. Kolda and B. W. Bader, "Tensor Decompositions and Applications," SIAM Review **51,** 455–500 (2009).

[5] E. Bossy, M. Talmant, M. Defontaine, F. Patat, and P. Laugier, "Bidirectional axial transmission can improve accuracy and precision of ultrasonic velocity measurement in cortical bone: a validation on test materials," IEEE Trans. Ultrason. Ferroelectr. Freq. Control. **51**, 71–79 (2004).

[6] P. Moilanen, P.H.F. Nicholson, V. Kilappa, S. Cheng, and J. Timonen, "Measuring guided waves in long bones: Modeling and experiments in free and immersed plates," Ultrasound in Medecine and Biology **32,** 709–719 (2006).

[7] V. C. Protopappas, I. C. Kourtis, L. C. Kourtis, K. N. Malizos, C. V. Massalas, and D. I. Fotiadis, "Three-dimensional finite element modeling of guided ultrasound wave propagation in intact and healing long bones," J. Acoust. Soc. Am. **121**, 3907–3921 (2007).

[8] D. Ta, W. Wang, Y. Wang, L. H. Le, and Y. Zhou, "Measurement of the dispersion and attenuation of cylindrical ultrasonic guided waves in long bone," Ultrasound in Medecine and Biology **35,** 641–652 (2009).

[9] M Sasso, M. Talmant, G. Haiat, S. Naili, and P. Laugier, "Analysis of the most energetic late arrival in axially transmitted signals in cortical bone," IEEE Trans. Ultrason. Ferroelectr. Freq. Control **56**, 363-2470 (2009).

[10] M. Davy, J. G. Minonzio, J. de Rosny, C. Prada, and M. Fink, "Influence of noise on subwavelength imaging of two close scatterers using time reversal method: Theory and Experiments," Progress in Electromagnetic Research **98**, 333–358 (2009).

[11] M. Cheney, "The linear sampling method and the MUSIC algorithm," Inverse Problems **17**, 591–595 (2001).

[12] J.-G. Minonzio, M. Talmant, and P. Laugier, "Guided wave phase velocity measurement using multi-emitter and multi-receiver arrays in the axial transmission configuration," J. Acoust. Soc. Am. 127, 2913-2919 (2010).

Journal of Physics: Conference Series **269** (2011) 012004 doi:10.1088/1742-6596/269/1/012004

# Coherent acoustic response of a screen containing a random distribution of scatterers: comparison between different approaches

**J. Dubois, C. Aristégui, O. Poncelet, A. L. Shuvalov**

Université de Bordeaux, CNRS, UMR 5469, Laboratoire de Mécanique Physique, 351, Cours de la Libération, Talence, F-33405, France

E-mail: `j.dubois@lmp.u-bordeaux1.fr`

**Abstract.** Theoretical models underlying the ultrasonic study of suspensions and bubble swarms in liquids often rely on the concept of a coherent wave response, within which the given medium is viewed as an effective homogeneous medium. More specifically, the coherent response can be formulated in a straightforward manner via the effective wave number and the effective impedance. These in turn can be expressed through the actual material properties of the given scattering medium. The derivation of the effective wave number and impedance is the issue of a number of different approaches existing in the multiple-scattering theory.

The present work deals with a coherent response of acoustic wave impinging on a screen of cylindrical inclusions randomly distributed in a fluid. The reflection and transmission coefficients have been expressed via the effective wave number and impedance that are provided by three different models due to Foldy, Waterman & Truell, and Linton & Martin. The obtained expressions have been analyzed with a view to illuminate what is the quantitative difference between those approaches as revealed in the coherent response, and how this difference depends on the basic parameters of the problem such as the frequency, the concentration of scatterers and their contrast relatively to the fluid matrix. Another aspect of this work is to compare the above analytical results with the numerical data. It has been obtained by means of a deterministic computational code which delivers the coherent wave field through averaging the outputs for various samplings of the positions of scatterers. Knowing this numerical benchmark allows us to specify the validity domains for each of the three analytical methods under study.

## 1. Introduction

Propagation of waves in many types of inhomogeneous media, e.g. of seismic waves in the crust, ultrasonic waves in composites, waves in bubbly fluids or in suspended particles, is accompanied by multiple scattering. If the exact location of each scatterer is known, it is theoretically possible to solve the deterministic problem, but the solution becomes not reachable when the scatterers are too numerous. Moreover, in most of the cases, only a statistical description of the medium is accessible.

In his seminal paper [1], Foldy studied multiple scattering of scalar waves by a random distribution of isotropic point-like scatterers in a fluid matrix. By averaging the field over all possible configurations of the positions of scatterers, he determined the effective wave number $k_{\text{eff}}$ of the coherent wave in the form $k_{\text{eff}}^2 = k_0^2 - 4\,\mathrm{i}\,g n_0$, where $k_0$ is the wave number in the matrix, $g$ is the scattering coefficient for an individual scatterer and $n_0$ is the number of scatterers per unit volume. Extending the model to anisotropic point-like scatterers, Waterman & Truell [2] obtained the effective wave number with the quadratic term in $n_0$: $k_{\text{eff}}^2 = k_0^2 + \delta_1 n_0 + \delta_2 n_0^2$, where $\delta_{1,2}$ are computable quantities (see § 3.1). Taking into account the real size of anisotropic scatterers, Lloyd & Berry [3] derived a formula, which differs from that of [2] by the form of the coefficient $\delta_2$. Using the assumption of [3], Linton & Martin [4] provided a similar formula for $k_{\text{eff}}$ in the two-dimensional case. Other expressions for the effective wave number may be found e.g. in [5–7].

Experiments show that a coherent wave, which is created by a plane wave impinging on a random medium, is also plane. Therefore a random medium can be replaced by an equivalent homogeneous medium from the viewpoint of the coherent wave. Kuster & Toksöz [8] expressed the effective properties of such media for the long-wavelength limit by considering only simple scattering. Aristégui & Angel [9] developed a procedure for evaluating the mechanical properties of an effective medium in the framework of Waterman & Truell, while Luppé & Conoir [10] proceeded in a similar way but in the framework of Linton & Martin. The effective properties obtained by either of these methods are complex-valued and frequency-dependent.

The aim of this work is to compare the theoretical predictions of the homogenization techniques [9, 10], which are based on the Foldy (F), Waterman & Truell (WT) and Linton & Martin (LM) models of multiple scattering theory (MST), with our numerical simulations performed by means of a finite difference time domain (FDTD) code. For this purpose, a fluid matrix containing fluid cylindrical scatterers will be considered. The paper is organized as follows. First, the statement of the problem and the numerical scheme for simulating the coherent reflected and transmitted waves are presented. Second, we recapitulate the explicit results of [9, 10], which provide the theoretical predictions for the reflection and transmission coefficients on an infinite screen of scatterers. Then the numerical results are presented and compared with the above theoretical estimations at different concentrations and at different contrasts between the properties of the matrix and of the scatterers.

## 2. Statement of the problem

We consider a screen of fluid cylindrical scatterers, immersed in a nonviscous fluid matrix, see Fig. 1. The scatterers of radius $a$, which are parallel and identical to each other, are uniformly and randomly distributed with an area fraction $\phi$. The infinite screen of thickness $e$ is insonified by a normally incident plane wave.

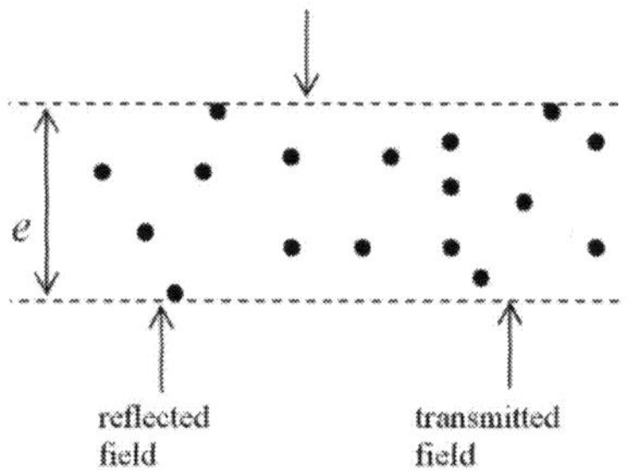


**Figure 1.** Screen of scatterers immersed in a matrix.

Multiple scattering gives rise to a reflected field and distorts the transmitted one. These signals consist of a coherent and incoherent part. The incoherent part vanishes due to averaging over a sufficiently large number of configurations of positions of the scatterers. In practice it is not easy to access the coherent-wave properties with only one sample of the random medium, for this requires to perform plenty of local measurements along the sample to restore the averaging procedure. Therefore, numerical simulations prove to be a good alternative.

Fig. 2 shows the geometrical parameters that were used in the FDTD scheme. The horizontal extent of the screen is restricted to a finite length with periodical boundary conditions imposed at each lateral side in order to ensure a plane front of the incident wave. This length is taken to be large enough with respect to the scatterer radius ($200a$) for making the effect of the periodical boundary conditions negligible. For each simulation, a new random configuration of the positions of scatterers is generated. The only condition imposed for any distribution is that at least one scatterer is tangent to each of the virtual interfaces of the screen. The reflection and transmission coefficients are defined by dividing the coherent reflected and transmitted spectra by the incident one. We have made between 90 and 210 simulations (depending on the parameters of the system) in order to achieve the convergence of the averaged field.

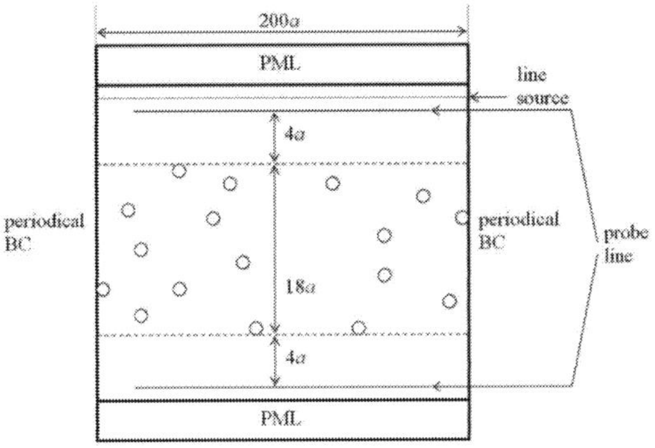


**Figure 2.** Geometry of the FDTD scheme (PML are perfectly matched layer).

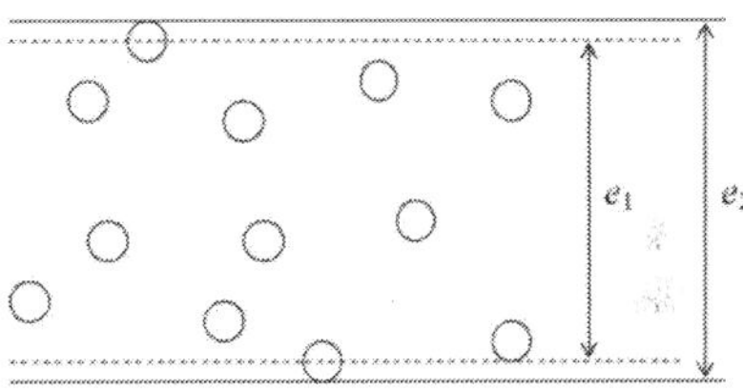


**Figure 3.** Definitions of the thickness $e_1$ for MST of the thickness $e_2$ for FDTD.

The thickness $e$ of the screen may admit two different values. According to the standard anzats [1] of the MST, the thickness $e_1$ is defined by the centers of the extreme scatterers, see Fig. 3. The corresponding concentration $\phi_1$ is given by $\phi_1 = \frac{N\pi a^2}{A_{e_1}}$, where $N$ is the number of scatterers inside the screen and $A_{e_1}$ is the area of the screen determined by the thickness $e_1$. On the other hand, actually the screen is limited by the tangents to the extreme scatterers, which define the thickness $e_2$ (Fig. 3). From this perspective, the concentration $\phi_2$ of scatterers inside the screen is $\phi_2 = \frac{N\pi a^2}{A_{e_2}}$, where $A_{e_2}$ is the area of the screen determined by the thickness $e_2$. For the simulations $e_2(e_2 = e_1 + 2a)$ is taken to be $18a$.

## 3. Theoretical background (the effective-medium models)

### *3.1. Multiple scattering theory: expressions of the effective wavenumber*

Under the assumption $\frac{n_0 Q_s}{k_0} \approx (k_0 a)^3 << 1$, where $Q_s$ is the scattering cross-section and $n_0$ is the number of scatterers per unit area, the coherent wave number provided by Waterman & Truell [2] is

$$\frac{k_{\text{eff}}^2}{k_0^2} = 1 - \frac{4\,\mathrm{i}\,n_0}{k_0^2} f\,(0) - \frac{4n_0^2}{k_0^4}\left[f^2\,(0) - f^2\,(\pi)\right], \tag{1}$$

where

$$f\,(\theta) = \sum_{n=0}^{\infty} \epsilon_n C_n \cos\,(n\theta) \tag{2}$$

is the angular shape function of a cylinder (in the three-dimensional case, the angular shape function of a sphere is used), $C_n$ is the scattering amplitude factors of the $n$th mode, and $\epsilon_n$ is the Neumann symbol equal to 1 if $n = 0$ and to 2 for $n > 0$.

For isotropic scatterers, $f \equiv f_0$ is constant, hence $f(\pi) = f(0)$, and so Eq. (1) simplifies to the form:

$$\frac{k_{\text{eff}}^2}{k_0^2} = 1 - \frac{4\,\mathrm{i}\,n_0}{k_0^2} f_0. \tag{3}$$

This expression is similar to the result of Foldy, which, however, takes into account only the first term of the modal series (2).

Fikioris & Waterman [7] introduced the "hole correction" to avoid overlapping of the scatterers by means of applying a pair correlation function in the calculation of the mean field. Using this, Linton & Martin [4] determined the low-frequency limit of the coherent wave number for the case of cylindrical scatterers as:

$$\frac{k_{\text{eff}}^2}{k_0^2} = 1 - \frac{4\,\mathrm{i}\,n_0}{k_0^2} f(0) - \frac{8n_0^2}{k_0^4} \sum_{n=-\infty}^{\infty} \sum_{m=-\infty}^{\infty} |m-n|\, C_n C_m. \tag{4}$$

Note that Eqs. (1), (3) and (4) are identical to the first order in $n_0$.

### *3.2. Homogenization techniques: expression of the effective impedance*

For the purpose of subsequent comparison with the numerical results, the reflection and transmission coefficients $R$ and $T$ shall be evaluated by way of the classical formulas for a homogeneous isotropic slab but with the effective mechanical properties, i.e.:

$$R = -\frac{Q\mathrm{e}^{-\mathrm{i}\,k_0 e}}{1 - Q^2\mathrm{e}^{2\,\mathrm{i}\,k_{\text{eff}} e}} \left(1 - \mathrm{e}^{2\,\mathrm{i}\,k_{\text{eff}} e}\right), \tag{5}$$

$$T = \frac{1 - Q^2}{1 - Q^2\mathrm{e}^{2\,\mathrm{i}\,k_{\text{eff}} e}} \mathrm{e}^{\mathrm{i}(k_{\text{eff}} - k_0)e}, \tag{6}$$

with

$$Q = \frac{Z_0 - Z_{\text{eff}}}{Z_0 + Z_{\text{eff}}},$$

where $Z$ is the impedance and the subscripts "0" and "eff" indicate the matrix and the effective medium, respectively.

By analyzing the reflection and transmission for a screen of scatterers immersed in a nonviscous fluid matrix (Fig. 1), Aristégui & Angel [9] identified the effective impedance as

$$\frac{Z_{\text{eff}}}{Z_0} = \frac{k_0}{k_{\text{eff}}} \left(1 - \frac{2\,\mathrm{i}\,n_0}{k_0^2} \left[f(0) - f(\pi)\right]\right). \tag{7}$$

For isotropic scatterers, Eq. (7) becomes

$$\frac{Z_{\text{eff}}}{Z_0} = \frac{k_0}{k_{\text{eff}}}, \tag{8}$$

which in view of the definition $Z = \rho\omega/k$ means that the effective mass density $\rho_{\text{eff}}$ is equal to the mass density of the matrix.

Luppé & Conoir [10] identified the effective impedance in the LM framework as:

$$\frac{Z_{\text{eff}}}{Z_0} = \frac{k_0}{k_{\text{eff}}} \left(1 - \frac{2\,\mathrm{i}\,n_0}{k_0^2} \left[f(0) - f(\pi)\right] + C\frac{4n_0^2}{k_0^4}\right), \tag{9}$$

where

$$C = \begin{bmatrix} \frac{1}{2}\left(f(0)^2 - f(\pi)^2\right) + \frac{f(\pi) - C_0 - f(0)}{f(0)} J_{(0)} + \frac{C_0}{f(0)} J_{(\pi)} \\ + C_0 \frac{f(0) - C_0}{f(0)} I_{(\pi)} - C_0 \frac{2f(\pi) - f(0) - C_0}{f(0)} I_{(0)} \end{bmatrix},$$

$$J_{(\alpha)} = \sum_{n=-\infty}^{\infty} \sum_{m=-\infty}^{\infty} |m-n| \, C_n C_m e^{\mathrm{i}n\alpha},$$

$$I_{(\alpha)} = \sum_{n=-\infty}^{\infty} |n| \, C_n e^{\mathrm{i}n\alpha}.$$

## 4. Numerical results and their comparison with the theory

The following simulations utilize water scatterers and one of the two other fluids as the matrix (Table 1). The ratios between the impedance of these two fluids and of the water are relatively close to each other, while the density and velocity contrasts are weak for the oil/water case, and strong for the FC-72 (perfluorohexane)/water case.

**Table 1.** Density $\rho$ and velocity $v$ of the fluids

| | | |
|---|---|---|
| water | $\rho = 1000\ kg.m^{-3}$ | $v = 1500\ m.s^{-1}$ |
| oil | $\rho = 800\ kg.m^{-3}$ | $v = 1200\ m.s^{-1}$ |
| FC-72 | $\rho = 1680\ kg.m^{-3}$ | $v = 512\ m.s^{-1}$ |

### *4.1. Reflection and transmission coefficients*

On calculating the reflection and transmission coefficients, we have taken into account a possibility of two different definitions of the screen thickness $e_1$ and $e_2$ (Fig. 3) and hence of the concentration of scatterers. It was observed that choosing one or another definition has little effect on the transmission coefficient but affects noticeably the reflection, especially near its resonances. This is highlighted on the low-frequency zooms (Figs. 5, 7, 11), while all other diagrams are referred to the thickness $e_1$.

#### *4.1.1. Oil matrix and water scatterers*

Figs. 4-7 present modulus of reflection coefficient $|R|$ versus adimensional frequency $k_0a$ that has been numerically calculated for two different concentrations $\phi_1 = 4.42\%$ and 34.4%. These data are compared to the theoretical predictions obtained by [9] in the F and WT frameworks and by [10] in the LM framework. For the low contrast case in hand, the results of [9] in the WT framework and of [10] are too close to be discerned at the scale of the figures.

It is seen that the theoretical predictions of the reflection coefficient obtained by [9] in the F framework do not fit with the numerical data for both concentrations. At low concentration ($\phi_1 = 4.42\%$), the other predictions are in a qualitatively good agreement with the numerical reflection coefficient up to $k_0a = 3$, although they underestimate it. Fig. 5 shows the impact of the optional definitions $e_1$ or $e_2$ of the screen thickness. It is seen that the numerical data lie in between the predictions assuming the two different thickness values. For the concentration $\phi_1 = 13.2\%$ (not shown here), the results are similar, except that it is only the first lobe that

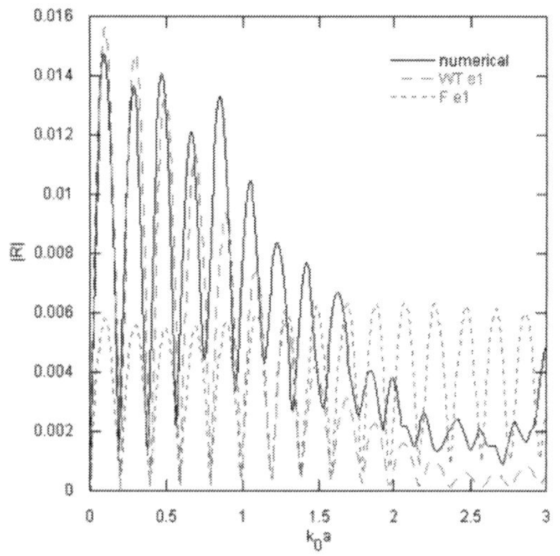


**Figure 4.** Reflection coefficient for oil matrix and water scatterers, $\phi_1 = 4.42\%$.

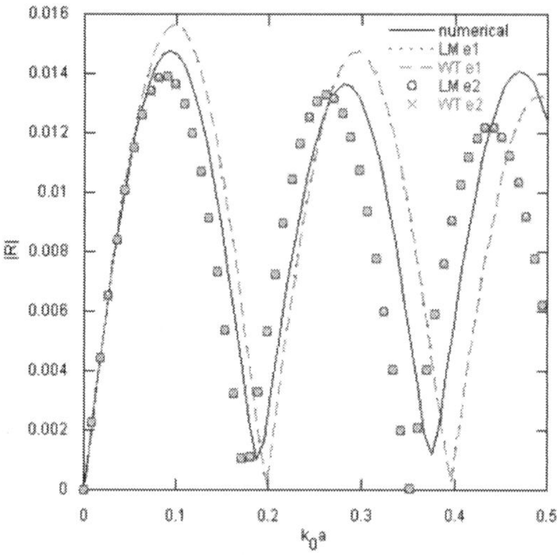


**Figure 5.** Low frequency zoom of Fig. 4.

lies in between the predictions assuming $e_1$ and $e_2$. For the high concentration $\phi_1 = 34.4\%$ (Figs. 6, 7), the theories still have good agreement at low frequencies, and the first lobe remains in between the predictions for $e_1$ and $e_2$. However, the theoretical reflection coefficient decreases with the frequency, whereas its numerical value first increases and after decreases.

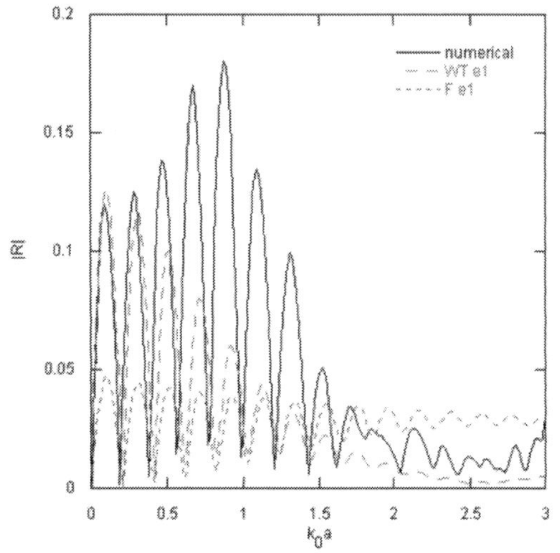


**Figure 6.** Reflection coefficient for oil matrix and water scatterers, $\phi_1 = 34.4\%$.

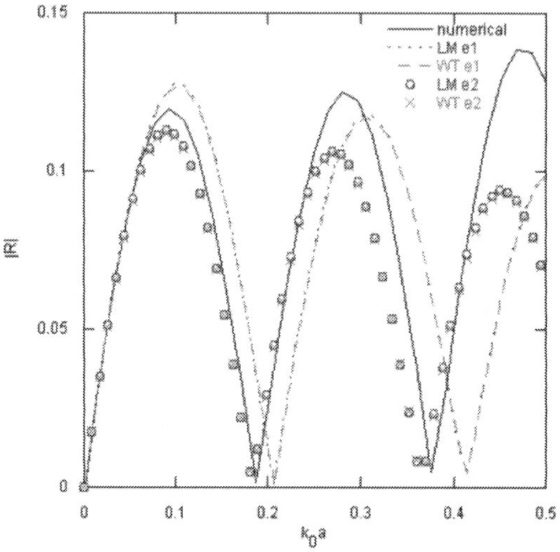


**Figure 7.** Low-frequency zoom of Fig. 6.

The numerical results for the transmission-coefficient modulus $|T|$ and their comparison with the theoretical predictions of [9, 10] are shown in Figs. 8,9. It is seen that the results of [9, 10] provide a similar evaluation of the transmission at low contrast. The relative error is less than

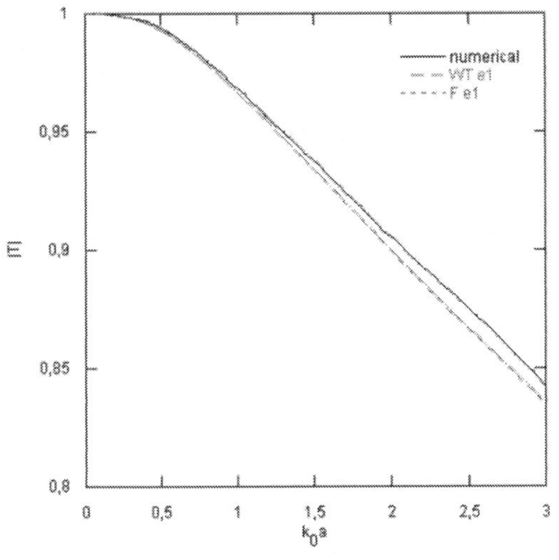


**Figure 8.** Transmission coefficient for oil matrix and water scatterers, $\phi_1 = 4.42\%$.

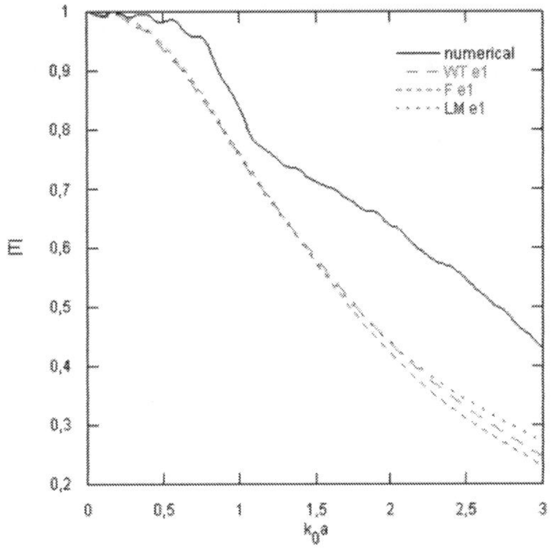


**Figure 9.** Transmission coefficient for oil matrix and water scatterers, $\phi_1 = 34.4\%$.

1% for $\phi_1 = 4.42\%$ and increases to 7% for $\phi_1 = 13.2\%$. For the high concentration $\phi_1 = 34.4\%$, the error reaches 45% at $k_0a = 3$ but does not exceed 5% for $ka \leq 0.5$.

We have also performed simulations for the case of oil scatterers in a water matrix which yields similar results except that the relative errors are slightly higher.

#### *4.1.2. FC-72 matrix and water scatterers*

This case is characterized by stronger density and velocity contrasts between the two phases than in the previous setting. For the reflection coefficient, this leads to a quantitatively greater difference between [9] in the WT framework and [10], which is most noticeable at low frequencies and high concentration, see Figs. 10, 11. This difference increases as concentration grows: it is equal to 1.8% and 8% for $\phi_1 = 4.42\%$, and 31.4% respectively. It is seen that the numerical data are closer on the first lobe to the theoretical prediction given by [10] than to the other predictions. The theoretical prediction of [9] in the F framework is found to be no longer valid for the high-contrast case: even for the low concentration $\phi_1 = 4.42\%$, it is five times greater than the numerical data.

Contrary to the low-contrast case of § 4.1.1, the three theories predict slightly different values for the transmission coefficient, which though are still inferior to the numerical data (Fig. 12). The difference between the three theories becomes noticeable only for the high concentration. The relative error is greater than for the low-contrast case: it is 2.3%, 18% and 76% (12% if $k_0a$ <0.5) for $\phi_1 = 4.42\%$, 13.2% and 31.4% respectively.

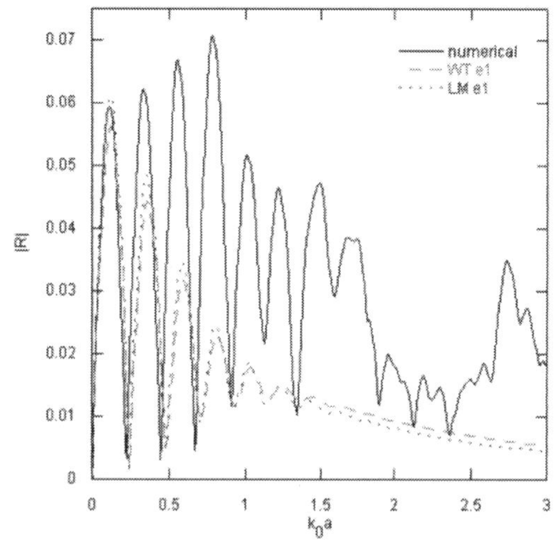


**Figure 10.** Reflection coefficient for FC72 matrix and water scatterers, $\phi_1 = 31.4\%$.

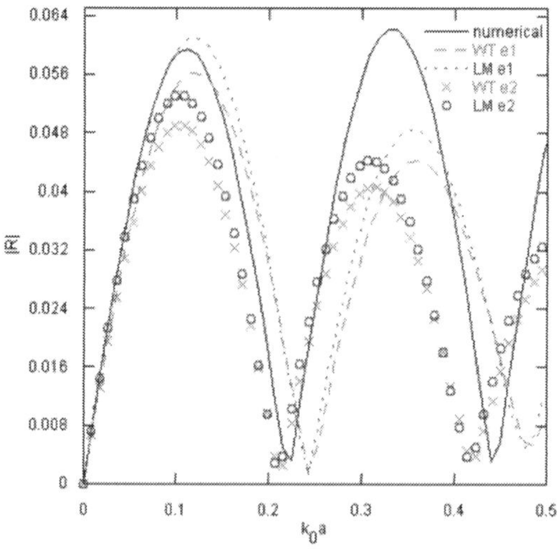


**Figure 11.** Low-frequency zoom of Fig. 10.

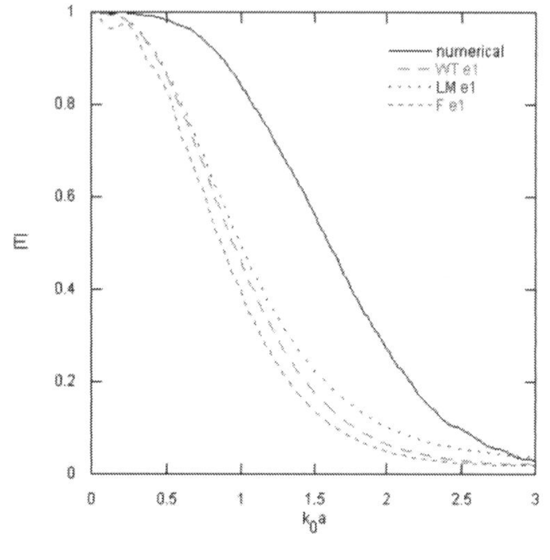


**Figure 12.** Transmission coefficient for FC72 matrix and water scatterers, $\phi_1 = 31.4\%$.

*4.2. The effective thickness*

As remarked in § 2, the definition of the screen thickness is ambiguous. The thickness-resonances calculated for the two different thickness values $e_1$ and $e_2$ do not coincide with the numerical ones, see Figs. 5, 7. Let us introduce in Eq. (5) the effective thickness $e_{\text{eff}}$, for which the first theoretical and numerical resonance fit together. The concentration is adapted to this thickness similarly to the case of $e_1$ and $e_2$ explained in § 2.

The so-defined effective thicknesses $e_{\text{eff}}$ is equal to $16.82a$, $16.64a$ and $17.54a$ for the concentrations $\phi_1 = 4.42\%$, 13.2%, 31.4%, respectively. Even though the resonances are actually not periodic in $k_0a$ due to a dispersive $k_{\text{eff}}$, it is observed that the effective thickness enables the coincidence of theoretical and numerical resonances up to $k_0a = 1.5$ and 0.5 for $\phi_1 = 4.42\%$ and 31.4%, respectively, see Figs. 13, 14. This observation confirms that, within these frequency ranges, the effective velocity $v_{\text{eff}} = \omega/\Re(k_{\text{eff}})$ is well defined (here $\Re$ means real part).

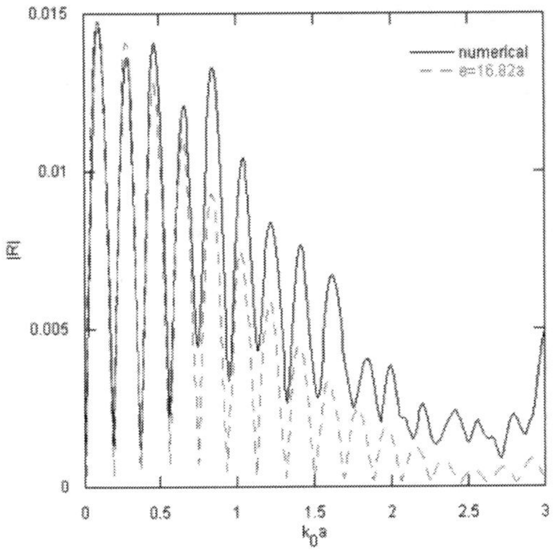


**Figure 13.** Reflection coefficient for oil matrix and water scatterers, $\phi_1 = 4.42\%$.

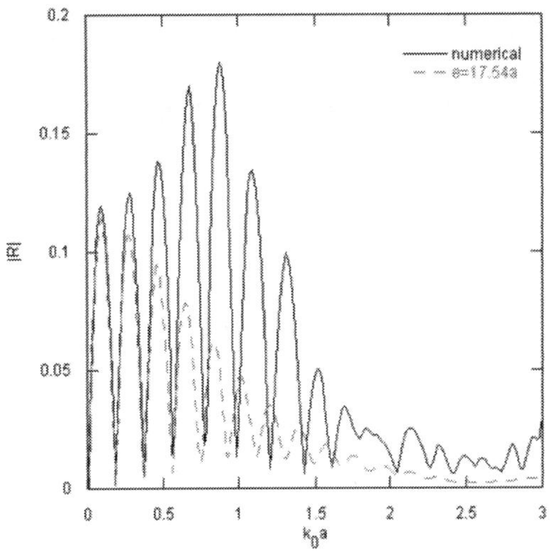


**Figure 14.** Reflection coefficient for oil matrix and water scatterers, $\phi_1 = 34.4\%$.

## 5. Conclusion

The coherent wave transmitted through a random medium is numerically reachable with not so many simulations (less than thirty). On the other hand, access to the coherent reflected wave is more effortful and time-consuming, because one needs between one and two hundreds of simulations to achieve the convergence to the average.

The performed numerical simulations and their comparison with the theoretical results of [9, 10], obtained within three different theoretical frameworks, yield the following observations. The transmission coefficient predicted by the three theories differs only for high contrast and high concentration. Thus, if only the transmission is dealt with, the results of [9] in the Foldy framework, which is the simplest model, are well-suited for the purpose. At the same time, this model is too "simple" in that it leads to the effective density being equal to the mass density of the matrix and, as a result, does not agree with the numerical data for the reflection coefficient.

It is seen that for low contrast between the matrix and the scatterers, the predictions obtained by [9] in the framework of Waterman & Truell and by [10] are similar. For high contrast, a noticeable difference appears for high concentrations of scatterers. In this case the model of [10] is closer to the numerical data for the reflection coefficient than the model of [9].

The overall conclusion is that the homogenization techniques in the context of multiple scattering theory remain valid at low frequencies even for high concentrations.

## Acknowledgment

We are grateful to F. Luppé and J.-M. Conoir for making the result of their paper [10] available to us. This work has been supported by the grant ANR-08-BLAN-0101-01 from the ANR (Agence Nationale de la Recherche) and also the project SAMM (Self-Assembled MetaMaterials) supported by the cluster AMA (Advanced Materials in Aquitaine).

## References

[1] Foldy L L 1945 *Phys. Rev.* **67** 107–19

[2] Waterman P C and Truell R 1961 *J. Math. Phys.* **2** 512–37

[3] Lloyd P and Berry M V 1967 *Proc. Phys. Soc.* **91** 678–88

Journal of Physics: Conference Series **269** (2011) 012004 doi:10.1088/1742-6596/269/1/012004

[4] Linton C M and Martin P A 2005 *J. Acoust. Soc. Am.* **117** 3413–23
[5] Lax M 1951 *Rev. Mod. Phys.* **23** 287–310
[6] Twersky V 1952 *J. Acoust. Soc. Am.* **24** 42–6
[7] Fikioris J G and Waterman P C 1964 *J. Math. Phys.* **10** 1413–20
[8] Kuster G T and Toksöz M N 1974 *Geophysics* **39** 587–606
[9] Aristégui C and Angel Y C 2007 *Wave Motion* **44** 153–64
[10] Luppé F and Conoir J M 2010 *J. Phys.: Conf. Ser.* (this issue)

Journal of Physics: Conference Series **269** (2011) 012005 doi:10.1088/1742-6596/269/1/012005

# Experimental determination of plate parameters with an air coupled instrument

**Abdelhak El Mouhtadi, Jean Duclos and Hugues Duflo**

Laboratoire Ondes et Milieux Complexes, FRE CNRS 3102, Groupe Ondes Acoustiques, Université du Havre, Place R. Schuman, 76610, Le Havre, France

E-mail: `abdelhak.ilel@gmail.com`

**Abstract.** The present work deals with an experimental determination of acoustical properties of a viscoelastic homogeneous plate using an air coupled equipment. Usually, these properties are determined separately through several measurements, while our paper provides a quick inverse method which permits to measure them simultaneously and requires only the knowledge of the sound velocity in air and its density. The experimental transmission through a face parallel plate, at normal incidence, is investigated and compared with the theoretically predicted one. Measurements are conducted using a chirp signal having 1.2 $MHz$ central frequency with 800 $kHz$ bandwidth and 100 $\mu s$ duration. Two signals are being exploited : one only through air path and another one through the sample inserted between the transducers. The FFTs of these signals provide the complex transmission coefficient as a function of frequency. The identification of parameters is carried out using the real and imaginary parts of the transmission coefficient. Then, assuming the wave velocity and density of air are known, we deduce the four properties of the plate (thickness, density, longitudinal velocity and attenuation coefficient) as well as the attenuation coefficient of air. A variety of viscoelastic materials, whose impedances have weak values, has been studied such as Polyethylene, Plexiglas and carbon/epoxy composite. The physical properties for each plate are obtained with a good accuracy in the frequency range of investigation.

## 1. Introduction

For a long time, traditional ultrasonic non-destructive inspection has widely proved itself in the investigation and characterization of materials. Most of these inspection methods are based on the generation and detection of sound waves that propagate through a test sample which is placed between the source and detector of ultrasound. In most cases, a liquid coupling medium, such as water or gel is required to ensure the good transfer of acoustical energy between the transducers and the test material.

The most efficient and popular coupling method is the immersion one in which the test sample is being completely immersed in a water liquid [1] which has an acoustic impedance of the same order of magnitude as the sample. However, there are many cases for which the use of water may not always be suitable for instance in presence of porous and composite materials that could be easily damaged or polluted through absorption of the liquid.

In another method using contact transducers, a physical contact with the specimen is required and the inspection is done by applying a gel to the surface of the structure. This method has the inconvenience of maintaining the contact pressure between the transducer and the test

Journal of Physics: Conference Series **269** (2011) 012005 doi:10.1088/1742-6596/269/1/012005

piece and sometimes, the use of this type of inspection becomes impossible for example at high temperatures or if the test piece is inaccessible.

Among various techniques proposed to overcome such problems, techniques that employ air-coupled[2] ultrasonic inspections appear most promising, largely due to their ability to perform measurements directly in atmospheric air. However, air-coupled systems have also many disadvantages compared with liquid-coupled systems: First, because of the large difference in acoustic impedance between air and most solid materials, the ultrasound is almost entirely reflected at the air/sample boundary. Second, the amplitude of the transmitted wave is extremely reduced due to the high absorption of ultrasound in air, especially at high frequencies. Finally, additional difficulties arise from the properties of air which are found to be considerably dependent on temperature, pressure and humidity environment.

Fortunately, to overcome these drawbacks, our equipment enables the generation of ultrasonic waves with high levels and allows their detection with sufficient sensitivity in order to improve the transfer of acoustic energy in air. Moreover, all measurements are made in a constant temperature, pressure and humidity environment.

This paper presents a non-contact ultrasonic inspection method for characterizing viscoelastic, homogeneous materials using piezoelectric air-coupled transducers. This method consists in studying the ultrasonic, plane wave transmission through an isotropic face parallel plate, in the same way as in the immersion method, but replacing the water medium with air. The aim is to determine the velocity and attenuation of ultrasonic waves in the material and the density and thickness of the sample, using a non-linear least-squares algorithm which minimizes the difference between experimental and calculated transmission coefficient data.

Kinra and Iyer[3] have developed a new technique for ultrasonic non-destructive evaluation of the acoustic properties of a thin plate through normal transmission coefficient measurement. The analysis was performed for the determination of any one of the four acoustic quantities characterizing the plate, assuming the other three are known by using the spectrum of normally transmitted ultrasonic signal and without any prior knowledge of the ambient medium properties.

In this work, we describe a novel inverse method for simultaneous determination of all properties of the plate: density, thickness, and longitudinal velocity and attenuation using two ultrasonic measurements at normal incidence: with and without the sample. This method requires only the knowledge of the sound velocity in air and its density. In addition, A preliminary analysis of the effect of variations in parameters on the model has shown that the sensitivity of identification to parameters is higher for the real and imaginary parts than for the modulus. Therefore, in the algorithm used for the identification, we use the complex (real and imaginary parts) form of the transmission coefficient instead of using its modulus which is inadequate.

## 2. Theory

Ultrasonic through transmission measurements are frequently used for the characterization of the acoustical properties of a thin plate. The purpose of this section is to review briefly the theoretical plane wave transmission coefficient for a homogeneous, isotropic plate.

Figure 1 shows schematically an incident longitudinal plane wave $p_{inc}$ at normal incidence, propagates in the $x$ direction through a parallel face plate of thickness $E$ and density $\rho_2$, placed in air whose density is $\rho_1$. The reflected and transmitted waves are denoted as $p_r$ and $p_t$, respectively. The diagram shows also the waves propagating downwards $p_-$ and upwards $p_+$ inside the plate.

To describe the dispersion and attenuation behavior of waves one can define a complex longitudinal propagation constant $\gamma_j(\omega)$, where the real part is an attenuative part related to the frequency dependent attenuation, $\alpha_j(\omega)$, while the imaginary part is a dispersive part related to the frequency dependent phase velocity, $V_j(\omega)$. These relations can be written as

$$\gamma_j(\omega) = \alpha_j(\omega) + i\frac{\omega}{V_j(\omega)} \quad (j = 1 \text{ } or \text{ } 2) \tag{1}$$

where $w$ is the circular frequency of the incident wave and the subscripts 1 and 2 refer to air and the plate, respectively. The attenuations of longitudinal waves in air and the plate are assumed to be frequency dependent as follows[4]

$$\alpha_1(f) = \mu_1 f^2 \tag{2}$$

$$\alpha_2(f) = \mu_2 f \tag{3}$$

where $\mu_1$ and $\mu_2$ are the longitudinal attenuation coefficients in air and the plate, respectively. $f = \frac{\omega}{2\pi}$ is the frequency.

For a viscoelastic homogeneous plate, the harmonic plane wave transmission coefficient, at normal incidence, can be expressed in complex form as

$$T_{th}(f) = \frac{2e^{\gamma_1 E}}{\left(\frac{Z_1}{Z_2} + \frac{Z_2}{Z_1}\right)\sinh(\gamma_2 E) + 2\cosh(\gamma_2 E)} \tag{4}$$

where

$$\gamma_1 E = \frac{Ef}{V_1}(\mu_1 V_1 f + i2\pi) \tag{5}$$

$$\gamma_2 E = \frac{Ef}{V_2}(\mu_2 V_2 + i2\pi) \tag{6}$$

$$\frac{Z_1}{Z_2} = \frac{\rho_1 V_1}{\rho_2 V_2} \times \frac{\mu_2 V_2 + i2\pi}{\mu_1 V_1 f + i2\pi} \tag{7}$$

$Z_1$ and $Z_2$ are the complex acoustic impedances of air and the plate and are given by

$$Z_1 = \frac{i\rho_1\omega}{\gamma_1} \tag{8}$$

$$Z_2 = \frac{i\rho_2\omega}{\gamma_2} \tag{9}$$

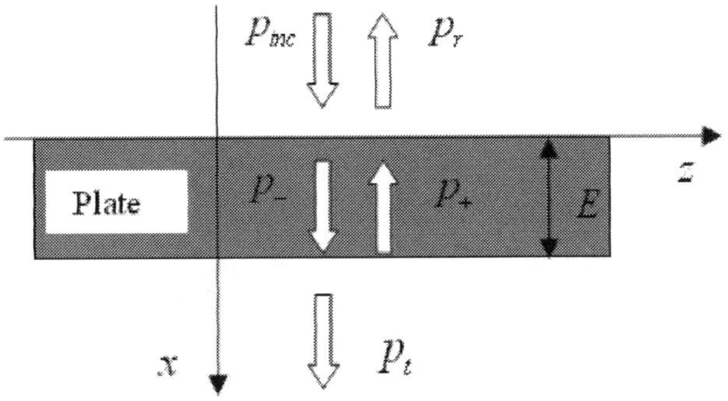


**Figure 1.** A schematic diagram of a sound wave $p_{inc}$ normally incident on a plate of thickness $E$. This wave propagates in the $x$ direction and gives rise to reflected and transmitted waves denoted as $p_r$ and $p_t$, respectively.

In addition, it is assumed that air behaves in a non dispersive manner. This means that the phase velocity $V_1$ is a constant. It depends only on temperature, humidity and atmospheric pressure. Whereas, the dispersion in the plate is taken into account by using the following dispersion relationship[5,6] linking the phase velocity and attenuation of the plate

$$\frac{1}{V_2(f)} = \frac{1}{V_0} + \frac{\mu_2}{\pi^2} ln\left(\frac{f_0}{f}\right) \tag{10}$$

where $V_0$ is the phase velocity at a reference frequency $f_0$.

## 3. Experiment

In this work, only normally incident waves are used for the characterization of the acoustical properties of a thin plate. A simplified drawing of the experimental set-up is shown in figure 2. Two identical ultrasonic air-coupled transducers (Ultran) are placed to face each other in air and are separated by a distance of $L = 8\ cm$. In order to perform normal through transmission measurements, the sample is mounted between the transducers so that it is perpendicular to the incident wave as shown in figure 2*b*. The transducers are piezoelectric devices with a nominal central frequency of 1 $MHz$ and an active diameter of 25 $mm$. The sample is a parallelepipedal plate of nominal thickness $E$, made up of a homogeneous, isotropic viscoelastic material. In the present work, three samples are tested. The first is a Polyethylene plate of nominal thickness 6 $mm$, the second is a 3-$mm$-thick Plexiglas plate and also tested is a carbon fiber epoxy sample of thickness 2.35 $mm$.

The transmission coefficient can be obtained by measuring the through transmission signals without and with the sample using the same experimental set-up. Therefore, two successive independent measurements are carried out. In the first, without the sample, the chirp waveform produced by the transmitter propagates through air and is directly acquired by the receiver. This received signal is called the reference signal $f_r(t)$. Next, with the specimen placed between the transducers, the ultrasonic signal transmitted through the plate is acquired as well and recorded as the sample signal $f_s(t)$.

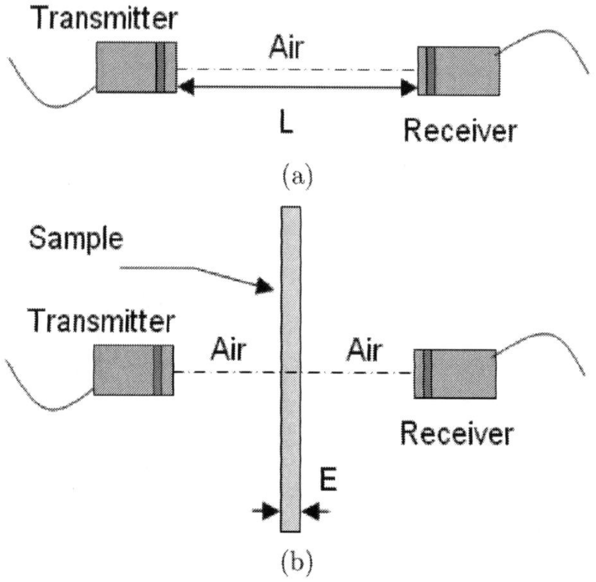


**Figure 2.** Measurement configuration: (a) through air path; (b) through the sample.

For each measurement, the transmitting transducer is excited by a chirp signal having 1.2 $MHz$ central frequency with 800 $kHz$ bandwidth and 100 $\mu s$ duration, produced by a transmitting/receiving generator (SIA 7-VN Instruments). As the transmitted signal is extremely weak, the receiving transducer is connected to an amplifier in order to increase the signal level. The amplified signal is then digitized using a digital oscilloscope (Accura 100

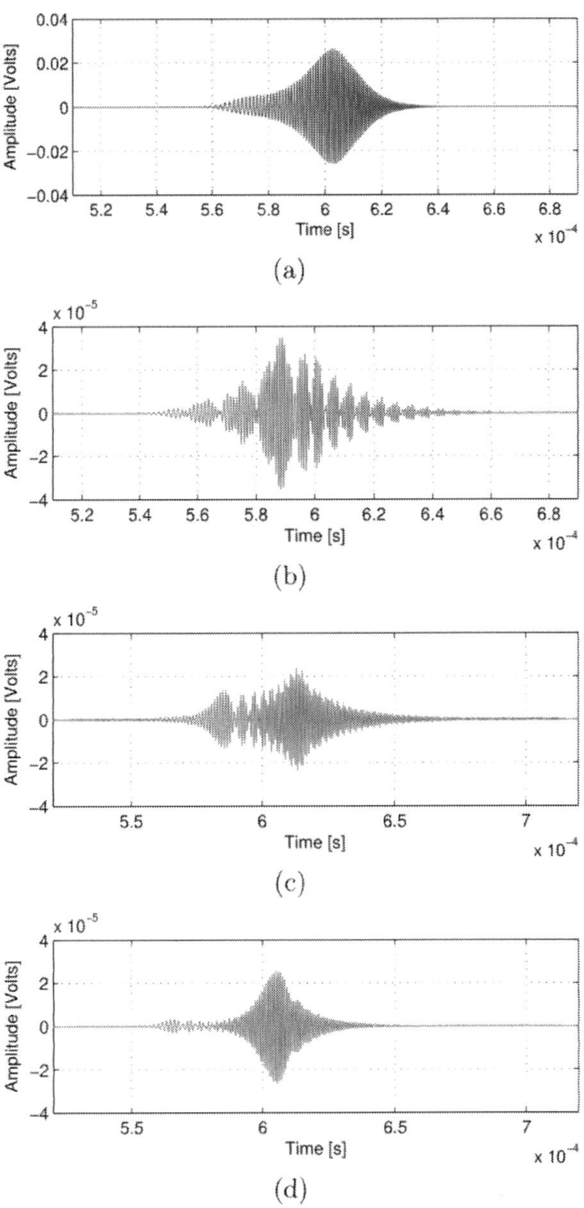


**Figure 3.** Time domain signals using 1 $MHz$ air-coupled transducers: (a) reference signal; signals transmitted through a (b) 6-$mm$-thick Polyethylene; (c) 3-$mm$-thick Plexiglas; (d) 2.35-$mm$-thick composite plate.

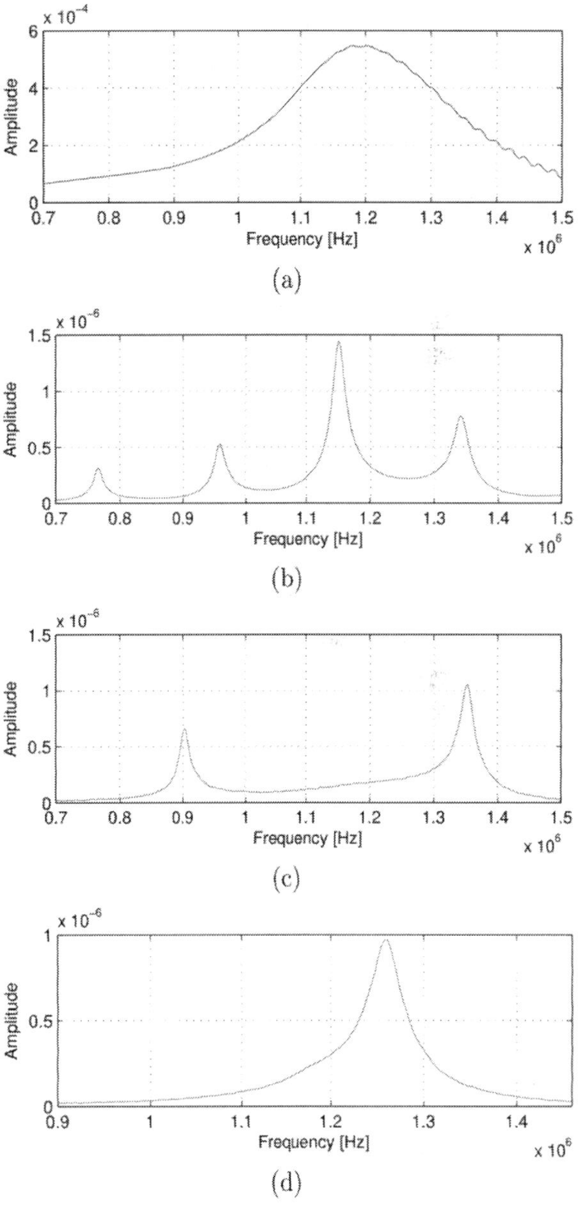


**Figure 4.** (a) FFT modulus of the reference signal; FFT modulus of signals transmitted through a (b) 6-$mm$-thick Polyethylene; (c) 3-$mm$-thick Plexiglas; (d) 2.35-$mm$-thick composite plate.

Journal of Physics: Conference Series **269** (2011) 012005 doi:10.1088/1742-6596/269/1/012005

Nicolet-12 bits), and averaged the required number of times to improve the signal-to-noise ratio (almost 30 $dB$ of gain) before being transferred to a computer for further processing. Note that, compared to the reference signal, the sample signal level drops considerably, with a lower signal-to-noise ratio that requires a higher amplification and more signal averaging to obtain a useful signal. Frequency spectral analysis of these signals using Fast Fourier transform (FFT) algorithms then allows the calculation of the transmission coefficient as a function of frequency. Indeed, if the FFTs of $f_r(t)$ and $f_s(t)$ are denoted as $F_r(f)$ and $F_s(f)$, respectively, the measured transmission coefficient is given by

$$T_{exp}(f) = \frac{F_s(f)}{F_r(f)} \tag{11}$$

Corresponding to the reference signal shown in figure 3$a$, the transmitted waves through three plates made of Polyethylene, Plexiglas and composite material with nominal thickness of 6, 3 and 2.35 $mm$ are shown, respectively, in figures 3$b$, $c$, $d$. The corresponding frequency spectra (modulus) are shown in figures 4$a$, $b$, $c$ and $d$.

As the velocity of sound in air depends significantly on temperature, all measurements mentioned above were made in a constant temperature environment in order to avoid the sound velocity fluctuations in air caused by small ambient temperature changes. Consequently, the temperature as well as humidity and pressure of air have been controlled and continuously recorded, during experiments, to assure the stability of the sound velocity in air.

## 4. Identification

In principle, the objective is to determine the four unknown real quantities characterizing the plate (thickness $E$, mass density $\rho_2$, longitudinal phase velocity $V_0$ and attenuation coefficient $\mu_0$). Substitution of Eq. (10) into Eqs. (6) and (7) yields

$$\gamma_2 E = \frac{E}{V_0} f \left\{ \mu_2 V_0 \left[ 1 + \frac{2i}{\pi} \ln\left(\frac{f_0}{f}\right) \right] + i2\pi \right\} \tag{12}$$

$$\frac{Z_1}{Z_2} = \frac{\rho_1 V_1}{\rho_2 V_0} \times \frac{\mu_2 V_0 \left[1 + \frac{2i}{\pi}\ln\left(\frac{f_0}{f}\right)\right] + i2\pi}{\mu_1 V_1 f + i2\pi} \tag{13}$$

Then by introducing Eqs. (5), (12) and (13) in Eq. (4), it can be observed that seven physical quantities appear,

$$E,\ V_0,\ \mu_2,\ \rho_2,\ V_1,\ \mu_1,\ \rho_1. \tag{14}$$

Since the wave velocity in air $V_1$ and its density $\rho_1$ will be considered as known, the number of physical properties to be found can be reduced and only five combinations of them denoted as

$$a_1 = \frac{E}{V_1},\ a_2 = \mu_1 V_1,\ a_3 = \frac{\rho_1 V_1}{\rho_2 V_0},\ a_4 = \frac{E}{V_0},\ a_5 = \mu_2 V_0, \tag{15}$$

are used for the optimization as shown in the following equations

$$\gamma_1 E = a_1 f \left(a_2 f + i2\pi\right) \tag{16}$$

$$\gamma_2 E = a_4 f \left\{ a_5 \left[ 1 + \frac{2i}{\pi} \ln\left(\frac{f_0}{f}\right) \right] + i2\pi \right\} \tag{17}$$

$$\frac{Z_1}{Z_2} = a_3 \times \frac{a_5\left[1 + \frac{2i}{\pi}\ln\left(\frac{f_0}{f}\right)\right] + i2\pi}{a_2 f + i2\pi} \tag{18}$$

where the constants $a_m$ ($m = 1$ to 5) linking the different acoustic quantities of both the plate and air, are the parameters which are to be identified using Eq. (4). The five unknown physical quantities of the set (14) are defined from the set of parameters (15) by

$$V_0 = \frac{a_1 V_1}{a_4},\ \rho_2 = \frac{\rho_1 a_4}{a_1 a_3},\ \mu_2 = \frac{a_4 a_5}{a_1 V_1},\ E = a_1 V_1,\ \mu_1 = \frac{a_2}{V_1} \tag{19}$$

One can observe that the precision and accuracy of the plate properties are related to those of the wave velocity in air; hence the interest in measuring it with a good precision.

To determine the acoustical properties of the plate, it is necessary to know two of the three physical quantities characterizing air. Therefore, an additional measurement of wave velocity in air is performed by moving the receiver through known distances from the transmitter and measuring the time of flight. Concerning density we refer to tables giving the density of air as a function of temperature.

### *4.1. Effect of parameters on transmission coefficient*

For the identification problems, the sensitivity analysis can be an essential help to know the behavior of the model and to predict the effect of each parameter on this model in order to find the optimal parameters values. This effect is investigated by determining the relative variation of the complex transmission coefficient due to a few per cent change in each parameter using a relative error function

$$\eta = \sqrt{\frac{\sum_{n=1}^{N} \left[RT_1(f_n) - RT_0(f_n)\right]^2 + \sum_{n=1}^{N} \left[IT_1(f_n) - IT_0(f_n)\right]^2}{\sum_{n=1}^{N} \left[RT_0(f_n)^2 + IT_0(f_n)^2\right]}}, \tag{20}$$

where $N$ is the total number of frequency data points taken for comparison. $RT_0(f_n)$ and $IT_0(f_n)$ are, respectively, the real and imaginary parts of the theoretical transmission coefficient for the frequency $f_n$, calculated with standard parameters. $RT_1(f_n)$ and $IT_1(f_n)$ are, respectively, the real and imaginary parts of the transmission coefficient when the value of a parameter is changed.

In the case of the 6-mm-thick polyethylene plate, this study has demonstrated that the variations for the complex theoretical transmission coefficient due to a 1% change in each of parameters $a_1$, $a_2$, $a_3$, $a_4$ and $a_5$ are 108%, 0.16%, 1%, 72% and 0.69%, respectively. Thus, one can observe that the transmission coefficient is strongly dependent on parameters $a_1$ and $a_4$, while it is weakly dependent on parameters $a_2$, $a_3$ and $a_5$.

### *4.2. Parameters exploration*

For the inverse algorithm, we begin by comparing the experimental $T_{exp}$ and calculated $T_{th}$ transmission coefficients. By sweeping all parameters one by one, the relative error between $T_{exp}$ and $T_{th}$ is investigated using Eq. (20) in which $T_0$ and $T_1$ are replaced by $T_{exp}$ and $T_{th}$, respectively. The goal is to obtain approximate values of the initial guesses with upper and lower bounds.

For parameters $a_2$, $a_3$ and $a_5$, the shape of the error function presents only a minimum while it contains many minima for the case of parameters $a_1$ and $a_4$. Therefore, the study of the location of these minima permits to distinguish between the true minimum and the false minima so that they can be easily avoided during the inverse process.

### *4.3. Inverse algorithm*

The parameters are identified by solving an inverse identification problem which consists in optimizing the parameters values such that the theoretical transmission formula becomes as close as possible to the experimental data in the useful frequency domain. This numerical process is done by minimizing the following objective function

$$g(a_m) = \sum_{n=1}^{N} \left[ \left[RT_{th}(f_n, a_m) - RT_{exp}(f_n)\right]^2 + \left[IT_{th}(f_n, a_m) - IT_{exp}(f_n)\right]^2 \right] \qquad (21)$$

$RT_{th}(f_n, a_m)$ and $IT_{th}(f_n, a_m)$ are, respectively, the real and imaginary parts of the theoretical transmission coefficient, computed with parameters $a_m$ representing the material properties which are to be optimized. $RT_{exp}(f_n)$ and $IT_{exp}(f_n)$ are, respectively, the real and imaginary parts of the experimental transmission coefficient measured for frequency $f_n$. The optimization of parameters $a_m$ is conducted by using the Matlab® routine *lsqnonlin* which consists in solving non-linear least squares problems.

The inverse algorithm is supplied with an initial guess for which lower and upper bounds are specified for all the parameters. The tolerances in parameters and function values are set at $1.0 \times 10^{-6}$ for all estimations. The parameters are estimated for $N = 11700$ frequency points, which are included in the bandwidth of the transducers.

Because the transmission coefficient is weakly dependent on parameters $a_2$, $a_3$ and $a_5$, the simultaneous research for all parameters in a five-dimensional space is not easy for finding the optimal solution.

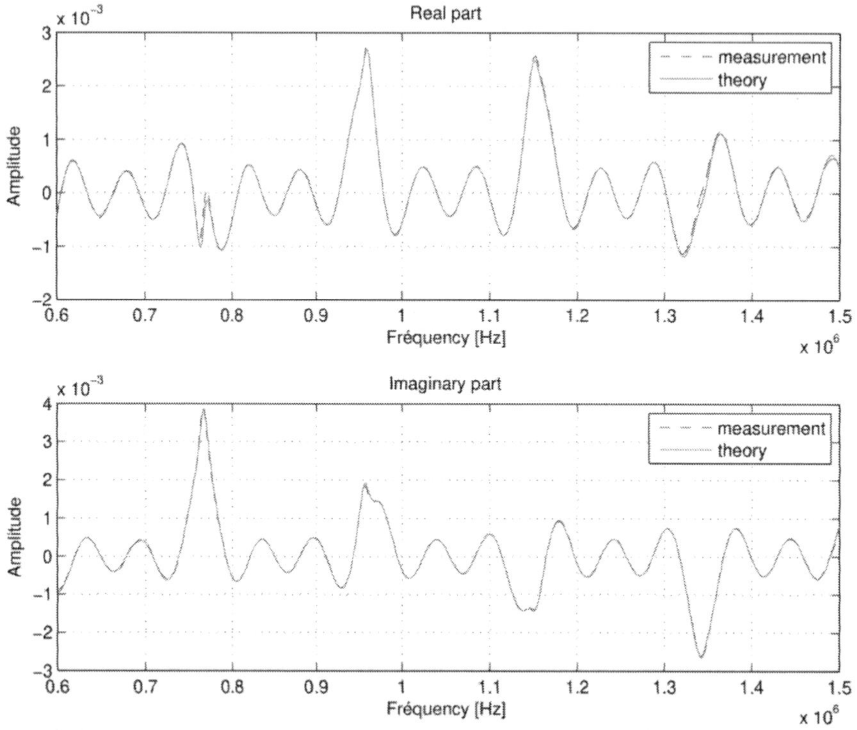


**Figure 5.** Complex transmission (real and imaginary parts) of a plate of polyethylene of nominal thickness 6 *mm*: blue dotted line represents experiments and red continuous line represents the modelling transmission computed with optimized parameters values.

Consequently, a strategy is defined for optimizing parameters. First of all, we determine two parameters $a_1$ and $a_4$ considering $a_2$, $a_3$ and $a_5$ are known. Then, for the optimal values obtained, the three other parameters are optimized together by running the algorithm again. This procedure is repeated until the minimum possible between the calculated and experimental transmission coefficient is found. This gives good estimates of parameters.

Figure 5 shows an example of identification for a 6-mm-thick polyethylene plate and table 1 gives the optimized acoustical properties values for the three tested samples assuming the wave velocity in air and its density are known. The standard deviation between the measured and theoretical curves is estimated at $4 \times 10^{-5}$.

| Sample | $V_1(m/s)$ | $\rho_1(kg/m^3)$ | $V_0(m/s)$ | $\rho_2(kg/m^3)$ | $\mu_2(N_p m^{-1} Hz^{-1})$ | $E(mm)$ | $\mu_1(N_p m^{-1} Hz^{-2})$ |
|---|---|---|---|---|---|---|---|
| Polyethylene | 344.9 | 1.19 | 2291 | 942 | $2.26 \times 10^{-5}$ | 5.974 | $2.16 \times 10^{-11}$ |
| Plexiglas | 345.1 | 1.187 | 2739 | 1185 | $1.90 \times 10^{-5}$ | 3.034 | $2.25 \times 10^{-11}$ |
| Composite | 344.7 | 1.19 | 3073 | 1511 | $2.68 \times 10^{-5}$ | 2.439 | $2.00 \times 10^{-11}$ |

**Table 1.** The optimized acoustical properties values identified for the three tested samples. We choose $f_0 = 1\ MHz$.

## 5. Conclusion

In this article, we describe an ultrasonic method for simultaneous determination of four acoustical quantities (thickness, mass density, longitudinal velocity and attenuation coefficient) of a viscoelastic plate embedded in a known fluid (air). Measurements are made locally in areas potentially inaccessible (center of the plate). This method can be applied to study the transmission (reflection) coefficient of a plate immersed in a liquid (water).

The results are found to be in good agreement with additional independent measurements (average thickness, average mass density and longitudinal wave velocity). Moreover, the values of the acoustical properties are very consistent with those obtained with a classical, ultrasonic, contact technique.

Measuring the wave velocity in air with a good accuracy, during experiments, permits to obtain uncertainties in thickness and velocity, clearly lower than 1 per cent, while attenuation coefficient and mass density are estimated with an accuracy of 1 and 2 per cent. The value of attenuation coefficient in air is in accordance with standard measurements[7].

Further work is to investigate a theoretical model taking into account the weak diffraction effects in the transmission coefficient in order to improve the accuracy of the parameters determination. The accuracy and reliability of these results justify the further study of materials with inhomogeneity in density, sound speed or thickness.

## References

[1] A. Marini G. Guidarelli and L.Palmieri. Ultrasonic method for determining attenuation coefficients in plate-shaped materials. *J. Acoust. Soc. Am*, 94 (3):1476–1481, 1993.

[2] D.W.Schindel and D.A.Hutchins. Through-thickness characterization of solids by wideband air-coupled ultrasound. *Ultrasonics*, 33 (1):11–17, 1995.

[3] V. K. Kinra and V. R. Iyer. Ultrasonic measurement of thickness, phase velocity, density or attenuation of a thin-viscoelastic plate. *Ultrasonics*, 33 (2):95–109, 1995.

[4] P. He. Direct measurement of ultrasonic dispersion using a broadband transmission technique. *Ultrasonics*, 37:67–70, 1999.

[5] Thomas L.Szabo. Causal theories and data for acoustic attenuation obeying a frequency power law. *J. Acoust. Soc. Am*, 97 (1):14–24, 1995.

[6] P. He. Experimental verification of models for determining dispersion from attenuation. *IEEE Trans. Ultrason., Ferroelect., Freq. Contr.*, 46 (3):706–714, 1999.

[7] L. J. Bond. Absorption of ultrasonic waves in air at high frequencies. *J. Acoust. Soc. Am*, 92 (4):2006–2045, 1992.

# Plate models and hidden informations.

**Philippe Destuynder**[1]

[1]Conservatoire National des Arts et Métiers,
Department of Mathematics, 292 rue saint Martin 75003 Paris

E-mail: philippe.destuynder@cnam.fr

**Abstract.** A plate or a shell structure is mainly characterized by the fact that the thickness is small compared to the other dimensions. This allows some kinematical and constitutive approximations which can be justified in a mathematical framework as far as the energy of the model is concerned. But in most cases, the three dimensional local behaviour is not represented by the plate or the shell model. For instance, interlayer stresses in a multilayered plate can be singular and lead to a delamination. But they are not contained in a standard plate model. The local waves trapped in one of these layers are not represented in these structural models. The goal of this paper is to give few possibilities to overcome these difficulties.

## 1. Introduction

A large number of papers and books have been devoted to the plate theory. But the most important contribution is certainly the one given by Kirchhoff and Love [17], [12]. It is recalled hereafter in a first part. The ideas based on a more accurate approximation of the transverse shear strains introduced by Mindlin [18] and Reissner [21] have also contributed to meaningful improvements of the models, in particular for the numerical aspects. Concerning the validation of these models, the natural question was to derive an upper bound of the error between the three dimensional solution and the one of the plate model. This was first obtained using an asymptotic method. The idea of the asymptotic strategy is due to Friedrichs and Dressler [9] but the mathematical framework which enables one to derive a full justification came latter (see the books by P.G. Ciarlet [4] and its very complete bibliography which resumes the various contributions and [7] which focus on linearly elastic plates). The main error bounds are recalled in the second section of this paper. Nevertheless, the boundary-layer which plays an important role in the delamination of multilayered plates contains strange but precious informations. The goal of the third fourth and fifth sections is to explain how these informations can be derived from the plate model.

This paper, which is an extension of the Kirchhoff-Love's method to the boundary-layer effect, corresponds to a twisting free boundary condition on a part of the lateral edge of the plate.

## 2. The almost usual Kirchhoff-Love's model

Let us start with the notations. The plate that we consider occupies in space the open set $\Omega^\varepsilon$, the small parameter $\varepsilon$ denoting half the thickness. The middle surface is $\omega$ with a boundary $\gamma$ and the other geometrical notations are explicited on figure 1. The lateral boundary is $\Gamma^\varepsilon = \Gamma_0^\varepsilon \cup \Gamma_1^\varepsilon$. The plate is clamped on $\Gamma_0^\varepsilon = \gamma_0 \times ]-\varepsilon, \varepsilon[$ and the twisting is free on $\Gamma_1^\varepsilon = \gamma_1 \times ]-\varepsilon, \varepsilon[$. The displacement field is $u = \{u_i\}$ and the linearized strain field is: $\gamma(u) = \{\gamma_{ij}(u)\}$ where

$\gamma_{ij}(u) = \frac{1}{2}(\partial_i u_j + \partial_j u_i)$ where $i, j \in \{1, 2, 3\}$ and the notation $\partial_i$ is the partial derivative with respect to the coordinate $x_i$. The $2^{nd}$ Piola-Kirchhoff's stress is denoted by $\sigma = \{\sigma_{ij}\}$. The

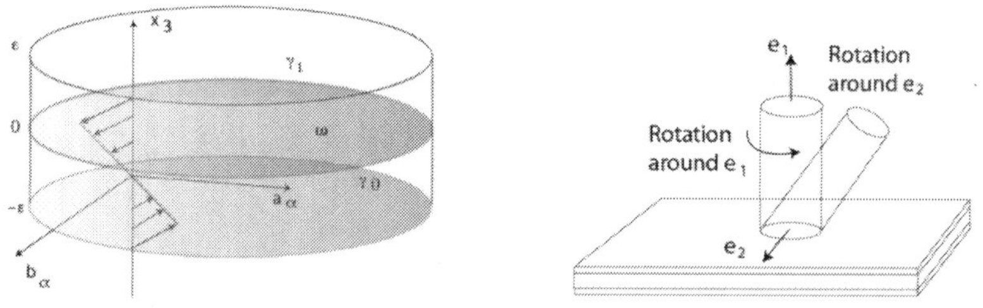


**Figure 1.** The 3D plate and the boundary conditions (left) an example (right)

Hooke's law [8] is (one uses the implicit summation convention from 1 to 3 for latin's indices and fom 1 to 2 for greek's indices):

$$\sigma = R\gamma(u), \quad \text{or else } \forall i, j, k, l \in \{1, 2, 3\}, \; \sigma_{ij} = R_{ijkl}\gamma_{kl}(u), \quad \text{where } R_{ijkl} = R_{klij} = R_{jikl}, \tag{1}$$

and it is commonly admitted that the materials used in a plate are monoclinic with respect to the thickness direction so that:

$$\forall \alpha, \beta, \lambda \in \{1, 2\}, \; R_{\alpha\beta\lambda 3} = 0.$$

This property enables to decouple the inplane stresses and the transverse shear stresses which is a basic point in plate and shell models. Furthermore the coefficient $S_{ijkl}$ are assumed to be constant with respect to the in plan coordinates $x_\alpha$ (even constant) and piecewise constant with respect to $x_3$. It is convenient to use the compliance $S = R^{-1}$ instead of the stiffness $R$. Therefore, the constitutive relationship can also be written:

$$S_{ijkl}\sigma_{kl} = \gamma_{ij}(u) \text{ and } S_{\alpha\beta\lambda 3} = 0. \tag{2}$$

Furthermore, the ellipticity-continuity condition is always satisfied:

$$\exists c_0 > 0, \; c_1 > 0, \; \text{s.t. } \forall \tau = \{\tau_{ij}\}, c_0 \sum_{i,j=1,3} |\tau_{ij}|^2 \leq S_{ijkl}\tau_{ij}\tau_{kl} \leq c_1 \sum_{i,j=1,3} |\tau_{ij}|^2 \tag{3}$$

Let us now introduce the principle of virtual work. First of all the admissible space for the displacements is defined by (see figure 1 for a geometrical representation, $r$ being a rotation along $\gamma_1$):

$$V = \{v = \{v_i\}, \; v_i \in H^1(\Omega^\varepsilon), \; v_i = 0 \text{ on } \Gamma_0^\varepsilon, \; v_3 = 0, \; v_\alpha b_\alpha = x_3 r \text{ on } \Gamma_1^\varepsilon\} \tag{4}$$

Thus, the principle of virtual work for a dynamical model without external load but with initial conditions for instance, can be stated as follows (the upper double dot is the second order time derivative):

$$\forall v \in V, \; \int_{\Omega^\varepsilon} \varrho \ddot{u}_i v_i + \int_{\Omega^\varepsilon} \sigma_{ij}\gamma_{ij}(v) = 0 \tag{5}$$

$$u(x, 0) = u_d(x) \; \dot{u}(x, 0) = u_v(x). \tag{6}$$

The set of equations (2), (5) and (6) is the three dimensional dynamical model for the plate that we consider in this paper (with the boundary conditions chosen). A classical result based on Korn's inequality [8] enables one to prove the classical following statement.

**Theorem 1** *Let $u_d \in V$ and $u_v \in [L^2(\Omega^\varepsilon)]^3$. The coefficients $S_{ijkl}$ of the compliance tensor are supposed to satisfy (3). For any $T > 0$, there exists a unique solution $u \in \mathcal{C}^0([0,T];V) \cap \mathcal{C}^1([0,T];[L^2(\Omega^\varepsilon)]^3)$ to (2)-(5)-(6). If the initial data are smooth enough additional regularity can also be obtain by derivating in time the model or/and using the classical regularity results for elliptic operator (see P. Grisvard [11]).*

Our goal now is to derive a surface model which would be a nice approximation of the three dimensional one. But a cornerstone is also to be able to derive hidden informations from the plate solution using few mathematical analysis between the 3D and the 2D. Concerning the error estimate, the basic trick is to use Prager and Synge's inequality [20] following the initial idea of P. Ladevèze [14] for static problems and with some additional and important mathematical modifications concerning the dynamical aspect and the boundary-layer effects. The strategy that we develop in this paper is mathematically similar to the one presented in [4] and [7]. But in our opinion, it is more efficient for practical applications and a better mechanical understanding of what happens in a thin structure.

## 3. The Kirchhoff-Love's theory

Let us introduce *a priori* assumptions which can be fully justified by an asymptotic method based on the smallness of the thickness of the plate [4] [7]. There are two sets of such hypothesis: one concerning an approximation of the kinematics, the other concerns the energy.

### *3.1. The Kirchhoff-Love's kinematical assumption*

Due to the smallness of the thickness compared to the other dimension of a plate, one can suggest to neglect the necking strain $\gamma_{33}(u)$ but also (less justified but acceptable) the transverse shear strain $\gamma_{\alpha 3}(u)$. Hence one introduces a subspace of $V$ defined by:

$$V^{KL} = \{v = \{v_i\} \in V,\ \partial_3 v_3 = 0,\ \partial_\alpha v_3 + \partial_3 v_\alpha = 0,\ \alpha = 1,2\}, \tag{7}$$

and this space can be identified to $V_t \times V_3$ defined by:

$$\begin{cases} V_t = \{v = \{v_\alpha\},\ v_\alpha \in H^1(\omega),\ v_\alpha = 0 \text{ on } \gamma_0\} \\ V_3 = \{v \in H^2(\omega),\ v = 0 \ \text{ on } \gamma_0 \cup \gamma_1\}. \end{cases} \tag{8}$$

In fact, there is an isomorphism between $V^{KL}$ and $V_t \times V_3$ defined by:

$$v = (\underline{v}_\alpha, v_3) \in V_t \times V_3 \to j_{KL}(v) = (\{\underline{v}_\alpha - x_3\partial_\alpha v_3\}, v_3) \in V^{KL}. \tag{9}$$

The kinematical Kirchhoff-Love's hypothesis can be stated as follows:

*Hypothesis 1 (kinematics) The displacement $u$ solution to three dimensional model can be approximated by a function $u^{KL}$ in the space $V^{KL}$.*

### *3.2. The implicit Koiter's assumption*

In a different formulation W.T. Koiter [13] underlined that the Kirchhoff-Love's kinematical hypothesis was not sufficient for defining a consistent plate or shell model. Hence he stated that the elastic energy (or free energy) was only dependent on the inplate stresses (or strains). This is an assumption. But in a mathematical framework this sentence is not sufficientely precise. And one controverse was that Koiter omitted to mention explicitely the possibility of transverse and necking stresses. This is why it appeared necessary to reformulate his statement in a more precise way as follows.

Journal of Physics: Conference Series **269** (2011) 012006 doi:10.1088/1742-6596/269/1/012006

*Hypothesis 2 (energy) The elastic energy due to the transverse shear and the necking stresses are negligible compared to the one of the in plan stresses.*
This second hypothesis is a little bit difficult to handle completely and it is at the origin of many misunderstandings in plate and shell models. In fact, it contains two informations: the first one is that in the expression of the complementary energy:

$$\sigma \in [L(\Omega^\varepsilon)]^9_s \to E(\sigma) = \int_{\Omega^\varepsilon} S_{ijkl}\sigma_{ij}\sigma_{kl},$$

the contribution due to tansverse shear stresses $\sigma_{\alpha 3}$ and the necking stress $\sigma_{33}$ can be neglected. Hence, one sets:

$$E(\sigma) \simeq \int_{\Omega^\varepsilon} S_{\alpha\beta\lambda\mu}\sigma_{\alpha\beta}\sigma_{\lambda\mu}.$$

This hypothesis doesn't mean that $\sigma_{\alpha 3} = \sigma_{33} = 0$, but only that their energy can be neglected. In fact, the transverse and the necking stresses will be defined from the equilibrium equations as functions which are only distributions in many cases. Therefore a regularity condition must be be satisfied in order to ensure that their energy is negligible compared to the one of the inplane stresses $\sigma_{\alpha\beta}$.

### *3.3. The Kirchhoff-Love's model*
The plate's model that we are dealing with in this paper is defined as follows:

$$\left\{\begin{array}{l} \text{Find } (u^{KL}, \sigma^{KL}) \in V^{KL} \times [L^2(\Omega^\varepsilon)]^9_s, \ (s \text{ for symmetry}), \text{ s.t.:} \\ \forall \tau \in [L^2(\Omega^\varepsilon)]^9_s, \ \displaystyle\int_{\Omega^\varepsilon} S_{\alpha\beta\lambda\mu}\sigma^{KL}_{\alpha\beta}\tau_{\lambda\mu} = \int_{\Omega^\varepsilon} \gamma_{\alpha\beta}(u^{KL})\tau_{\alpha\beta}, \\ \forall v \in V^{KL}, \ \displaystyle\int_{\Omega^\varepsilon} \sigma^{KL}_{ij}\gamma_{ij}(v) + \int_{\Omega^\varepsilon} \varrho\ddot{u}.v = 0, \\ u^{KL}(0) = P^1_{KL}(u_d), \ u^{KL}_v(0) = P^0_{KL}(u_v), \end{array}\right. \tag{10}$$

where $P^1_{KL}$ (resp. $P^0_{KL}$) is the euclidian projection from $V$ onto $V_{KL}$ (resp. $[L^2(\Omega^\varepsilon)]^3$). Let us point out that the space for the virtual displacement field is $V^{KL}$ and not $V$. This is a cornerstone in plate modeling. Let us give the first basic result (see for instance H. Brezis [3] or G. Duvaut-J.L. Lions [8]).

**Theorem 2** *The model (10) has a unique solution $\underline{u}^{KL}_t \in \mathcal{C}^0([0,T]; V_t) \cap \mathcal{C}^1([0,T]; [L^2(\omega)^2)$ and $u^{KL}_3 \in \mathcal{C}^0([0,T]; V_3) \cap \mathcal{C}^0([0,T]; [L^2(\omega)]^2)$ as soon as $u_d \in V$ and $u_v \in [L^2(\Omega^\varepsilon)]^3$. This solution can be characterized as follows:*

$$\left\{\begin{array}{l} \textit{first of all let us set:} \\ R^{2m}_{\alpha\beta\lambda\mu} = \displaystyle\int_{-\varepsilon}^{\varepsilon} [S_{\alpha\beta\lambda\mu}]^{-1}, \ R^{2f}_{\alpha\beta\lambda\mu} = \int_{-\varepsilon}^{\varepsilon} x_3^2[S_{\alpha\beta\lambda\mu}]^{-1}, \ \varrho_a = \int_{-\varepsilon}^{\varepsilon} \varrho, \ I_3 = \int_{\omega} x_3^2\varrho, \\ \textit{and } \nabla \textit{ is the inplane gradient operator.} \end{array}\right. \tag{11}$$

$$
\left\{
\begin{array}{l}
\textit{The displacements } \underline{u}_t^{KL}(t) \in V_t \quad u_3^{KL}(t) \in V_3 \textit{ are solution of:} \\
\forall \underline{v}_t = \{\underline{v}_\alpha\} \in V_t \ \displaystyle\int_\omega R^{2m}_{\alpha\beta\lambda\mu}\gamma_{\alpha\beta}(\underline{u}_t^{KL})\gamma_{\lambda\mu}(\underline{v}) + \int_\omega \varrho_a \ddot{\underline{u}}_\alpha^{KL}\underline{v}_\alpha = 0, \\
\forall v_3 \in V_3, \ \displaystyle\int_\omega R^{2f}_{\alpha\beta\lambda\mu}\partial_{\alpha\beta}u_3^{KL}\partial_{\lambda\mu}v_3 \\
\qquad\qquad + \displaystyle\int_\omega \varrho_a \ddot{u}_3^{KL} v_3 + \int_\omega I_3 \nabla \ddot{u}_3^{KL} ; \nabla v_3 = 0.
\end{array}
\right. \tag{12}
$$

**Remark 1** *It is convenient to introduce the in plan stress resultant and the bending moment by:*

$$
\left\{
\begin{array}{l}
n_{\alpha\beta} = \displaystyle\int_{-\varepsilon}^{\varepsilon} \sigma_{\alpha\beta} = R^{2m}_{\alpha\beta\lambda\mu}\gamma_{\lambda\mu}(\underline{u}_t), \\
m_{\alpha\beta} = \displaystyle\int_{-\varepsilon}^{\varepsilon} x_3\sigma_{\lambda\mu} = -R^{2f}_{\alpha\beta\lambda\mu}\partial_{\lambda\mu}u_3.
\end{array}
\right. \tag{13}
$$

*Therefore, the two dimensional equilibrium equations are respectively for the membrane and the bending behaviour:*

$$
\left\{
\begin{array}{l}
\forall \underline{v}_\alpha \in V_t, \ \displaystyle\int_\omega n_{\alpha\beta}\partial_\alpha \underline{v}_\beta + \int_\omega \varrho_a \ddot{\underline{u}}_\alpha \underline{v}_\alpha = 0, \\
\forall v_3 \in V_3, \ -\displaystyle\int_\omega m_{\alpha\beta}\partial_{\alpha\beta}v_3 + \int_\omega \varrho_a \ddot{u}_3 v_3 + \int_\omega I_3 \nabla \ddot{u}_3 . \nabla v_3 = 0.
\end{array}
\right. \tag{14}
$$

*Finally, the in plan stress $\sigma^{KL}_{\alpha\beta}$ is given by the following expression (the upper index $^{-1}$ denotes the inverse):*

$$
\sigma^{KL}_{\alpha\beta} = [S_{\alpha\beta\lambda\mu}]^{-1}\gamma_{\lambda\mu}(u^{KL}) = [S_{\alpha\beta\lambda\mu}]^{-1}[R^{2m}_{\lambda\mu\eta\xi}]^{-1}n^{KL}_{\eta\xi} - x_3[S_{\alpha\beta\lambda\mu}]^{-1}[R^{2f}_{\lambda\mu\eta\xi}]^{-1}m^{KL}_{\eta\xi}. \tag{15}
$$

*For instance, in a multilayered plate, this in plan stress is piecewise linear through the thickness of the plate. Another important point concerns the boundary conditions on $\Gamma_1^\varepsilon$. In the three dimensional model, the free twisting movement implies that one has $\sigma_{\alpha\beta}a_\alpha b_\beta = 0$. From integrations by parts on the equations (14) and because of the boundary conditions satisfied by the displacements in $V^{KL}$, one has (see for instance [7]):*

$$
m^{KL}_{\alpha\beta}b_\alpha b_\beta = 0, \ \partial_s(m^{KL}_{\alpha\beta}a_\alpha b_\beta) + \partial_\alpha m^{KL}_{\alpha\beta}b_\beta = 0. \tag{16}
$$

*Hence, the 3D torsion-free condition is not satisfied by the plate solution. This phenomenon is at the origin of a torsion boundary-layer that we describe in the following.*

## 4. The error bound

In order to obtain a validation of the plate model it is useful to construct a 3D stress field which satisfy exactly the principle of virtual work (PPW). This strategy has been first observed by P. Ladevèze [14], but it is also necessary that this 3D stress field could be connected to an admissible displacement field belonging to the space $V$. This is the most technical part which was overcomed in [7] using a space interpolation method due to J.L. Lions [15]. The point is that no matter the method you are using, (asymptotic expansion or *a priori assumptions*) the error bound strategy is the same: i) define a 3D stress field which satisfies the PPW, ii) apply an interpolation result. Let us sketch the method. First of all, let us consider the 3D in plan equilibrium equation in which $\sigma_{\alpha\beta}^{KL}$ and $u_{\alpha}^{KL}$ are known (but not yet the transverse shear stress $\sigma_{\alpha 3}^{KL}$):

$$\left\{ \forall v_t \in V_t, \int_{\omega} \sigma_{\alpha 3}^{KL} \partial_3 \underline{v}_{\alpha} = -\{\int_{\Omega^{\varepsilon}} \varrho \ddot{u}_{\alpha}^{KL} v_{\alpha} + \int_{\Omega^{\varepsilon}} \sigma_{\alpha\beta}^{KL} \partial_{\beta} v_{\alpha}\}. \right. \tag{17}$$

One can make explicit the local conditions contained in this weak formulation:

$$\begin{cases} \partial_3 \sigma_{\alpha 3}^{KL} = \varrho[\ddot{\underline{u}}_{\alpha}^{KL} - x_3 \partial_{\alpha} \ddot{u}_3^{KL}] - \partial_{\beta} \sigma_{\alpha\beta}^{KL}, \\ \sigma_{\alpha 3} = 0 \text{ for } x_3 = \pm\varepsilon. \end{cases} \tag{18}$$

One can compute $\sigma_{\alpha 3}^{KL}$ explicitely from these relations (the two boundary conditions are compatible because of (14)):

$$\sigma_{\alpha 3}^{KL} = \ddot{\underline{u}}_{\alpha}^{KL} \int_{-\varepsilon}^{x_3} \varrho - \partial_{\alpha} \ddot{u}_3^{KL} \int_{-\varepsilon}^{x_3} x_3 \varrho - \int_{-\varepsilon}^{x_3} \partial_{\beta} \sigma_{\alpha\beta}^{KL}. \tag{19}$$

Under classical regularity assumptions one can claim that $\sigma_{\alpha 3}^{KL} \in \mathcal{C}^0([0,T]; L^2(\Omega^{\varepsilon}))$. But there is a compatibilty relation which should be satisfied by the Kirchhoff-Love's solution in order to ensure that the equation (17) is fully satisfied. It is obtained by an integration by part on the right handside of (17) and is due to the fact that $v_{\alpha} a_{\alpha} \neq 0$ on the boundary $\Gamma_1$. It is the twisting free edge condition:

$$\sigma_{\alpha\beta} a_{\alpha} b_{\beta} = 0 \text{ and unfortunately } \sigma_{\alpha\beta}^{KL} a_{\alpha} b_{\beta} \neq 0. \tag{20}$$

This is the starting point of the boundary-layer theory for the thin structure and it is discussed in the next section. For the time being, let us go on with the computation of the necking stress $\sigma_{33}^{KL}$. From (14) one should have:

$$\forall v_3, (0, v_3) \in V, \int_{\Omega^{\varepsilon}} \sigma_{33}^{KL} \partial_3 v_3 = -\{\int_{\Omega^{\varepsilon}} \sigma_{\alpha 3}^{KL} \partial_{\alpha} v_3 + \int_{\Omega^{\varepsilon}} \varrho \ddot{u}_3^{KL} v_3\}. \tag{21}$$

The solution is:

$$\sigma_{33}^{KL} = \ddot{u}_3^{KL} \int_{-\varepsilon}^{x_3} \varrho - \int_{-\varepsilon}^{x_3} \partial_{\alpha} \sigma_{\alpha 3}^{KL}. \tag{22}$$

If the plate solution is smooth enough ($\underline{u}_{\alpha} \in \mathcal{C}^2([0,T], H^3(\omega))$ and $u_3 \in \mathcal{C}^2([0,T]; H^4(\omega))$), one can claim that both $\sigma_{33}^{KL}$ and $\sigma_{\alpha 3}^{KL} \in \mathcal{C}^0([0,T]; L^2(\Omega^{\varepsilon}))$. Because the stress field $\sigma^{KL}$ coupled with the displacement field $u^{KL}$ doesn't satisfy exactly the PPW (due to (20)). Therefore one can construct a boundary-layer model which takes into account an additional term $(d_i^{BL}, \chi_{ij}^{BL})$ (displacement and stress field) such that $\sigma_{ij}^{KL} + \chi_{ij}^{BL}$ would be close to satisfy the PPW.

### *4.1. The boundary-layer*

First of all, only the torsion effect is involved. Therefore the local model which is parametrized by the curvilinear coordinate along the boundary $\gamma_1$ is set over a semi-infinite strip $S^{\varepsilon}$ as shown on figure 2. The torsion displacement the magnitude of which is denoted here by $\varphi$, is transverse to the strip. It is solution to the following equations where the shear modulus in each layer - assumed to be the same in $s\xi$ and $sx_3$ ($R_{s3s3} = R_{s\xi s\xi}$) for sake of simplicity- is denoted by $G$ and $\nabla_t$(resp. $div_t$) is the gradient (resp. the divergence) with respect to the coordinates $(\xi, x_3)$) as shown on figure 2:

$$\begin{cases} -div_t(G\nabla_t \varphi) = 0 \text{ in } S^{\varepsilon}, \\ \dfrac{\partial \varphi}{\partial x_3} = 0 \text{ for } x_3 = \pm\varepsilon, \; G\dfrac{\partial \varphi}{\partial \xi} = -\sigma_{\alpha\beta}^{KL} a_{\alpha} b_{\beta} \text{ for } \xi = 0, \lim_{\xi \to \infty} \varphi = 0. \end{cases} \tag{23}$$

The solution of this model has several remarkable properties that are discussed in the following sections. The existence and uniqueness of a solution is classical. It is obtained analytically from

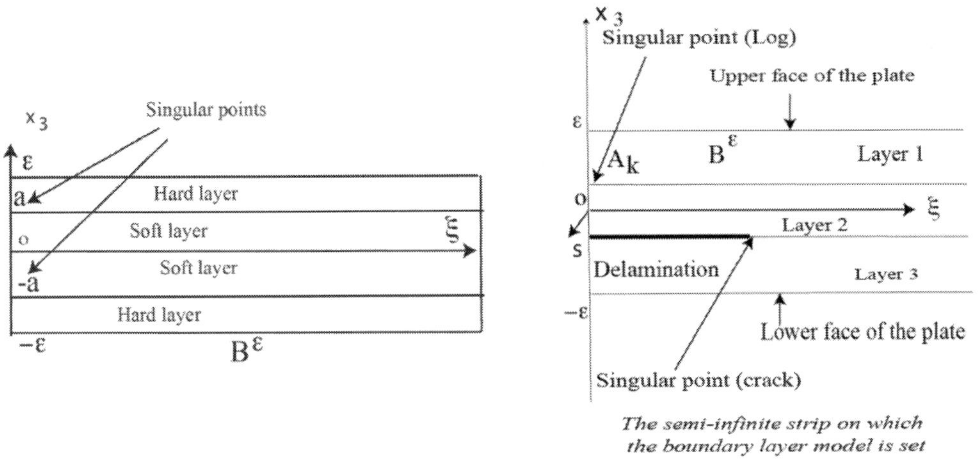


**Figure 2.** The strip $S^\varepsilon$ on the left and few singular points on the right

a Fourier series by (one has $A_0 = 0$ because the average of $G\dfrac{\partial\varphi}{\partial\xi}(0, x_3)$ is zero):

$$\varphi(\xi, x_3) = \sum_{k=1,\infty} A_n e^{-\dfrac{n\pi\xi}{2\varepsilon}} \cos(\frac{n\pi x_3}{2\varepsilon}),\ A_n = -\frac{1}{\varepsilon}\int_{-\varepsilon}^{\varepsilon} \frac{\sigma^{KL}_{\alpha\beta} a_\alpha b_\beta}{G} \sin(\frac{n\pi x_3}{2\varepsilon}). \tag{24}$$

In fact the previous series converges in the space $H^1(S^\varepsilon)$. The corrector term is then defined as follows in the local frame $(a, b, e_3)$ (see figure 1) where $\eta$ is a truncation function equal to 1 in the vicinity of $\xi = 0$:

$$\Phi_i = \varphi a_i \eta(\xi),\ a_3 = 0,\ \ \chi_{ij} = \eta((\xi) R_{ijkl}(\partial_k \Phi_l + \partial_l \Phi_k)/2. \tag{25}$$

But the point is to describe the local behaviour of this solution in the neighbourhood of a point like $A_k$. A singular behaviour appears due to the discontinuity of the coefficient $G$ from one layer to the other. Thus one considers the following model the solution of which describes the one of $\varphi$ in the neighbourhood of the boundary between the two layers (see point $A_k$ on figure 3 for the notations). Therefore another basis of harmonic functions can be used.

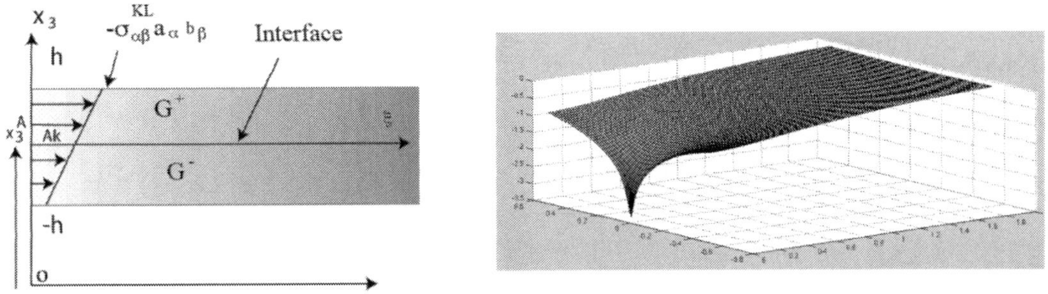


**Figure 3.** Strip for the singularity computation and the logarithm behaviour

Journal of Physics: Conference Series **269** (2011) 012006 doi:10.1088/1742-6596/269/1/012006

One can write in a semi-infinite strip $H^\varepsilon = ]0,\infty[\times]-h,h[$:

$$\varphi(s,\xi,x_3) = \frac{A_0}{2} + \sum_{n\geq 1}[C_n\cos(\frac{n\pi x_3}{h}) + D_n\sin(\frac{n\pi x_3}{h})]e^{-\frac{n\pi\xi}{h}}], \tag{26}$$

the coefficients $C_n$ and $D_n$ are defined in order to satisfy the boundary condition at $\xi = 0$ by:

$$C_n = \frac{1}{n\pi}\int_{-h}^{h}\frac{\sigma^{KL}_{\alpha\beta}a_\alpha b_\beta}{G}\sin(\frac{n\pi x_3}{h}),\quad D_n = -\frac{1}{n\pi}\int_{-h}^{h}\frac{\sigma^{KL}_{\alpha\beta}a_\alpha b_\beta}{G}\cos(\frac{n\pi x_3}{h}). \tag{27}$$

In fact, only the series with the sine functions leads to a singularity. Let us set:

$$\chi(x_3) = \begin{cases} 1 \text{ if } x_3 > x_3^{A_k}, \\ -1 \text{ if } x_3^{A_k}. \end{cases}$$

One can split $-\dfrac{\sigma^{KL}_{\alpha\beta}a_\alpha b_\beta}{G}$ as follows (where $g$ is continuous on $]-h,h[$):

$$-\frac{\sigma^{KL}_{\alpha\beta}a_\alpha b_\beta}{G} = \kappa\chi(x_3) + g(x_3), \tag{28}$$

where $\kappa$ is a coefficient which takes into account the discontinuity of the material properties (details are given in J.L. Davet and Ph. D [6]). The contribution of the first term on the right hanside of (28), to the series representing $\varphi$ can be explicited analytically and the second term is smoother (derivatives are $\mathcal{C}^0$). A simple computation leads to (let us set $x_3^\delta = x_3 - x_3^A$):

$$\frac{\partial\varphi^S}{\partial x_3}(s,\xi,x_3^\delta) = \frac{2\kappa}{\pi}\mathrm{Log}[\frac{1+2\cos(\pi x_3^\delta)e^{-\xi\pi/h}+e^{-2\xi\pi/h}}{1-2\cos(\pi x_3^\delta)e^{-\xi\pi/h}+e^{-2\xi\pi\xi}}]. \tag{29}$$

This singular function represented on figure 3. It is a singular peeling stress which can be at the origin of a delamination.

### *4.2. The mathematical result*

The basic point in plate or shell theory is to justify that the the 3D and 2D solutions are closed enough. But this can be obtained in an energy norm (excepted if the two solutions are very smooth in oder to be able to apply the implicit function theorem for linear or nonlinear models [5]), and no pointwise error bounds have been obtained for singular behaviours. Therefore, all the conclusions drawn from the plate model should be justified from an energetical point of view. The fundamental result is the following one and its proof is quite standard from the classical methods used in variational methods and from the results given in [4] and [7].

**Theorem 3** *Let us introduce the error variables by:*

$$\begin{cases} \overline{\sigma}^\varepsilon_{ij} = \sigma^\varepsilon_{ij} - \sigma^{KL}_{ij} - \chi_{ij}, \\ \overline{u}^\varepsilon_i = u^\varepsilon_i - u^{KL}_i - \Phi_i. \end{cases} \tag{30}$$

*There exists two constant $c_0$ and $c_1$ which are independent on $\varepsilon$ and can be computed from the plate solution, such that:*

$$\forall t > 0,\ \frac{1}{2}\int_{\Omega^\varepsilon}\varrho|\dot{\overline{u}}^\varepsilon(t)|^2 + \frac{1}{2}\int_{\Omega^\varepsilon}\overline{\sigma}^\varepsilon_{ij}\gamma_{ij}(\overline{u}^\varepsilon)(t) \leq (E(0) + c_0\varepsilon)e^{c_1 t}.$$

*Hence the error is in $\sqrt{\varepsilon}$ in the energy norm as soon as the error on the initial conditions (term $E(0)$) is small enough. The energy norm is:*

$$||u(t)||_E = \sqrt{\frac{1}{2}\int_{\Omega^\varepsilon} \varrho|\dot{\overline{u}}^\varepsilon|^2 + \frac{1}{2}\int_{\Omega^\varepsilon} \overline{\sigma}^\varepsilon_{ij}\gamma_{ij}(\overline{u}^\varepsilon)}.$$

It is possible to upgrade Theorem 3 in different ways by adding new terms. This can be done in two differents ways. One consists in using the asymptotic method based on the fact the thickness of the plate compared to its other dimensions, is a small parameter. This is explained in [4] and [7]. Another one consists in enlarging the approximation spaces for both the displacements and the stress fields. This is the case of the Reissner and Mindlin models [18][21].

## 5. Few aspects contained in the TBL

The energy contained in the boundary-layer model is meaningful because of Theorem 3. The logarithmic singularity leads an energy term (the hidden regularity of $m_{\alpha\beta}$ and $n_{\alpha\beta}$ (see J.L. Lions [16]) enables one to claim this property. A numerical simulation of this effect is represented on figure 4 for different time in the bending case of the plate.

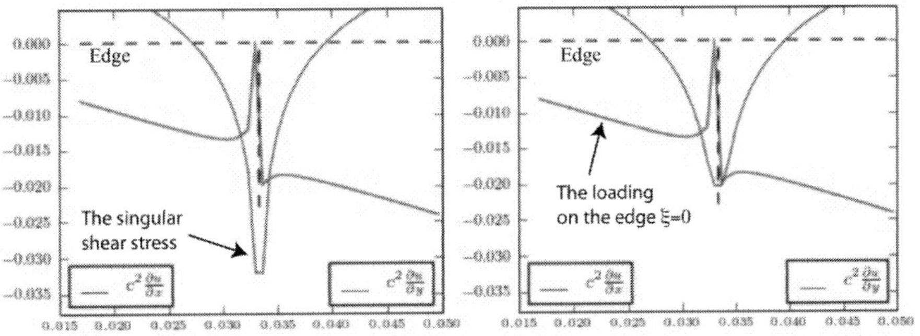


**Figure 4.** The singular stress (continuous) in the vicinity of the singular point $A_k$ at two different time steps along the free edge ($\xi = 0$) and the loading applied (piecewise linear)

### *5.1. Delamination in the boundary-layer*

Due to the singular peeling stress in the boundary-layer, a debonding between two layers can occur. It is called a delamination. As far as its length is small enough, one can justify the use of the plate model which doesn't take into account the presence of this small crack. But the boundary-layer model does. And the energy release rate which is the opposite of the energy derivative with respect to crack length, is an energy stable quantity. Therefore it can be evaluated from the solution of the boundary-layer term. Let us consider that a small (*length* $\simeq \varepsilon$) delamination crack has appeared at the interface between two layers. The local behavior of $\varphi$ near the crack tip can be computed following the standard methods used in fracture mechanics (see H.D. Bui [2]). In fact $K$ is the so-called stress intensity factor and $\theta$ represents a virtual movement of the crack tip (parallel to it). In fact its support is included in a neighbourhood $B^\varepsilon$ of the crack tip and the energy release rate only depends on its value at this point. The advantage of this second expression is that it is perfectly stable from the energetical point of view, ($D_t\theta$ is the Jacobian matrix of $\theta$, (.,.) is the scalar product).

Journal of Physics: Conference Series **269** (2011) 012006 doi:10.1088/1742-6596/269/1/012006

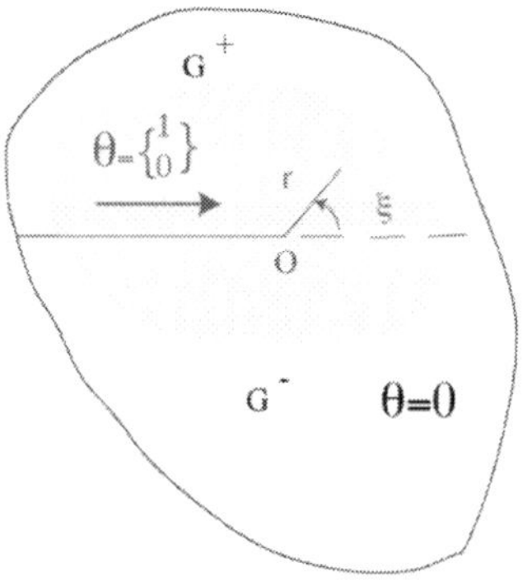


*Analytical computation:*

$$\begin{cases} \varphi = \dfrac{K}{G}\sqrt{r}\sin(\dfrac{\xi}{2}) + \varphi^R(\xi, x_3), \\ K^2 = -\dfrac{8}{\pi(G^+ + G^-)}\dfrac{\partial E}{\partial l}, \end{cases}$$

*where the energy release rate is defined by:*

$$-\frac{\partial E}{\partial l} = \frac{1}{2}\int_{B^\varepsilon} [|\dot{\varphi}|^2 - G|\nabla_t \varphi|^2] div_t(\theta) + \int_{B^\varepsilon} G(D_t\theta\nabla_t\varphi, \nabla_t\varphi)$$

*Analytical expression of the singularity at the crack tip and the energy release rate.*

### *5.2. Love-Stoneley's local waves*

The boundary-layer model can also be (for highest frequencies) modeled by a dynamical model (wave equation). The dynamical term can be neglected for low frequency terms, but can include meaningful energy contributions for hight frequencies. Even if the pate is vibrating at a low frequency, the discontinuity of material (sandwich structure) can be at the origin of a spill over process on highest frequencies because of the local waves. For instance, the reflection of incident waves on a crack is more efficient than on an interface where a part of the energy is swallowed by the Love-Stoneley waves mainly localized in the weakest part of the sandwich plate. Because of the simplicity of the wave operator for the twisting movement, all the eigenmodes are known analytically [10]. Let us show one of them on figure 5 which is computed for the half upper part of the infinite strip without crack. The model consists in finding $\varphi$ such that:

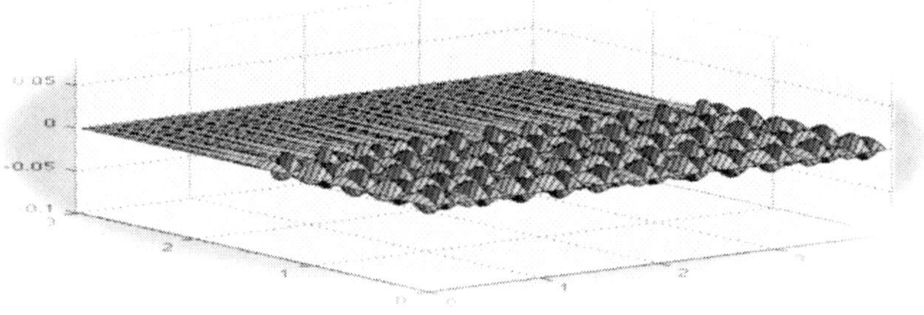

**Figure 5.** Love-Stoneley local wave in the twisting model (half strip and the coordinates $\xi$ varies from left to right

$$\begin{aligned} &\varrho\ddot{\varphi} - G\frac{\partial^2\varphi}{\partial\xi^2} - \frac{\partial}{\partial x_3}(G\frac{\partial\varphi}{\partial x_3}) = 0, \\ &G\frac{\partial\varphi}{\partial\xi}(t, 0, x_3) = -\sigma^{KL}_{\alpha\beta}a_\alpha b_\beta, \;\; G\frac{\partial\varphi}{\partial x_3} = 0 \text{ for } x_3 = \pm\varepsilon, \\ &\text{for instance: } \frac{\partial\varphi}{\partial\xi} = 0 \;\text{ at } \xi = L >> \varepsilon \text{ and initial conditions.} \end{aligned} \tag{31}$$

The numerical simulation (figure 6) shows that the energy measured on the hard part is much smaller without a crack (left) than when there is one (right). The loadind case is the one defined in the boundary layer from the plate solution in dynamics.

**Figure 6.** Density of total energy: left without crak (important absorption) and right (weak absorption) with a crack at the upper interface. The loading case is the same as on figure 4.

## 6. Conclusion

This paper gives few ideas concerning plate theory and focus on two important points:

- how to construct a plate (or a shell) model which is accurate when the thickness is small,
- how to derive additional informations using a boundary layer model.

## Acknowledgments

The author thanks Olivier Wilk who has performed part of the numerical tests.

## References

[1] Adams R. 1975 *Sobolev spaces,* (Academic press, New York)
[2] Bui H.D. 1978 *Mécanique de la rupure fragile,* (Masson, Paris)
[3] Brezis H. 1983 *Analyse Fonctionnelle,* collection mathématiques appliquées pour la maîtrise (Masson, Paris)
[4] Ciarlet P.G. 1997 *Mathematical elasticity* vol. 2, Theory of plates, (North-Holland, Amsterdam)
[5] Ciarlet P.G. and Destuynder Ph. 1979 A justification of a nonlinear model in plate theory, *Comp. meth. Appl. Mechs. Eng.*, vol. **17/18**, pp. 227-258
[6] Davet J.L. and Destuynder Ph. 1986 Free-edge stress concentration in composite laminates: a boundary layer approach, *CMAM*, vol. **59**, pp. 129-140
[7] Destuynder Ph. 1986 *Une théorie asymptotique des plaques minces en élasticité linéaire,* (Masson, Paris)
[8] Duvaut G. and Lions J.L. 1972 *Les inéquations en mécanique et en physique,* (Dunod, Paris)
[9] Friedrichs K.O. and Dressler R.F. 1961 A boundary-layer theory for elastic plates, *Comm. Pure Appl. Math.* vol. **14**, pp. 1-33
[10] Fung Y.C. 1965 *Foundations of solid mechanics, p.182,* (Prentice-Hall Inc., Englewood Cliffs, USA)
[11] Grisvard P. 1992 *Singularities in boundary value problems,* (Masson, Paris)
[12] Kirchhoff G. 1850 Über das Gleichgewiicht und die Bewegnung einer elastichen Shribe, *J. Reine Angew. Math.*, vol. **40**, pp. 51-58
[13] Koiter W.T. 1970 On the foundations of the linear theory of thin elastic shells, Proc. Kon. Ned. Akad. Wetensch., Part 1: **B**, vol. **13**, pp. 169-182
[14] Ladevèze P. 1975 *Comparaison de modèles de mécanique des milieux continus,* (Thèse d'Etat, Université Pierre-et-Marie-Curie, Paris)
[15] Lions J.L. 1973 *Perturbations singulières dans les problèmes aux limites en contrôle optimal,* Lecture notes in mathematics, vol. **323**, (Springer-Verlag, Berlin)
[16] Lions J.L. 1988 *Contrôlabilité exacte, perturbations et stabilisation de systèmes distribués,* (Masson, Paris)
[17] Love A.E.H. 1924 *A treatise on the mathematical theory of elasticity,* Fourth edition, (Cambridge University press and Dover edition in 1944)
[18] Mindlin R. 1951 Influence of rotary inertia and shear on flexural motions of isotropic plate, *J. Appl. Mech.*, vol. **18**, pp. 31-58
[19] Necas J. 1967 *Les méthodes directes en théorie des équations elliptiques,* (Masson, Paris)
[20] Prager W. and J.L. Synge J.L. 1947 Approximation in elasticity based on the concept of function space, *Quat. Appl. Math.* vol. **5**, $n^3$), pp. 241269
[21] Reissner E. 1945 The effect of tansverse shear deformations on the bending of elastic plates, *J. App. Mech.* vol. **12**, **A**, pp. 69-77

Journal of Physics: Conference Series **269** (2011) 012007 doi:10.1088/1742-6596/269/1/012007

# The effective wavenumber of a pre-stressed nonlinear microvoided composite

**W.J. Parnell and I.D. Abrahams**

School of Mathematics, Alan Turing Building, University of Manchester, Oxford Road, Manchester M13 9PL, UK

E-mail: William.Parnell@manchester.ac.uk

**Abstract.** By using nonlinear elasticity and a modified version of classical multiple scattering theory we derive an explicit form for the effective wavenumber for horizontally polarized shear (SH) elastic waves propagating through a pre-stressed inhomogeneous material consisting of well-separated cylindrical voids embedded in a neo-Hookean rubber host phase. The resulting effective (incremental) antiplane shear modulus is thus also derived.

## 1. Introduction

Multiple scattering in linear elastic inhomogeneous materials such as fibre reinforced or particulate composites (without pre-stress) in the quasi-static regime is relatively well understood. Much work has centred on the derivation of the effective wave number (and resulting effective elastic properties) given a random distribution of inhomogeneities (see e.g. [2], [22]). Additionally, the influence of nonlinear pre-stress on subsequent incremental linear wave propagation in elastic media has been studied extensively over the last few decades using the theory of *small-on-large* [9], [18]. In this theory, a linearization is performed about the nonlinear equilibrium state in order to determine the incremental wave propagation characteristics of the pre-stressed medium. However, to the authors' knowledge, in the literature interest has centred almost exclusively on the influence of *homogeneous* stretch distributions (and hence induced anisotropy) on subsequent wave propagation, see e.g. [4], [10]. When the medium in question is *inhomogeneous* (for example a fibre-reinforced, or particulate composite material), where the host phase is nonlinear-elastic, pre-stress will almost always lead to non-homogeneous stretch distributions, except in very special cases (see e.g. [20]). Degtyar et al. [5] analysed the case of stressed composites where residual stresses occur in the vicinity of inhomogeneities, in the context of linear elasticity and a self consistent scattering formulation of the effective wavenumber was used. We emphasise that this analysis took place in the linear elastic (small displacement gradient) regime.

Of interest however, is how an initial *nonlinear* pre-stress affects subsequent multiple wave scattered fields in heterogeneous media. In the low frequency regime this information would allow us to determine the incremental behaviour of the material, the effective incremental homogenized properties and subsequently the overall effective strain energy function of the nonlinear inhomogeneous material. Recently, the so-called *second-order homogenization method* was applied to determine such homogenized behaviour in the static regime for fibre reinforced composites [3]. Nonlinear pre-stress can be extremely useful in practice, allowing us to *tune* materials in order to permit or restrict waves of specific frequency ranges to propagate through the structure. This property was described by Parnell [20] and discussed further in subsequent articles in different contexts [7], [1].

Although a significant amount of work exists regarding multiple scattering in stress-free configurations, there is no existing theory which can deal with the pre-stress problem where the host phase is nonlinear-elastic (e.g. rubber). A canonical scattering problem regarding pre-stress was analysed by Parnell and Abrahams [21]. In that article, the problem of elastic antiplane or horizontally-polarized shear (SH) wave scattering from a cylindrical void embedded in a host medium which is capable of finite deformation and is neo-Hookean in its constitutive behaviour, was studied. In particular it was shown that the initial pre-stress has *no effect* on the scattering coefficients of the scattered field due to an incident line source. Therefore in the far-field it is not possible to see the effect of the pre-stress: its effect is felt only within the local vicinity of the void. Since standard multiple scattering theory usually describes the effective wavenumber in terms of the far-field scattering pattern, this latter result suggests that the effective wavenumber is unaffected by the pre-stress. However, the modified theory that we develop in this paper suggests that this is not the case.

In section 2 we summarize the pre-stress problem regarding an isolated cylindrical void embedded in a neo-Hookean elastic material. We then derive an exact solution for incremental SH waves in this pre-stressed state in section 3 by using a mapping corresponding to the finite deformation associated with the pre-stress. The *multiple* scattering problem is then considered in section 4, assuming that each void is well separated. In this instance, each void sees the pre-stress as if it were isolated. By taking configurational averages and assuming a random distribution of voids in the pre-stressed state, we then derive an explicit expression for the effective wavenumber incorporating a parameter $P$ which characterizes the level of pre-stress within the material. We conclude in section 6 indicating where further work is required.

## 2. Pre-stress

Let us first consider the pre-stress problem associated with the finite deformation which ensues when a body of infinite extent, except for the presence of a cylindrical void of radius $A$, is subjected to hydrostatic pressure in the far-field, modifying its radius to $a$. We shall assume that the host phase is isotropic and nonlinear-elastic so that it is capable of finite deformation. Its constitutive behaviour is described by a strain energy function $W_{SE} = W_{SE}(I_1, I_2, I_3)$ where $I_j$ are the principal strain invariants of the deformation [9], [18]. We assume incompressibility of this phase, which is a very good approximation for materials of a rubber type. This means that the strain invariant $I_3 = 1$ and thus $W_{SE} = W_{SE}(I_1, I_2)$. Furthermore in this article we shall restrict attention to a material whose strain energy function is of the incompressible *neo-Hookean* type, whose strain energy function is $W_{SE} = \mu(I_1 - 3)/2$, where $\mu$ is the linear-elastic shear modulus of the material [18].

Let us define the cylindrical polar (Lagrangian) coordinate system $(R, \Theta, Z)$ with origin at the centre of the cylindrical void in the undeformed configuration. Additionally we define the cylindrical polar (Eulerian) coordinate system $(r, \theta, z)$ associated with the *deformed* configuration. In the far-field we impose a hydrostatic pressure $\sigma^{rr} = -p_\infty$ (per unit area in the deformed configuration, corresponding to a Cauchy stress) as $r \to \infty$ (uniform in the longitudinal direction), as shown in figure 1. The material is also held at a fixed stretch $L$ in the longitudinal $Z$ direction. The ensuing deformation is therefore described by the following relationships

$$R = rQ(r), \qquad \Theta = \theta, \qquad Z = \frac{z}{L}, \tag{1}$$

where $Q(r)$ is a function which must be determined from the incompressibility condition and equilibrium equations. We shall use the convention that upper case variables correspond to the *undeformed* configuration whilst lower case variables correspond to the *deformed* configuration.

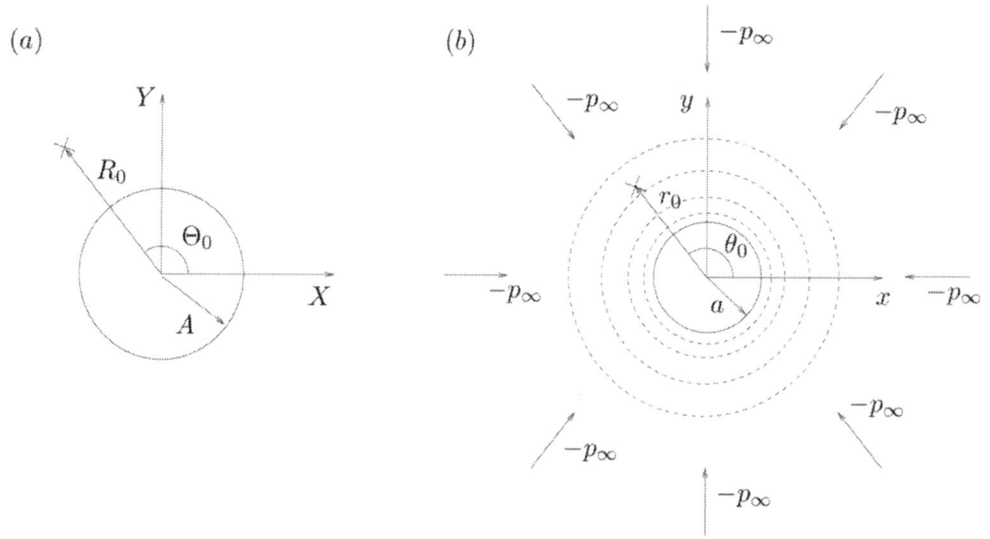


**Figure 1.** (a) Shows the undeformed cavity, with a line source indicated by crossed lines at the position $(R_0, \Theta_0)$. In (b) we show how the cavity deforms due to the hydrostatic pressure $-p_\infty$, successive dotted circles indicating decreasing strain as we move away from the cavity. Under this deformation the line source moves to the location $(r_0, \theta_0)$, noting that $\theta_0 = \Theta_0$ due to symmetry.

With reference to [21] we note that incompressibility gives $Q(r) = (\sqrt{r^2 + M})/r$ where $M = A^2/L - a^2$ and where we remind the reader that $A$ and $a$ are the undeformed and deformed cylinder radii. Furthermore by consideration of the induced stress field, in [21] it was shown that when $r \to \infty$, given that $\sigma_{rr} \to -p_\infty$,

$$\frac{p_\infty}{\mu} = \frac{1}{2L}\left[\frac{A^2}{La^2} - 1 + \log\left(\frac{A^2}{La^2}\right)\right], \tag{2}$$

and we see therefore that this is a (nonlinear) equation for the determination of the deformed to undeformed radius ratio $a/A$ as a function of the ratio $p_\infty/\mu$ and the stretch $L$, both of which are assumed specified. We plot $a/A$ as a function of $p_\infty/\mu$ for three different prescribed values of $L = 0.7, 1, 1.3$ in figure 2. The exhibited behaviour is as one would expect.

## 3. Incremental deformation

In order to set up the multiple scattering problem in the pre-stressed medium, we consider the problem of incremental wave scattering from the finitely deformed medium described in section 2. We use the theory of *small-on-large*, i.e. linearization about a nonlinear deformation state [9]. The total displacement field may therefore be represented by

$$\hat{\mathbf{u}} = \mathbf{u} + \eta\mathbf{u}' \tag{3}$$

where $\mathbf{u} = u^i\mathbf{g}_i = u_i\mathbf{g}^i$ is the displacement field derived from the finite deformation (1) and $\eta \ll 1$ is a small parameter associated with the magnitude of the incremental displacement.

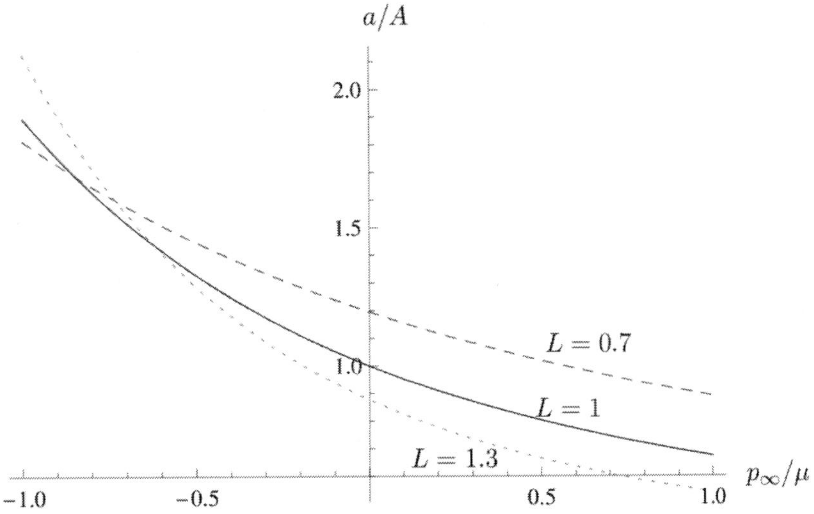


**Figure 2.** Plot of the deformed radius $a/A$ as a function of $p_\infty/\mu$ for the three prescribed values of $L = 0.7, 1, 1.3$.

Note that $\mathbf{g}_i$ and $\mathbf{g}^i$ are covariant and contravariant basis vectors associated with the deformed configuration [21]. Let us assume that the incremental displacement is of an anti-plane nature, i.e. of the form

$$\mathbf{u}' = u_3'\mathbf{g}^3 = u^{3'}\mathbf{g}_3 = \Re\left[w(r,\theta)\exp(-i\omega t)\right]\mathbf{g}_3 \tag{4}$$

so that the time-harmonic wave is a horizontally-polarized shear (SH) wave, polarized in the $z$ direction and propagating in the $r\theta$ plane.

The initial finite deformation described in section 2 leads to the following modified wave equation for $w$ [21]

$$\frac{\mu}{L}\left[\frac{1}{r}\frac{\partial}{\partial r}\left(r + \frac{M}{r}\frac{\partial w}{\partial r}\right)\right] + \frac{1}{(r^2+M)}\frac{\partial^2 w}{\partial \theta^2} + \rho\omega^2 w = 0, \tag{5}$$

where we remind the reader that $M = A^2/L - a^2$.

In [21], it was shown, by mapping back to the undeformed configuration, that (5) can be solved analytically and its (outgoing) solution is

$$w(r) = \sum_{n=-\infty}^{\infty} i^n a_n \mathrm{H}_n\left(k\sqrt{r^2+M}\right)e^{in\theta}, \tag{6}$$

where $H_n(x) = H_n^{(1)}(x) = J_n(x) + iY_n(x)$ is the Hankel function of the 1st kind of order $n$, where $J_n$ and $Y_n$ are Bessel functions of the first and second kind, of order $n$. Furthermore, we have introduced the notation $k^2 = L\rho\omega^2/\mu$, for the square of the modified wavenumber (modified due to pre-stress), noting that the original wavenumber $K$ is defined by $K^2 = \rho\omega^2/\mu$ . The coefficients $a_n$ depend upon the type of forcing. In [21] a line source forcing was assumed and it was shown explicitly that the coefficients $a_n$ *do not depend upon the pre-stress* if the *force* of the source strength remains unchanged during the finite deformation. In the section to follow we shall describe the multiple scattering problem of interest. This problem is, in fact, an eigenvalue problem in the sense that there is no assumed forcing since we take the limit where scatterers fill all space. The problem of an incident wave field is technically more difficult and leads to the same expression for the effective wavenumber and hence, for now at least, does not need to be considered.

## 4. Multiple scattering

Consider now the configuration consisting of $N$ circular cylindrical voids of infinite extent in the $Z$ direction distributed in an unbounded neo-Hookean host phase such that the voids are well-separated. In particular suppose that in the $X,Y$ plane these voids occupy a finite domain $\mathcal{D}_0$, defined by $(X,Y) \in [-Q,Q] \times [-H,H]$ within unbounded space. This domain becomes $\mathcal{D}$ (defined by $(x,y) \in [-q,q] \times [-h,h]$) after the pre-stress described in section 2 is imposed in the far-field. The area of $\mathcal{D}_0$ is denoted by $|\mathcal{D}_0|$ and analogously with $\mathcal{D}$. In the $X,Y$ plane there are $n_0$ inclusions per unit undeformed area so that $N = n_0|\mathcal{D}_0|$ and furthermore there are $n$ inclusions per unit *deformed* area so that $N = n|\mathcal{D}|$. The *volume* fraction occupied by the voids is therefore denoted by $\phi_0 = n_0\pi A^2$ (per unit length) in the undeformed configuration which in the deformed configuration becomes $\phi = n\pi a^2/L$, noting the factor of $L$ in the latter.

Note that since the voids are well-separated we assume that under the pre-stress described in section 2, each void deforms as if it were isolated and embedded inside the homogeneous host phase. Thus its radius deforms from $A$ to $a$ and furthermore the local incremental field due to a superimposed horizontally polarized shear wave can be described by (5).

We wish to determine the wave field which arises due to multiple scattering from the $N$ voids in the deformed configuration described above and in particular we wish to assess how the *effective* propagating wave is modified by the pre-stress. We will consider the limit when $\mathcal{D}$ expands to fill the all space, which is achieved by letting $q, h \to \infty$ whilst taking the limit $N \to \infty$ with $n$ held fixed (Fig. 3). In the deformed configuration, the voids are located at $\mathbf{p}_j = (p_j, q_j)$, $j = 1, 2, \ldots, N$ with local polar coordinate system $(r_j, \theta_j)$ relating to the local Cartesian system $x_j = r_j \cos\theta_j$, $y_j = r_j \sin\theta_j$, see Fig. 4.

Ensuring only an outgoing (scattered) field from each void, using (6), we may pose a solution of the form

$$w = \sum_{j=1}^{N} \sum_{n=-\infty}^{\infty} a_n^j Z_n \mathrm{H}_n\left(k\sqrt{r_j^2 + M}\right) e^{in\theta_j} \tag{7}$$

in terms of the local coordinate system described above. We do not assume any incident forcing; we seek eigensolutions to the problem as we shall see shortly by taking the limit as $N \to \infty$. However, we note that if we had alternatively considered a half-space problem in spirit of, for example [13], the result for the resulting effective wavenumber in the pre-stressed medium would remain unchanged. The terms $a_n^j$ are coefficients associated with the $j$th inclusion and $n$th mode and the coefficients $Z_n$ are introduced for convenience as we shall show shortly. Locally to the $s$th inclusion we may write

$$w(r_s,\theta_s) = \sum_{n=-\infty}^{\infty} a_n^s Z_n \mathrm{H}_n(k\sqrt{r_s^2 + M}) e^{in\theta_s} + \sum_{j=1, j\neq s}^{N} \sum_{n=-\infty}^{\infty} a_n^j Z_n \mathrm{H}_n(k\sqrt{r_j^2 + M}) e^{in\theta_j} \tag{8}$$

and given that the inclusions are well-separated, we can approximate this as

$$w(r_s,\theta_s) \approx \sum_{n=-\infty}^{\infty} a_n^s Z_n \mathrm{H}_n(k\sqrt{r_s^2 + M}) e^{in\theta_s} + \sum_{j=1, j\neq s}^{N} \sum_{n=-\infty}^{\infty} a_n^j Z_n \mathrm{H}_n(k r_j) e^{in\theta_j}. \tag{9}$$

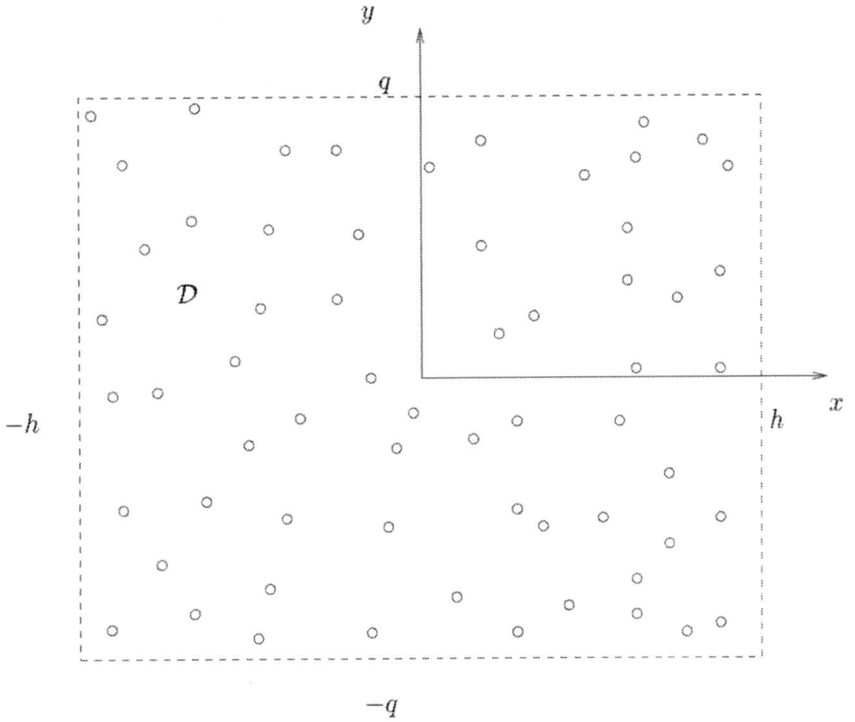


**Figure 3.** Inclusions reside inside a domain $\mathcal{D}$ in the deformed configuration, in an otherwise uniform and unbounded material. We consider the limit where $q, h \to \infty$, whilst letting the number of inclusions $N \to \infty$ with the volume fraction $\phi$ held fixed.

Referring to figure 4, we are now able to use Graf's addition theorem directly on the second term in (9) to find that

$$w(r_s, \theta_s) \approx \sum_{n=-\infty}^{\infty} a_n^s Z_n \mathrm{H}_n(k\sqrt{r_s^2 + M})e^{in\theta_s} + \sum_{j=1, j\neq s}^{N} \sum_{n=-\infty}^{\infty} a_n^j Z_n \sum_{m=-\infty}^{\infty} \mathrm{H}_{n-m}(kR_{js})e^{i(n-m)\theta_{js}} J_m(kr_s)e^{im\theta_s} \quad (10)$$

where $R_{js}$ is the distance between the $j$th and $s$th void and $\theta_{js}$ is the angle between the $j$th and $s$th void.

On applying the zero traction condition on $r_s = a$ and exploiting the orthogonality of $e^{in\theta_s}$, we find that

$$a_m^s + \sum_{j=1, j\neq s}^{N} \sum_{n=-\infty}^{\infty} a_n^j Z_n \mathrm{H}_{n-m}(kR_{js})e^{i(n-m)\theta_{js}} = 0 \quad (11)$$

where we have defined $Z_n$ as

$$Z_n = P\frac{J_n'(ka)}{\mathrm{H}_n'(Pka)}, \quad (12)$$

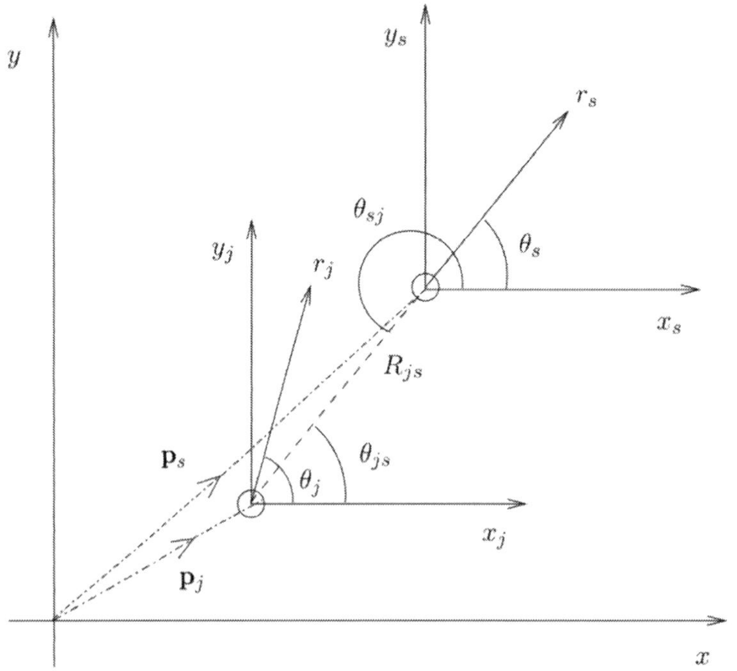


**Figure 4.** Geometry of the location of inclusions and their local coordinate systems.

and where we have introduced the notation $P = A/(a\sqrt{L})$. We also used the relationship $k = K\sqrt{L}$ in the above.

In particular note that in the low frequency limit, $ka \ll 1$, (once again using $k = K\sqrt{L}$) we find that

$$Z_0 \sim -\frac{\pi}{4i}P^2(ka)^2 = \pi\hat{Z}_0(ka)^2, \quad Z_{-1} = Z_1 \sim \frac{\pi}{4i}P^3(ka)^2 = \pi\hat{Z}_1(ka)^2, \quad Z_n \sim o(ka)^2. \tag{13}$$

## 5. Effective wavenumber

Let us introduce a probability density function, defining the statistics of the random distribution of voids [15], [22] in the deformed configuration. This is written as

$$p(\mathbf{p}_1, \mathbf{p}_2, ..., \mathbf{p}_N) = p(\mathbf{p}_s)p(\mathbf{p}_j|\mathbf{p}_s)p(\mathbf{p}_1, \mathbf{p}_2, ..., \not{\mathbf{p}}_j, ..., \not{\mathbf{p}}_s, ..., \mathbf{p}_N|\mathbf{p}_j, \mathbf{p}_s), \tag{14}$$

where $\not{\mathbf{p}}_j$ indicates that $\mathbf{p}_j$ is not present and $p(\mathbf{p}_j|\mathbf{p}_s)$ denotes the probability density of finding an inclusion at $\mathbf{p}_j$ given that there is already one located at $\mathbf{p}_s$. Thus, we can take the ensemble average of (11) as in Linton and Martin [13] and obtain

$$\langle a_m^1\rangle_1 + (N-1)\pi(ka)^2 \sum_{n=-1}^{1} \int_{-\infty}^{\infty}\int_{-h}^{h} p(\mathbf{p}_2|\mathbf{p}_1)\langle a_m^2\rangle_{12}\hat{Z}_n H_{n-m}(kR_{21})e^{i(n-m)\theta_{21}}\, dp_2 dq_2 = 0 \tag{15}$$

where we were able to take $s = 1, j = 2$ without loss of generality and we have retained only terms of $O((ka)^2)$, employing (13). We recall that

$$\langle f\rangle_s = \int_D\int_D \cdots \int_D p(\mathbf{p}_1, \mathbf{p}_2, .., \not{\mathbf{p}}_s, .., \mathbf{p}_N|\mathbf{p}_s) f \, d\mathbf{p}_1 d\mathbf{p}_2 ... d\not{\mathbf{p}}_s ... d\mathbf{p}_N, \tag{16}$$

$$\langle f\rangle_{js} = \int_D\int_D \cdots \int_D p(\mathbf{p}_1, \mathbf{p}_2, .., \not{\mathbf{p}}_j, .., \not{\mathbf{p}}_s, .., \mathbf{p}_N|\mathbf{p}_j, \mathbf{p}_s) f \, d\mathbf{p}_1 d\mathbf{p}_2 .. d\not{\mathbf{p}}_j .. d\not{\mathbf{p}}_s .. d\mathbf{p}_N, \tag{17}$$

are the ensemble averages of the field $f$ with the $s$th inclusion fixed and the $s$th and $j$th inclusions fixed respectively [13], [22].

Let us use the *hole correction* correlation function [6],[22], defined by

$$p(\mathbf{p}_2|\mathbf{p}_1) = \begin{cases} \dfrac{1}{|\mathcal{D}|}, & R_{21} > 2a, \\ 0, & R_{21} < 2a \end{cases} \tag{18}$$

which implies that there is no interaction between voids other than asserting that they cannot overlap. Finally, as with most multiple scattering theories, in order to avoid using higher order correlation functions, a closure assumption is invoked; we use the Lax Quasi-Crystalline Approximation (QCA) [11]

$$\langle a^2 \rangle_{12} = \langle a^2 \rangle_2. \tag{19}$$

In the limit as $N, q, h \to \infty$, we find that (15) becomes

$$\langle a_m^1 \rangle_1 + k^2 L\phi \sum_{n=-1}^{1} \oint \langle a_m^2 \rangle_2 \hat{Z}_n H_{n-m}(kR_{21}) e^{i(n-m)\theta_{21}} \, dp_2 dq_2 = 0 \tag{20}$$

where the notation on the integral sign indicates that the integral is over all space excluding the circular region of radius $2a$ centred at $(p_2, q_2) = (p_1, q_1)$ and is known as the hole correction.

We now pose the effective wave solution [13]

$$\langle a_m^1 \rangle_1 = i^m e^{i\beta q_1} C_m e^{i\gamma p_1}. \tag{21}$$

Under this assumption, (20) becomes (noting that $\theta_{21} = \theta_{12} + \pi$)

$$C_m e^{i\gamma p_1} + k^2 L\phi \sum_{n=-1}^{1} \hat{Z}_n (-i)^{n-m} e^{i\gamma p_1} C_n M_{n-m} = 0 \tag{22}$$

where

$$M_{n-m} = \oint_{-\infty}^{\infty} \oint \psi_{n-m}(p_{21}, q_{21}) \Psi(p_{21}, q_{21}) \, dp_2 dq_2 \tag{23}$$

and $p_{21} = p_2 - p_1, q_{21} = q_2 - q_1$. Furthermore, we have employed the same notation as Linton and Martin [13], i.e.

$$\psi_n(x, y) = H_n(kr) e^{in\theta}, \tag{24}$$

$$\Psi(x, y) = e^{i(\gamma x + \beta y)} = e^{i\tilde{\Gamma} r \cos(\theta - \zeta)}, \tag{25}$$

where

$$x = r\cos\theta, \qquad y = r\sin\theta, \tag{26}$$

$$\gamma = \tilde{\Gamma}\cos\zeta, \qquad \beta = \tilde{\Gamma}\sin\zeta. \tag{27}$$

Therefore since $\alpha^2 + \beta^2 = k^2, \gamma^2 + \beta^2 = \tilde{\Gamma}^2$, we have that $\gamma^2 - \alpha^2 = \tilde{\Gamma}^2 - k^2$.

We may follow the procedure of Linton and Martin [13] in order to evaluate the integrals in (23) using Green's theorem. As such these integrals become line integrals along the faces at $p_2 = 0$ and around the boundary of the circle of radius $2a$ centred at $(p_1, q_1)$ induced by the hole correction. Omitting this detail and comparing coefficients of $e^{i\gamma p_1}$ in the resulting equations, in the point scattering limit ($ka \to 0$) we find that

$$(1-\Gamma^2)C_m - L\phi \sum_{n=-1}^{1} \hat{Z}_n \mathcal{N}_{n-m}(\Gamma)C_n = 0, \qquad m = 0, \pm 1 \tag{28}$$

where we have introduced the notation $\Gamma = \tilde{\Gamma}/k$ and where

$$\mathcal{N}_n(\Gamma) = \lim_{\epsilon \to 0} 4\pi\epsilon[H_n'(2\epsilon)J_n(2\Gamma\epsilon) - (\Gamma\epsilon)H_n(2\epsilon)J_n'(2\Gamma\epsilon)] = 4i(\Gamma)^{|n|}. \tag{29}$$

Note the factor $L$ that arises in (28) since $\phi$ is the *volume fraction* of fibres whereas the probability density function describes the distribution of fibre locations in an *area* of the deformed configuration, see the discussion of the change in volume fraction at the start of section 4.

Given all of the above, the system (28) becomes

$$\begin{pmatrix} 1-\Gamma^2+L\phi\mathcal{M} & -L\phi\mathcal{D}\Gamma & L\phi\mathcal{M}\Gamma^2 \\ -L\phi\mathcal{M}\Gamma & 1-\Gamma^2-L\phi\mathcal{D} & L\phi\mathcal{M}\Gamma \\ L\phi\mathcal{M}\Gamma^2 & -L\phi\mathcal{D}\Gamma & 1-\Gamma^2+L\phi\mathcal{M} \end{pmatrix} \begin{pmatrix} C_{-1} \\ C_0 \\ C_1 \end{pmatrix} = 0 \tag{30}$$

where

$$\mathcal{D} = P^2, \qquad \mathcal{M} = P^3. \tag{31}$$

A nontrivial solution of (30) is only possible if the determinant of the matrix is zero which yields the following condition on $\Gamma$:

$$\Gamma^2 = (1 - L\phi\mathcal{D})\left(\frac{1+L\phi\mathcal{M}}{1-L\phi\mathcal{M}}\right) + o(1), \tag{32}$$

so that in dimensional form

$$\tilde{\Gamma}^2 = LK^2(1-LP^2\phi)\left(\frac{1+LP^3\phi}{1-LP^3\phi}\right) + o(1). \tag{33}$$

Furthermore, in the low frequency limit we may derive the effective incremental properties of the pre-stressed material,

$$\tilde{\Gamma}^2/K^2 = \frac{\rho_*}{\rho_0}\frac{\mu_0}{\mu_*} = L(1-LP^2\phi)\left(\frac{1+LP^3\phi}{1-LP^3\phi}\right) \tag{34}$$

and given that the effective density in the deformed configuration must have the form

$$\frac{\rho_*}{\rho_0} = (1-\phi) \tag{35}$$

this leaves the following expression for the effective incremental shear modulus

$$\frac{\mu_*}{\mu_0} = \frac{(1-\phi)(1-LP^3\phi)}{L(1-LP^2\phi)(1+LP^3\phi)}. \tag{36}$$

## 6. Conclusions

We have derived an explicit form for the effective wavenumber for SH elastic waves propagating through a pre-stressed inhomogeneous material consisting of well-separated cylindrical voids embedded in a neo-Hookean rubber host phase. The resulting effective (incremental) antiplane shear modulus has thus also been derived. Future work will include a study of the interaction of initial fields due to pre-stress and the error induced by the assumption which permitted the use of the standard Graf's addition theorem to be applied to (9). The eventual goal is the study of vector problems in higher dimensions (i.e. for particulate composites), compressible phases and the consideration of nonlinear elastic *inclusions* in addition to voids.

## References

[1] Bertoldi, K. and Boyce, M.C., 2008, Wave propagation and instabilities in monolithic and periodically structured elastomeric materials undergoing large deformations. Phys. Rev. B 78, 184107.

[2] Bose, S.K. and Mal, A.K., 1973, Longitudinal shear waves in a fibre-reinforced composite, Int. J. Solids Structures 9, 1075-1085.

[3] Brun, M., Lopez-Pamies, O. and Ponte-Castaneda, P. 2007, Homogenization estimates for fiber reinforced elastomers with periodic microstructures, Int. J. Solids Structures 44 5953-5979.

[4] Destrade, M. and Scott, N., 2004. Surface waves in a deformed isotropic hyperelastic material subject to an isotropic internal constraint. Wave Motion 40, 347-357.

[5] Degtyar, A.D., Huang, W. and Rokhlin, S.I., 1998, Wave propagation in stressed composites. J. Acoust. Soc. America 104, 2192-2199.

[6] Fikioris, J.G. and Waterman, P.C., 1964, Multiple scattering of waves II. Hole corrections in the scalar case. J. Math. Phys. 5, 1413-1420.

[7] Gei, M., Movchan, A.B. and Bigoni, D., 2009, Band-gap shift and defect-induced annihilation in prestressed elastic structures. J. Appl. Phys. 105, 063507.

[8] Graff, K.F., 1975, Wave Motion in Elastic Solids. Dover, New York.

[9] Green, A.E. and Zerna, W., 1992, Theoretical Elasticity, Dover, New York.

[10] Kaplunov, J.D. and Rogerson, G.A., 2002, An asymptotically consistent model for long-wave high frequency motion in a pre-stressed elastic plate. Math. Mech. Solids 7, 581-606.

[11] Lax, M., 1952, Multiple scattering of waves II. The effective field in dense systems, Phys. Rev. 85, 621-628.

[12] Leungvichcharoen, S. and Wijeyewickrema, A.C., 2004, Stress concentration factors and scattering cross-section for plane SH wave scattering by a circular cavity in a pre-stressed elastic medium. J. Appl. Mech. JSCE 7, 15-20.

[13] Linton, C.M. and Martin, P.A., 2005, Multiple scattering by random configurations of circular cylinders: Second order corrections for the effective wavenumber. J. Acoust. Soc. Am., 117, 3413-3423.

[14] Martin, P.A., 2006, Multiple Scattering. Interaction of time-harmonic waves with $N$ obstacles, Cambridge University Press, Cambridge.

[15] McQuarrie, D.A., 1976, Statistical Mechanics. Harper-Row.

[16] Ogden, R.W., 1982, Elastic deformations of rubberlike solids, in Mechanics of Solids, the Rodney Hill 60th Anniversary Volume, H.G. Hopkins and M.J. Sewell eds. Oxford, Pergamon Press, 499-537.

[17] Ogden, R.W., 1986, Recent advances in the phenomenological theory of rubber elasticity, Rubber Chem. Technol. 59, 361-383.

[18] Ogden, R.W., 1997, Nonlinear elasticity, Dover, New York.

[19] Pao, Y.-H. and Mow, C.-C., 1973, Diffraction of elastic waves and dynamic stress concentrations, Rand Coorporation, New York.

[20] Parnell, W.J., 2007, Effective wave propagation in a pre-stressed nonlinear elastic composite bar, IMA J. Appl. Math. 72, 223-244.

[21] Parnell, W.J. and Abrahams, I.D. 2010, Scattering from a cylindrical void in a pre-stressed neo-Hookean material. Submitted to Comm. Comp. Physics.

[22] Varadan, V.K. and Varadan, V.V. and Pao, Y-H, 1978, Multiple scattering of elastic Waves by cylinders of arbitrary cross section. I. SH Waves. J. Acoust. Soc. Am., 63, 1310-1319.

[23] Waterman, P.C. and Truell, R., 1961, Multiple scattering of waves, J. Math. Phys. 2, 512-537.

[24] Yang, R.-B. and Mal, A.K., 1996, Elastic waves in a composite containing inhomogeneous fibers, Int. J. Eng. Sci. 34, 67-79.

Journal of Physics: Conference Series **269** (2011) 012008 doi:10.1088/1742-6596/269/1/012008

# Ultrasonic NDT optimization using Randomized Adaptive Differential Evolution

**B Puel[1], S Chatillon[1], and D Lesselier[2]**

[1]CEA, LIST, Département Imagerie et Simulation pour le Contrôle, F-91191, Gif-sur-Yvette, France.

[2]Département de Recherche en Electromagnétisme, Laboratoire des Signaux et Systèmes (UMR8506, CNRS-Supélec-Univ. Paris Sud), 3 rue Joliot-Curie, 91192 Gif-sur-Yvette cedex, France

E-mail: benoit.puel@cea.fr

**Abstract**. Ultrasonic non-destructive testing (UT) enables to control the integrity of critical parts. The design of transducers and the optimization of inspection procedures might be involved tasks due to the possibly large number of parameters to consider. For complex configurations, experts generally have to perform a parametric study using simulation tools, which can be time-consuming and fastidious. Here, an automatic method for the optimization of inspection parameters (positioning, setting and/or designing the transducer) is proposed. It uses an evolutionary algorithm linked with forward modelling algorithms that are already implemented in the ultrasonic module of CIVA. The method is then successfully tested on a realistic UT application of nozzle inspection, and further extended to handle constraints and achieve specific accuracy in order to optimize the design of a transducer.

## 1. Introduction

Ultrasonic non-destructive testing (NDT) aims at the control of the integrity of critical components or structures by performing and interpreting more or less complex echograms. Because of the evolution of technologies in industry, ultrasonic testing (UT) has to be adapted to new materials or increasingly complicated geometries, which makes harder a proper design of the inspection tool and conditions of this inspection. Also, if phased-array techniques [1] improve performances and capabilities of UT, they also complicate the design of transducers themselves. So, UT experts have to carry out more and more parametric studies by means of simulation tools in order to optimize the transducer design, its displacement, and the overall setting.

Such a work can be performed by means of ultrasonic simulation tools as those available in CIVA (the NDT software platform developed by CEA LIST) which brings together various models and enables to simulate a wide range of configurations [2]. But such studies can still be fastidious and take a long time. Moreover, since only few solutions are tested at the end, there remains a risk to only find a local optimum design solution.

The aim of this contribution is to introduce and investigate in preliminary fashion an optimization tool which could solve a number of NDT problems by using forward models as those implemented within CIVA. In Section 2, one briefly describes the Evolutionary Algorithm (EA) that has been chosen as main tool and one tests it on a UT application. In section 3, one shows how one is able to solve constrained problems and to take into account the accuracy of variables.

## 2. Optimization method

Many EA have been introduced in the literature [3], so choosing one or another is difficult and rather hard to justify in practice unless thorough comparisons. In UT field, Yand *et al.* [4] have used a Genetic Algorithm to design a sparse array transducer by activating only 16 elements on a 32-element linear phased array. Lupien *et al.* [5] have proposed a software that optimizes the emission surface and the layout of a phased-array transducer in order to ensure the beam width at several depths using an Evolution Strategy. Yet, if EAs appear powerful to solve many kinds of problems with few adaptations, they also involve a high number of forward-model computations.

Here, one has focused onto the so-called Randomized Adaptive Differential Evolution (RADE) [6] in view of what appears to be its good performance, its good ability to solve engineering optimization problems, and also the small number of tuning parameters involved. RADE aims at the evolution of a population of candidate solutions using evolution theory metaphor in order to build up an optimal solution.

A candidate solution is a $n$-dimensional vector, where $n$ is the number of variables of the problem. Its ability is given by the objective function evaluated for this solution. In the present case, variables should correspond with CIVA parameters and objective functions should be computed using forward ultrasonic models implemented within it. Those obviously depend upon the problem at hand and will be given for all examples next. As is usual, the candidate solution associated with the least value of the objective function is sought.

### 2.1. Randomized Adaptive Differential Evolution

This algorithm is derived from Differential Evolution [7], an auto-regulation of the mutation factor being enforced. Standard evolutionary operators are used: mutation, crossover, and selection. The first step is the initialization of the population by a uniformly distributed random set within the search range. Then, mutated solutions are created by adding perturbations to parent solutions. Next, crossover combines variables between mutated and parent solutions. Last, the selection keeps solutions that improve on the objective function so CIVA simulation is run at this step. This evolution as described is performed until stopping criteria are fulfilled and each loop is called a generation.

Characteristic of the approach are three factors: population size (*NP*), a large value favouring robustness and a small value favouring speed (with the risk to only find a local optimum), choice of the mutation type (both implemented herein); *Rand* that favours diversity and *Best* that favours convergence (with the risk to prematurely converge); and crossover rate (*Cr*), a large value then meaning intensive exploration of the search space and a small value intensive exploitation of history.

### 2.2. UT application

This algorithm has been tested on a nozzle inspection (figure 2) to optimize the detection of one given plane flaw using a flexible 8×8-element probe operated at 2 MHz frequency. The objective is to maximize the amplitude of the corner echo of the breaking back-wall flaw. The variables of the problem are the position of the transducer: $y$ (vertical position), $\theta$ (angular position) and $\alpha$ (rotation on itself); and of the focal point: $x$, $y$ and $z$ (in Cartesian coordinates). The amplitude maximization depends upon three objectives regarding the beam: achievement of a small angular deviation (for the directivity), being as close as possible with the normal of the defect in the horizontal plane projection (in order to be specular to the defect), and positioning the probe close enough to the defect (in order to focus in the near field and to reduce attenuation loss). Improving on one objective might make others worsen because of the geometry. Yet the non-trivial solution that is found by RADE is 5dB better in terms of the objective function than the one found by the expert (figure 3).

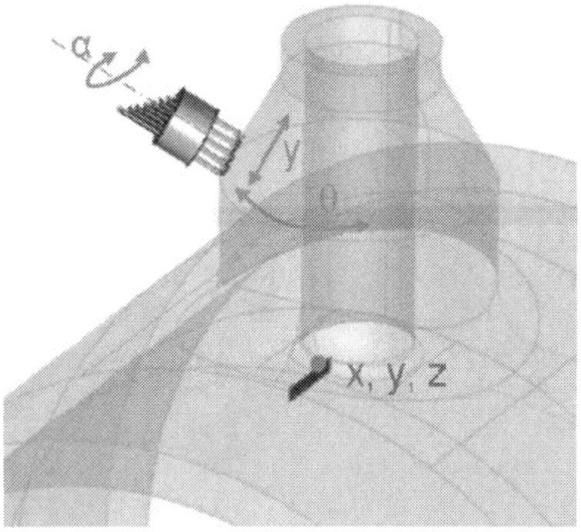


**Figure 2.** Definition of the nozzle inspection problem.

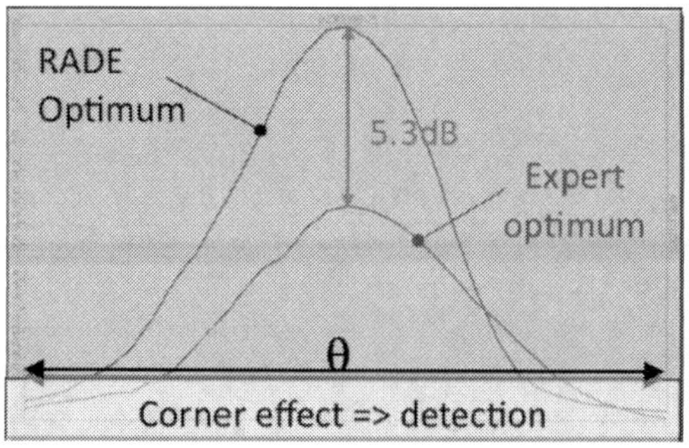


**Figure 3.** Comparison between the RADE optimal solution and the one proposed by the expert, displayed on echo-amplitude curves throughout a displacement following θ.

## 3. Constraint and accuracy handling (illustrated)

Some UT problems are more constrained than others (or, say, more *a priori* information is available) when an optimization is to be performed. In the first part of this section, such a constrained problem is described. Then, the constraint and accuracy handling as implemented in the proposed solution are presented and illustrated. Corresponding results are then reported.

### 3.1. UT application

The aim is to design a transducer which is able to detect a back-wall crack with pressure waves at 45° within a stainless steel pipe affected by structure noise. So, one has to maximize the signal-to-noise ratio by optimizing the focal characteristics of the beam with the additional constraint that the transducer should be displaced by hand, which means that it should not be too bulky. A dual element probe operated at 500 KHz (figure 4) has been chosen in order to reduce noise effects, grating lobe effects, and to improve focusing.

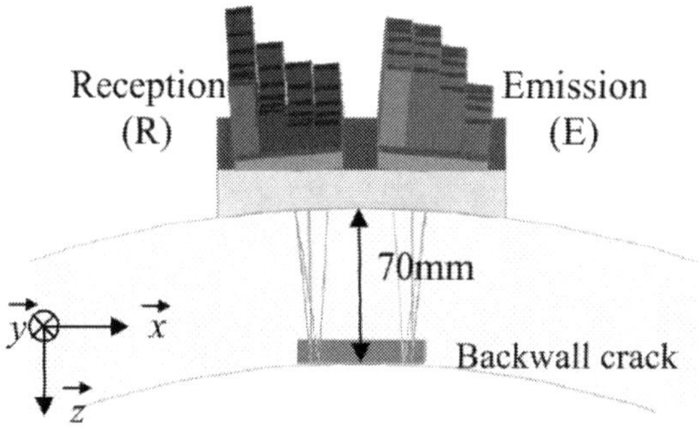


**Figure 4.** Definition of the nozzle inspection problem.

To optimize the transducer arrays (E) and (R), the following variables have been defined: the number of rows and columns of the arrays, the incident and orthogonal lengths of elements, the angle and the distance between arrays. Table 1 provides the range in which the solution has to lie and the accuracy below which the information is useless.

**Table 1.** Definition of the range and accuracy of variables for the transducer design optimization.

| | *Number of rows* | *Number of columns* | *Element incident length* | *Element orthogonal length* | *Roof angle* | *Distance ER* |
|---|---|---|---|---|---|---|
| **range** | [4,16] | [2,8] | [2,25] | [2,25] | [0,10] | [10,20] |
| **accuracy** | 1 | 1 | 1 | 1 | 0.5 | 0.5 |

Constraints that depend on several variables have also been defined: the number of elements should be less than 64, the incident dimension of E and R should be less than 71 mm, and their orthogonal dimension less than 51 mm.

### 3.2. Constraint handling

Two kinds of constraints should be taken into account: bounded and inequality constraints.

The bounded constraints guarantee that all solutions evaluated remain within the search range. If a variable is out, it is put back into it by axial symmetry on the closest boundary.

Inequality constraints are inequality relationships which involve several variables and have to be respected. Three are given in our application. For example, the constraint on the total number of elements is given by the relation: number of rows $\times$ number of columns $\times 2 \leq 64$. Then, according to Deb [8], two kinds of solutions are considered: feasible ones respect all constraints and unfeasible ones violate at least one. Consequently, feasible solutions are favoured *vs.* unfeasible ones.

### 3.3. Accuracy handling

Since the accuracy is of any use under a step given by the end-user, solutions that are too close to one another should not be computed. So, one has developed the following accuracy handling. First, the search space is sampled using the precision step. When a new trial solution is created, the algorithm finds the closest solution on the grid (one denotes it as "formatted solution"). Yet, one has to care for preservation of diversity since loss of diversity leads to being unable to generate new solutions. That is, if the formatted solution has already been computed, accuracy handling finds the closest non-computed solution on the grid within the neighbourhood.

As a conclusion, computation time is saved since solutions too close to one another are not computed, and the optimal solution found makes sense, in the design point of view, since it does not have unusable digits.

### 3.4. Optimization

To optimize the UT application introduced in section 3.1, the objective function has been computed by simulation of the defect response (figure 5) in a component containing four side-drilled holes (SDH). The goal is to guarantee that the transducer is focusing at 70 mm depth, so an objective function that maximizes the amplitude of this SDH is defined to that effect. To guarantee also that the amplitude of this SDH is maximum compared with the amplitude of other SDHs (figure 6) is achieved by using a penalization method for solutions which not respect it.

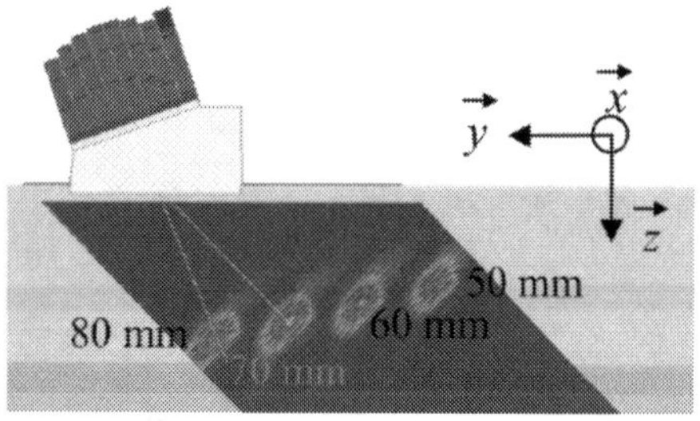


**Figure 5.** True BScan of the simulation performed to evaluate the objective function.

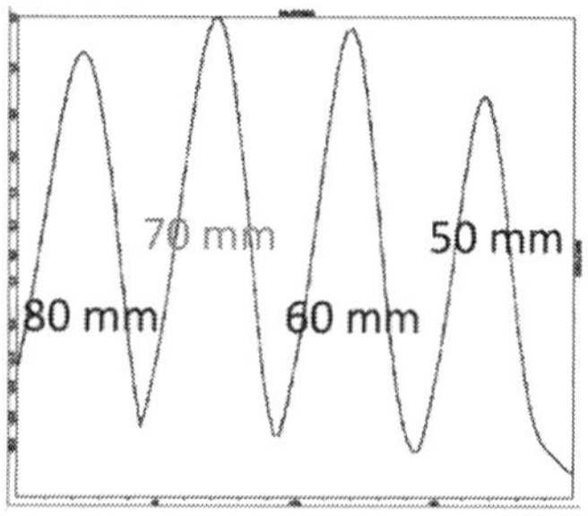


**Figure 6.** Echo-amplitude curve of the BScan from the simulation.

In Table 2 solutions obtained by RADE and an expert are compared. In terms of amplitude, RADE solution is 1 dB better than the one provided by the expert. This solution has been found for a population size $NP = 30$, a Rand mutation type, and a crossover rate $Cr = 0.2$. The evolution of the best solution's objective function for 200 generations is depicted in figure 7. It takes about 5 hours on a Core2Duo 2.66 GHz CPU with 3 Go of memory.

**Table 2.** Comparison of results obtained by RADE and an expert.

| | *Variables* | | | | | | *Constraints* | | |
|---|---|---|---|---|---|---|---|---|---|
| | *Nb of rows* | *Nb of columns* | *Elmt incid length* | *Elmt ortho length* | *Roof angle* | *Distance ER* | *nb of elmt* | *incid dim* | *ortho dim* |
| **RADE** | 6 | 4 | 11 | 12 | 8 | 10 | 48 | 71 | 51 |
| **Expert** | 8 | 4 | 8 | 12 | 5 | 16 | 64 | 71 | 51 |

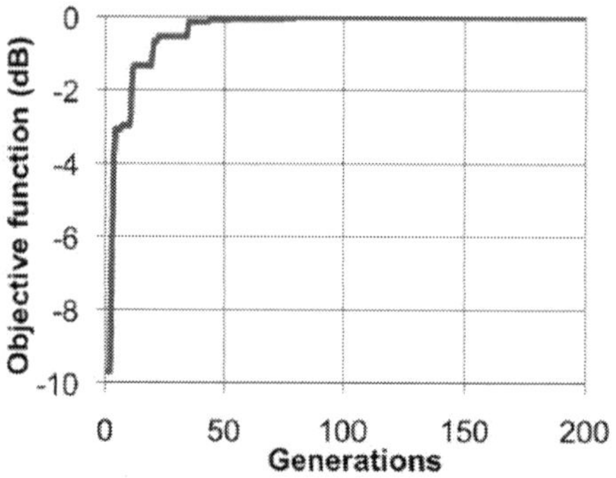


**Figure 7.** Evolution of the best solution's objective function for 200 generations.

## 4. Conclusion

In this paper, one has proposed to use forward modelling tools found within the ultrasonic module of CIVA in combination with the Randomized Adaptive Differential Evolution so as to optimize UT inspections. In addition, the algorithm has been improved in order to solve constrained optimizations problems which better describe realistic UT situations, and to account for accuracy in the variables that saves time of computation and yields realistic solutions. The approach has been successfully tested on two realistic applications. Future works will be on further testing for other UT applications, especially in the case of multi-objective problems, and on stopping test criteria as well.

**Acknowledgement**
This work has been supported by the Region Ile-de-France in the framework of the CAPVERS project.

**References**

[1] Chatillon S, Mahaut S and Dubois P 2008 Simulation of advanced UT phased array techniques with matrix probes and dynamic settings for complex component inspections *Review of Progress in QNDE* **27**, ed D O Thompson and D E Chimenti (Melville: AIP Conf. Proc. **1096**) 864
[2] CIVA software platform for simulating NDT techniques (UT, EC, RT) http://www-civa.cea.fr
[3] Rocca P, Benedetti M, Donelli M, Franceschini D and Massa A 2009 Evolutionary optimization as applied to inverse scatering problems *Inverse Problems* **25** 123003
[4] Yang P, Chen B and Shi K 2006 A novel method to design sparse linear arrays for ultrasonic phased array *Ultrasonics* **44** e717
[5] Lupien V, Hassan W and Dumas P 2006 Improved titanium billet inspection sensitivity trough optimized phased array design, part I : Design technique, modeling an simulation *Review of Progress in QNDE* **25**, ed D O Thompson and D E Chimenti (Melville: AIP Conf. Proc. **820**) 853
[6] Nobakhti A and Wang H 2008 A simple self-adaptive Differential Evolution algorithm with application on the ALSTOM gasifier *Applied Soft Computing* **8** 350
[7] Price K, Storn R and Lampinen J 2005 *Differential Evolution: a Practical Approach to Global Optimization* (Springer)
[8] Deb K 2000 An efficient constraint handling method for genetic algorithms *Comput. Methods Appl. Mech. Eng.* **186** 311

Journal of Physics: Conference Series **269** (2011) 012009 doi:10.1088/1742-6596/269/1/012009

# A 45° silicon mirror for acoustic propagation parallel to the plane of the substrate

Shengxiang Wang [a)], Julien Carlier [b)], Pierre Campistron, Weijiang Xu, Dorothée Callens-Debavelaere, Bertrand Nongaillard, Assane NDieguene

*Univ Lille Nord de France, F-59000 Lille, France*

*UVHC, IEMN, F-59313 Valenciennes, France*

*CNRS, UMR 8520, F-59650 Villeneuve d'Ascq, France*

Xingzhong Zhao

*Department of Electrical Science and Technology, Key Laboratory of Acoustic and Photonic Materials and Devices of Ministry of Education, Wuhan University, Wuhan 430072, China*

## Abstract

The interest of a thin silicon oxide layer on a {110} silicon crystal planes used as 45° acoustical mirrors for reflection has been demonstrated. Passive mirror-like planes (groove depth ~150 μm) were fabricated by wet chemical etching of (100) silicon. High frequency acoustical longitudinal waves were obtained thanks to thin film ZnO transducers on the bottom surface of the silicon wafer. 45° mirrors had been prepared on the top side of the wafer. The transmission signal was characterized between two transducers using two mirrors. The results of acoustical characterization show that thanks to a thin silicon oxide layer deposited on the mirror it is possible to improve the efficiency of the reflection of the longitudinal wave on the mirror by about one order of magnitude.

[a)] Email address : shengxiang.wang@iemn.univ-lille1.fr
[b)] Email address : jullien.carlier@univ-valenciennes.fr

Journal of Physics: Conference Series **269** (2011) 012009 doi:10.1088/1742-6596/269/1/012009

# I. Introduction

. High frequency (1 GHz) ultrasonic measurements may be sensitive to some properties on a micrometer scale [1-3]. Moreover, the miniaturization of fluidic systems has given rise to many exciting applications in biotechnology [4]. It is possible to integrate high frequency ultrasonic chips with microfluidic for application of cell biology and/or biotechnology. To make our system user friendly, one face of the silicon wafer is used to fabricate the microfluidic channels and the other face to integrate the ultrasonic devices. A microsystem is designed as to realize acoustic measurement in transmission mode through the microfluidic channel. It includes two 45° mirrors and a pair of ultrasonic transducers used as emitter/receiver. A schematic view of the mirrors and piezoelectric transducers fabricated on the silicon substrate is shown in Figure 1.

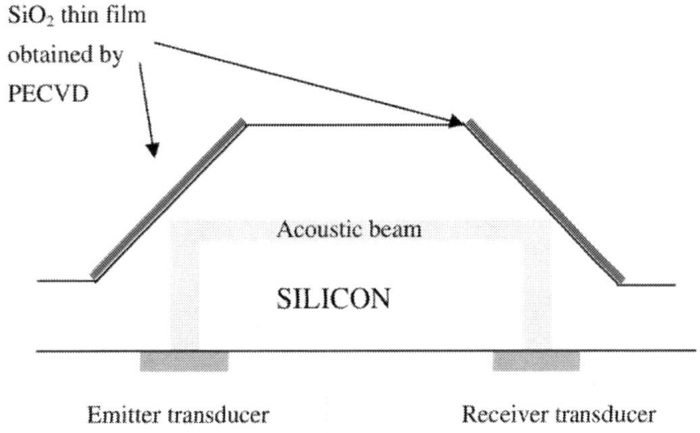


Figure 1. Scheme of the mirrors and piezoelectric transducers fabricated on silicon substrate.

The first face of the {1 1 0} silicon wafer is wet etched to obtain 45° micromachined mirrors as it can be done for optical applications [5-7]. The transducers can be fabricated by bonding single crystals or depositing thin films on a host wafer. The focus of this paper is to present the use of this kind of mirrors for acoustical signals transmitting through silicon in a direction parallel to the surface of the substrate. By means of such acoustical transmission, samples could be put on the path of the signals for characterization. Moreover the interest of coating a matching layer on these mirrors to improve the reflection of a longitudinal wave is demonstrated.

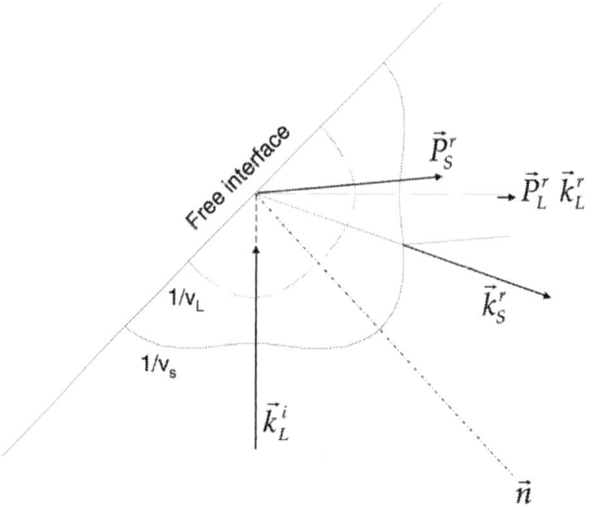


Figure 2. Propagation of acoustic wave in silicon. $v_L$ and $v_S$ velocity of longitudinal and shear wave; $\vec{k}_L^i$, $\vec{k}_L^r$ incident and reflected longitudinal wave vectors; $\vec{k}_S^r$ reflected shear wave vectors; $\vec{P}_L^r$, $\vec{P}_S^r$ reflected longitudinal and shear Poynting vectors.

The reflection of an acoustic longitudinal wave upon a free surface in oblique incidence produces a mode conversion, so that there are reflected waves of both the longitudinal and shear polarizations [8]. The directions of the wave vectors of these waves can be obtained using the Snell law $\frac{\sin\theta_i}{v_i} = \frac{\sin\theta_r}{v_r}$, with $v_i$ the velocity of the incident wave and $\theta_i$ the incident angle, $v_r$ the velocity of the reflected wave, which can support longitudinal or shear polarization, and $\theta_r$ the reflected angle. The longitudinal velocity $v_i$ in the {0 0 1} direction being the same as $v_r$ in the {0 1 0} direction ($v_l$ = 8432 m/s), the longitudinal wave is deflected by 90° ($\theta_i$ = 45°). It is not the same for the shear wave. Its slowness curve is computed (figure 2), from which the Snell law is solved in term of reflection angle ($\theta_{rs}$ = 27°) and velocity ($v_s$ = 5398 m/s). The direction of the shear energy propagation is given by the Poynting vector which is normal to the slowness curve. On can deduce that the energy of the shear wave propagate nearly horizontally with a deflection angle of about 95°. So a quasi shear wave and a longitudinal wave reach the second mirror. But this shear wave cannot be converted on the second mirror in a longitudinal wave. In fact, the Snell law clearly shows that for this range of incident angle, sin($\theta_{rs}$) is more than 1 and this shear wave is converted in an evanescent wave along the mirror. That is why, in this system, the conversion from shear to longitudinal wave on this mirror cannot be observed.

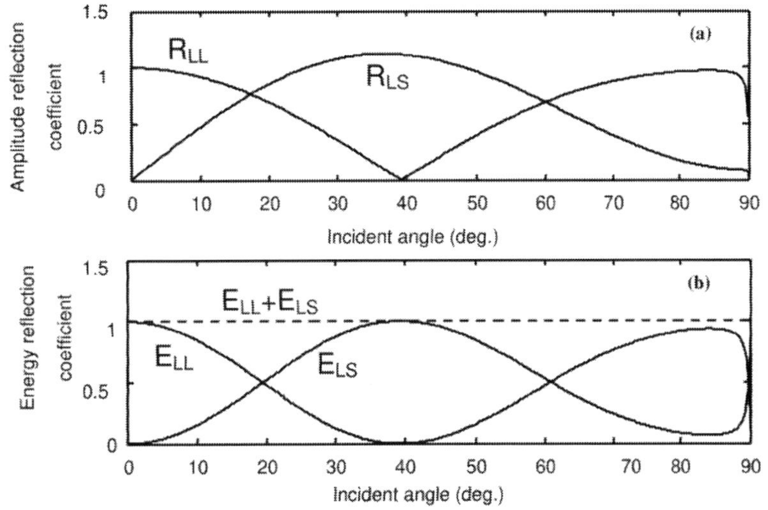


Figure 3. Reflection coefficient of longitudinal mode LL and shear mode TL for a longitudinal wave incidence at silicon-vacuum interface (a) in amplitude and (b) in energy. (ultrasonic characteristics of silicon : $V_L$= 8457 m/s; $V_T$ = 5887m/s; ρ= 2329 kg/m$^3$).

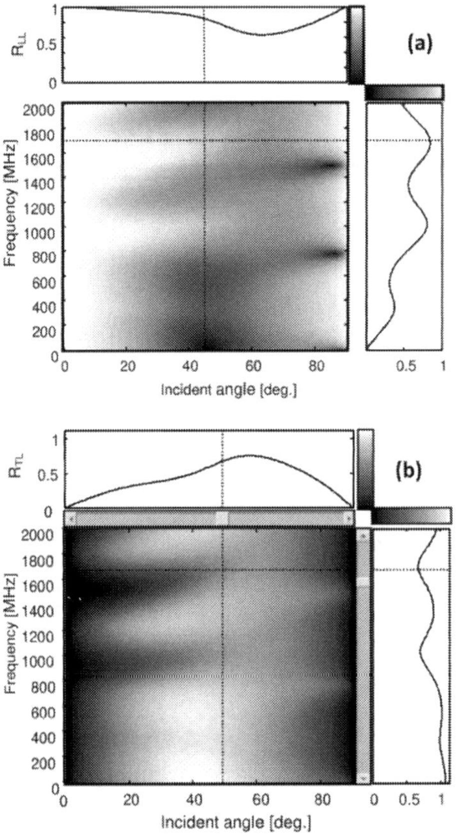


Figure 4. Reflection coefficient in amplitude of (a) longitudinal mode $R_{LL}$ and (b) transverse mode $R_{TL}$, for a longitudinal wave incidence at silicon-$SiO_2$ layer (4μm in thickness)-vacuum system (ultrasonic characteristics of $SiO_2$ : $V_L$= 6050m/s; $V_T$ = 4100 m/s; ρ= 2200 kg/m$^3$).

To estimate the wave reflection of different modes at the surface of the silicon mirror, we have taken a simplified isotropic model for silicon material and calculated the wave reflection coefficient in amplitude and in energy as a function of the incident angle for a longitudinal incident wave. A simplified model has been considered here because the use of the longitudinal wave is mainly interesting. Moreover, after reflected on a 45° mirror, the direction of propagation of such wave is equivalent to its propagation before the mirror due to the cubic crystalline structure of silicon. It is

Journal of Physics: Conference Series **269** (2011) 012009 doi:10.1088/1742-6596/269/1/012009

shown in Figure 3 that at 45° incidence, only less than 10% of energy is converted into the longitudinal wave.

Nevertheless the simulations have shown that the deposition of a thin film on the mirror can improve the reflection of the longitudinal wave on the mirror instead of converting it in a shear wave. It has been achieved with a thin $SiO_2$ thin film (about 4 μm). It can be observed in figure 4 that with this layer 90 % of the longitudinal wave is reflected at the frequency of 1.7 GHz.

## II. Experiment

The experimental work was performed on silicon wafer with ⟨100⟩ crystal orientation (*Siltronix, France*). Firstly, 1 μm thick $SiO_2$ masks were deposited on both sides of the wafer by PECVD (Plasma Enhanced Chemical Vapor Deposition) at 300 °C. Then rectangular grooves were aligned parallel to the $[0\bar{1}1]$ direction, perpendicular to the flat. $CHF_3$ and $CF_4$ (ratio of 1:1) were used to etch the opened $SiO_2$ after the patterns were obtained by S1828 (*Shipley Corporation*) positive photoresist. In order to obtain {110} plane 45° mirrors, wet micromachining employed anisotropic aqueous KOH solution was performed. Addition of isopropyl alcohol (IPA) was used to achieve smooth {110} plane [9]. Passive mirror-like planes (groove depth ~150 μm) were obtained finally. Figure 5 shows the cross-section scanning electron microscopy (SEM) image of 45° silicon mirror. After the {110} plane 45° mirrors were achieved on one side of the silicon wafer, ZnO piezoelectric transducers with thickness of about 1 μm were deposited on the other side of the silicon wafer using radio frequency magnetron sputtering [10]. As shown in Figure 1, the transducers were aligned to be under the mirrors. A silicon oxide thin film with thickness of 4 μm was deposited by means of the classic PECVD technique on the silicon 45° mirrors.

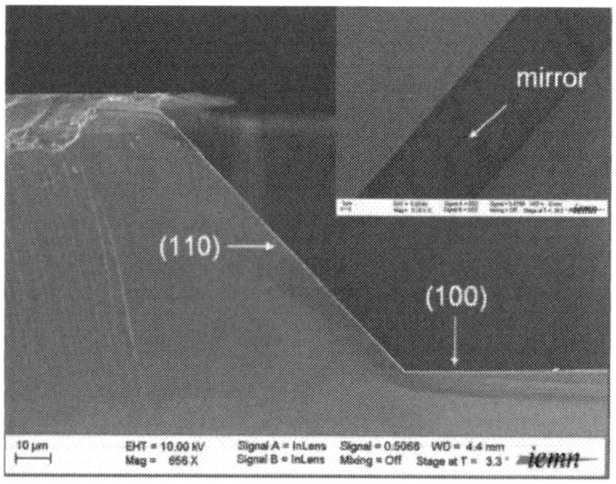


Figure 5. SEM image of the cross-section of the 45° mirror. The inset shows the top view of the mirror.

# III. Results and discussion

The scattering parameters were measured between the two transducers as shown in Figure 1 by a Rhode and Scharwtz Virtual Network Analyzer with a Karl Suss prober system. The central frequency of the transducers was characterized as about 1.75 GHz. In order to determine the acoustic velocity in silicon, $S_{11}$ impulse response of the system were obtained by the inverse Fourier transform. Longitudinal acoustic wave converting to shear wave was observed. The velocity of longitudinal wave and shear wave were calculated to be 8457 m/s and 5887 m/s, respectively, because the distances are known for normal incidence and reflection on the upper face of the substrate.

Figure 6 shows the impulse response of the system obtained by the inverse Fourier transform of the $S_{21}$ parameter without and with silicon oxide mirror coating. The distance between the mirrors $d_M$ is 390 μm and the total length of the path between the generator and receiver transducer $d_P$ is 1080 μm. From figure 6, it was found that the measured impulse response of longitudinal wave traveling from the generator transducer to the receiver transducer, named $t_{LLL}$, is 124.7 ns. Comparing to the expected delay for the longitudinal wave $t_{LLL}$ which is 127.7 ns, the error is of less than 3 %.

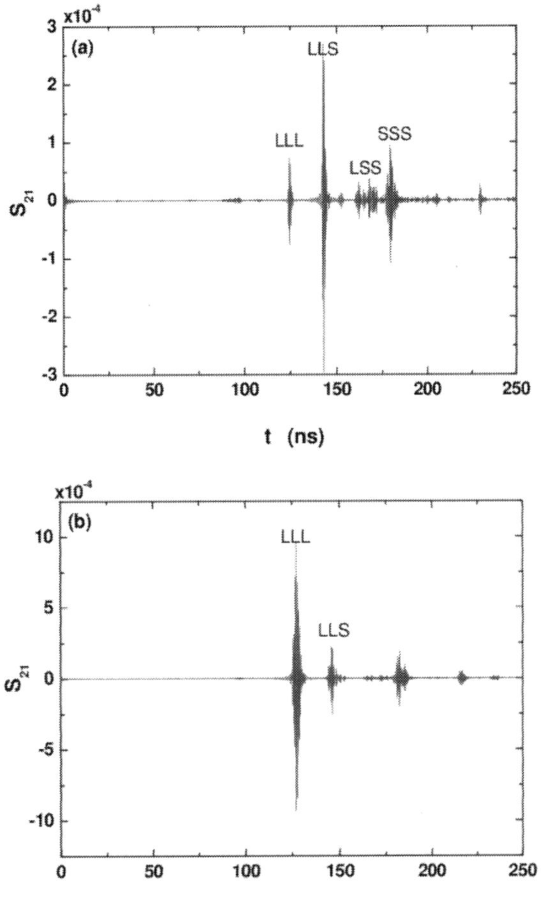


Figure 6. $S_{21}$ impulse response of the system vs. time (a) without silicon oxide on the mirror; (b) with a 4 μm silicon oxide matching layer on the mirrors.

In the system discussed, the longitudinal waves have two chances to convert to shear wave since there are two mirrors on the path of the signal from the first transducer to the second transducer. The impulse response of longitudinal wave converting to shear wave when the acoustic wave reached the second mirror, $t_{LLS}$, was observed to be 142.9 ns, which is of 2% error from the expected 145.5 ns. The comparison of the $S_{21}$ parameters from emitter to receptor with and without the silicon oxide layer on the mirror have shown an increase from 10 % to 80 % for amplitude as expected with the simulation. It can be seen in Figure 6 that the efficiency of the reflection of the longitudinal wave on the mirror can be improved by means of the deposition of a 4-μm thickness $SiO_2$ matching layer on the mirror. The increase of amplitude of the LLL signal is more than 20 dB.

## IV. Summary

A 45° mirror etched in silicon was presented for reflecting acoustic waves and integration of the mirrors with ZnO piezoelectric transducers in this paper. Scattering parameters of the system were characterized. The results show that the mirrors are efficient in reflecting acoustic waves. A thin silicon oxide layer deposited on the mirror can improve the efficiency of the reflection of the longitudinal wave on the mirror by about one order of magnitude.

## References

[1] R A Lemons and C F Quate 1974 *Appl. Phys. Lett.* **24** 163

[2] T Kundu, J Bereiter-Hahn, and I Karl 2000 *Biophys. J.* **78** 2270

[3] T Kundu, J P Lee, C Blase and J Bereiter-Hahn 2006 *J. Acoust. Soc. Am.* **120** 1646

[4] D Sobek, A M Young, M L Gray, and S D Senturia 1993 February 7-10 *Proc. IEEE Micro Electro Mechanical Systems* 219

[5] J D Sadler, M J Garter, C H Ahn, S Koh and A L Cook 1997 *J. Micromech. Microeng.* **7** 263

[6] M Hoffmann and E Vogaes 2002 *J. Micromech. Microeng.* **12** 349

[7] D Resnik, D Vrtacnik, U Aljancic, M Mozek and S Amon 2005 *J. Micromech. Microeng.* **15** 1174

[8] B A Auld 1990 Acoustic Fields and Waves in Solids, 2nd Edition, Vol II, Krieger Publishing Company

[9] C Strandman, L Rosengren, Hikan G A Elderstig, and Y Backlund 1995 *J. Microelectromech. S.* **4** 213

[10] S X Wang, P Campistron, J Carlier, D C -Debavelaere, B Nongaillard, A NDieguene, G Nassar, C Soyer and X Z Zhao 2009 *IEEE Transactions on Ultrasonics, Ferroelectrics, and Frequency Control* **56** 1483

Journal of Physics: Conference Series **269** (2011) 012010 doi:10.1088/1742-6596/269/1/012010

# Relation between ultrasonic scattering and microstructure of polycrystalline materials

**O Dupond[1], N Feuilly[1], B. Chassignole[1], T. Fouquet[2], J. Moysan[3] and G. Corneloup[3]**

[1] EDF R&D, Material and Mecanics of Components Department, Site des Renardières, 77818 Moret-sur-Loing, France
[2] EDF R&D, Neutronics simulation, information technologies and scientific calculations department, Clamart, France
[3] Laboratoire de Caractérisation Non-Destructive, Université de la Méditerranée, 13625 Aix-en-Provence, France

**Abstract** : Ultrasonic testing is a very common volumetric technique used in industry in order to detect and characterized defects. Within the framework of the maintenance of its nuclear power stations, Electricité de France (EDF) uses this technique on various components. In the case of polycrystalline materials, the scattering of the ultrasonic wave can result in an important attenuation of the signal and the appearance of structural noise. In the present study, we propose to analyze scattering noise in relation to the microstructure of the material. Both experimental and 2D finite elements modelling results are presented for different controlled structure of an homogeneous and isotropic Ni-based alloy.

## 1 – Introduction

Maintenance represents, for Electricité de France (EDF), a major issue to guaranty the safety of its nuclear power plants. In order to detect and characterize a potential defect, the use of ultrasonic inspection is widely spread. In this technique, the detection is mostly based on an amplitude criterion of the signal reflected from the defect. But while the ultrasonic wave is propagating inside the component, it interacts with the structure resulting in a random scattering of the wave. This phenomenon induces two main effects on the recorded echo: the attenuation of signal and a disturbing backscattered noise. In specific configurations, these effects are so important that the performances of ultrasonic techniques are not guaranteed. These conditions have been studied in previous works [1-4] and analytical expressions of the attenuation as a function of wavelength to grain size ratio ($\lambda/D$) have been proposed [2, 5].

For onsite nuclear inspection, the use of one non destructive testing (NDT) method is conditioned by a pre-qualification which has to demonstrate the performances of the chosen technique. In this context, modelling codes can be very helpful in the performance demonstration by parametric studies. Nonetheless, in order to get quantitative modelling results, all physical phenomena and particularly beam-to-structure interactions have to be taken into account.

In the literature, several modelling codes have been used to describe ultrasound wave propagation in solid media including analytical and semi-analytical approaches and Finite Element methods [1-4], [6-8]. EDF concentrates its efforts on one finite elements code called ATHENA 2D developed in collaboration with the French national Institute for research in Computer Science and Control (INRIA) [9]. In order to implement the wave scattering in the model, recent developments have been performed [10, 11]. The reproduction of the attenuation by the model can be implemented using two different approaches. The first one consists in modifications of the mathematical formulation of the equations,

Journal of Physics: Conference Series **269** (2011) 012010 doi:10.1088/1742-6596/269/1/012010

with the introduction of a tensor determined on the basis of experimental values of ultrasonic attenuation coefficients [11].
This development of the code has proven its efficiency in reproducing the attenuation of the acoustic wave during its propagation but can inherently not reproduce structural noise.
The second solution consists in coupling the conventional elastic model with a description of the material sophisticated enough to correctly reproduce the scattering phenomenon. The objective of the present study is to evaluate the relevance of a grain scale description of the material for the simulation of the ultrasonic beam scattering for a wide range of λ/D ratio. This paper is dedicated to the validation of this modelling approach on backscattered noise criteria. The first part introduces the experimental mock-ups, defects and measurements. In the second part, the ATHENA code and the modelling of the microstructure are introduced. Finally, experimental and modelling results are discussed with a view to validate this approach.

## 2. Experimental set-up

### *2.1. Material characterization*

The scattering of ultrasound in polycrystalline materials is complex as it involves interaction with microstructural features of the material (grain boundaries, precipitates, twins…). For simplification considerations, the study is conducted on forged bars of 135 mm diameter in Inconel 600®, a nickel based alloy used in the nuclear sector. Indeed, this material is purely austenitic and exhibits a homogeneous structure in terms of chemical composition and grain size.

Moreover, the metallographic examinations performed on this material (Figure 1) reveal an equiaxed structure characteristic of isotropic elastic properties. The mean grain size diameter ($\overline{D}$), estimated according to American Society for Testing and Materials (ASTM) standards, is equal to 220 microns. The presence of twins has furthermore been observed.

In order to study the scattering phenomenon on different microstructures in term of grain size, we have chosen to apply heat treatments at different temperatures on several samples. The consequence of those treatments is the increase of the mean grain size [12]. From 1150°C and a holding time of 90 minutes, the mean grain size increases and reaches a value of 450 microns (Figure 1b). After a treatment at 1200°C, the grain size is estimated at 1 mm (Figure 1c). Table 1 summarizes the characteristics of the various microstructures studied in this work.

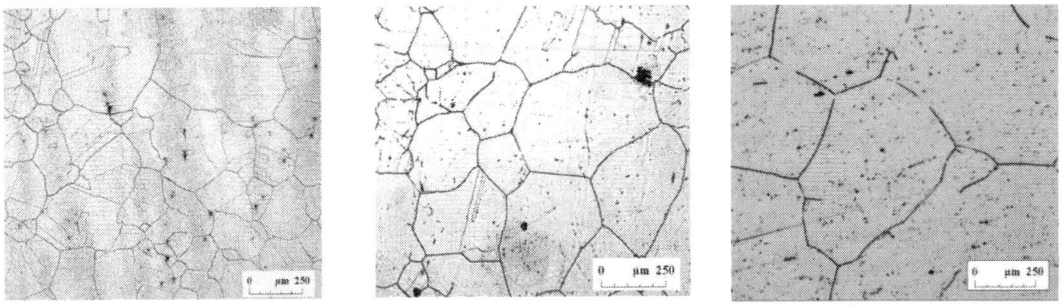


*(a)* *(b)* *(c)*

**Figure 1 :** (a) Initial state; (b) 1050°C; (c) 1200°C.

**Table 1.** Characteristics of microstructures studied.

| Sample | Heat treatment temperature (°C) | Heat treatment duration (mins) | Grain size (µm) |
|---|---|---|---|
| 1 | Initial state | | 220 |
| 2 | 1150 | 90 | 450 |
| 3 | 1200 | 360 | 1000 |

*2.2 Inspection configuration*

The objective of the experimental set-up is to produce some reference data to be compared to modelling results. Figure 2 presents the mock-up geometry. It consists of a section of the forged bar with a 100 mm thickness and a 135 mm diameter.

The reference defects used in this study are flat bottom holes of 30 mm height and respectively 3 mm, 5 mm and 8 mm diameter.

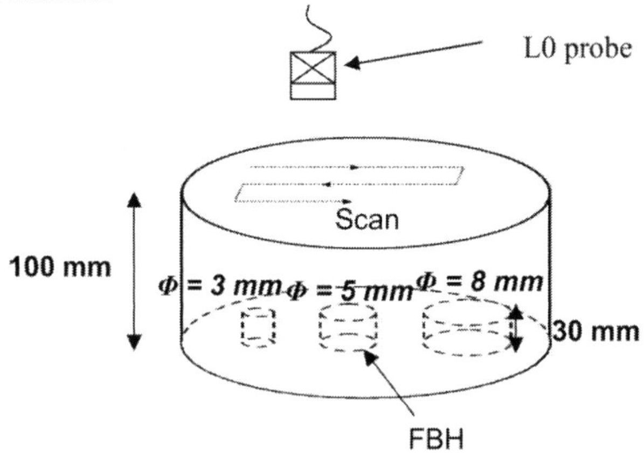


**Figure 2.** Schematic representation of the experimental set-up.

The ultrasonic inspection consists in an immersion inspection with longitudinal waves at 0° incidence. Three probes with different central frequencies (1 MHz, 2.25 MHz and 5 MHz) are selected in order to analyze the ultrasonic scattering for a large range of $\lambda$/D ratio.

The position of the probe from the upper surface of the mock-up is selected to perform a far field inspection. Once the distance from the upper surface is adjusted, the orientation of the probe is controlled by obtaining the maximum amplitude of the echo reflected by the upper surface in order to assure the orthogonality between the ultrasonic beam and the sample surface.

The machining respects the parallelity of the upper, backwall and the flat-bottom hole (FBH) surfaces.

*2.3. Estimation of the SNR*

The comparison between experiment and modelling is based on the signal to noise ratio estimation. The signal corresponds to the specular echo reflected by the flat bottom hole. The highest echo amplitude on the whole scan is kept in. The backscattered noise is defined as the maximum of echo amplitude out of defect localization and in a time gate of 1 μs located around the top of the FBH.

## 3. Modelling approach : ATHENA 2D Code

*3.1. ATHENA 2D Code*

ATHENA 2D is a finite elements code for elastodynamics developed by EDF Research and Development Division in collaboration with INRIA [9]. The model uses a regular mesh for the calculation zone. Contact and immersed transducers are implemented. Calculations can be performed for all type of structures, especially anisotropic and heterogeneous structures [13] and take into account wave attenuation due to grain scattering [10, 11]. The beam propagation and beam to defect interactions are calculated with all conversion modes. Furthermore, defects are represented with the fictitious domain method which gives the opportunity to separate the defect and the component meshes [14].

*3.2. Grain scale modelling compatible with ATHENA 2D*

A study of the attenuation parameter has estimated the judicious scale necessary to describe the structure [15, 16]. Indeed, it has been demonstrated that, if the first order parameter was grain boundaries, twins have also to be taken into account to correctly reproduce the scattering phenomenon. Indeed, twins are 2D cristallographic defects corresponding to a stacking fault of the dense planes, leading to a mirror symmetry of 60° relatively to the {111} plane. In first approximation, twins can be considered as grain boundaries. Consequently twin modelling can be achieved by reducing the mean grain size of the microstructure evaluated by a metallographic examination.

Microstructure modelling of polycrystalline materials can be computed using Voronoï diagrams. Voronoï polycrystals have already been used to study elastic wave scattering and provided a reliable description of a polycrystalline structure [8]. Figure 3 shows an example of a Voronoï microstructure. An example of the comparison between the simulated and the experimental grain size distribution is shown in Figure 4.

**Figure 3:** Microstructure described by Voronoï diagrams.

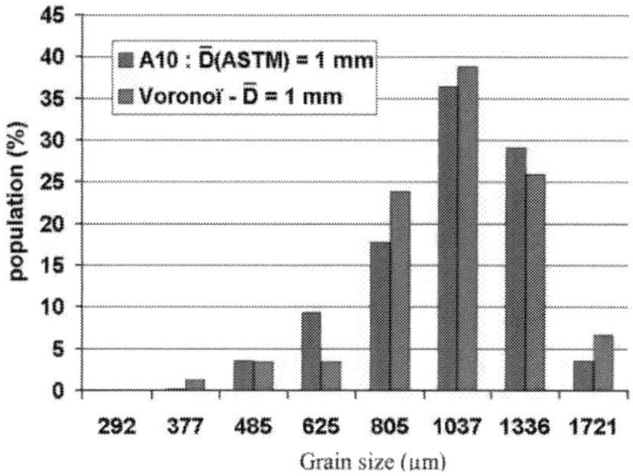


**Figure 4:** Histogram representing the grain size distribution estimated by metallographic examination of a metallography (sample A10) and of Voronoï description.

As the forged bar in Inconel 600® exhibits an isotropic structure, the determination of the orientation of one individual grain, represented by a Voronoï cell, is made through a random drawing on a thousand possible orientations. The matrix of elastic constants of each grain, expressed in the reference sample is then determined based on the single crystal elastic constants ($C_{11}$ = 234.6 GPa, $C_{12}$ = 145.4 GPa, $C_{44}$ = 126.2 GPa in the single crystal reference) [17] and the crystallographic orientation of each grain. Finally, since the 2D version of ATHENA software is used, only the four elastic constants whose indices correspond to the propagation plan are considered.

## 4. Results

Typical Bscan images are presented in the Figure 5 for both experimental and modelling results for the inspection at 5 MHz frequency of the sample with a mean grain size equal to 850 µm. These results demonstrate the relevance of finite elements and grain-scale modelling to reproduce the following scattering phenomena :

1. The presence of a backscattered noise in the form of random echoes whose amplitude decreases as the time of flight increases. This result is attributed to the attenuation of the backscattered signal and to its spatial diffusivity.
2. The poor signal to noise ratios for the FBH echoes due to both the attenuation of the defect echo and the high level of structural noise.

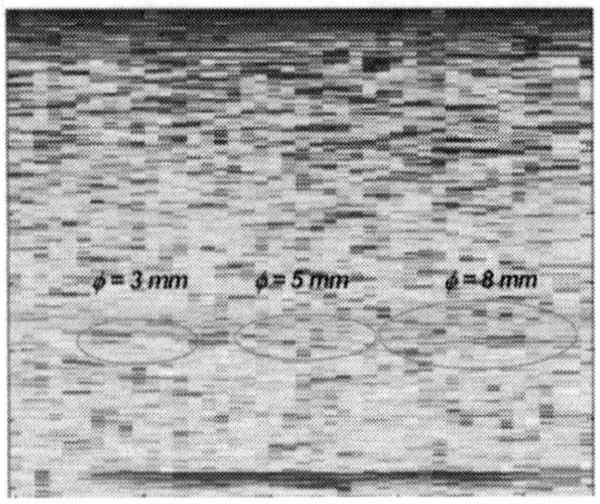


**Figure 5.a:** Experimental Bscan.

**Figure 5.b:** Voronoï description for the mean grain size equal to 850 μm.

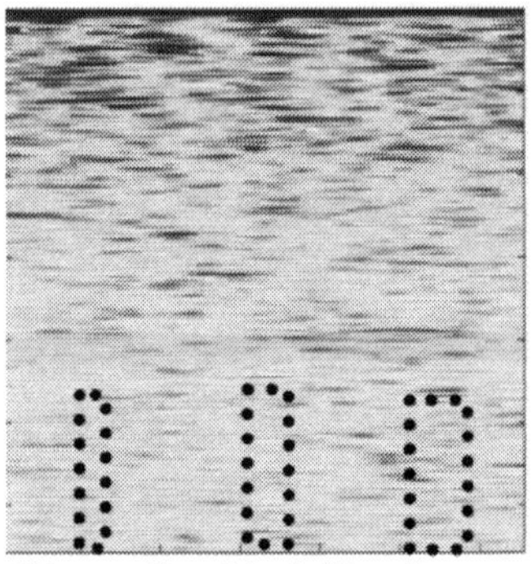

**Figure 5.c:** Modelling Bscan.

**Figure 5:** Experimental and modelling results for the 850 μm grain size mock-up with a 100 mm thickness.

To complete this validation study, quantitative values of signal to noise ratios were evaluated for every inspection configurations. The comparison between experimental and modelling SNR are presented in table 2 and table 3 for microstructures with a mean grain diameter respectively equal to 370 μm and 850 μm.

As the local microstructure around the defect is unknown, the amplitude of the FBH signal in modelling corresponds to the mean of echo amplitudes obtained from 10 different Voronoï descriptions. On the other hand, the backscattered noise in modelling corresponds to the maximum amplitude estimated for each 2D Voronoï description in order to be in agreement with the experimental procedure.

A limitation of the present approach is that 2D modelling overestimates the amplitude of the FBH echo for certain configurations (especially for 3 mm diameter FBH and low frequencies) because the fact that the beam width is higher than the defect diameter in the third dimension is not taken into account. To overcome this difficulty, the 2D effect is compensated using correction coefficients calculated for each defect and each probe. They are defined as the difference between experimental and modelling FBH to backwall echo amplitude ratio estimated in a structure with small grains corresponding to a very low attenuation.

**Table 2:** Experimental and modelling results in term of signal to noise ratio (dB) for a mean grain size equal to 370 μm.

| Frequency (MHz) | FBH 3 mm | | FBH 5 mm | | FBH 8 mm | |
|---|---|---|---|---|---|---|
| | Experiment | ATHENA 2D | Experiment | ATHENA 2D | Experiment | ATHENA 2D |
| 1 | 6.0 | 5.9 | 10.9 | 11.0 | 16.2 | 15.4 |
| 2.25 | 3.4 | 5.3 | 8.8 | 10.0 | 15.9 | 15.4 |
| 5 | Not detected | Not detected | 0.2 | 0.2 | 4.8 | 4.0 |

**Table 3:** Experimental and modelling results in term of signal to noise ratio (dB) for a mean grain size equal to 850 μm.

| Frequency (MHz) | FBH 3 mm | | FBH 5 mm | | FBH 8 mm | |
|---|---|---|---|---|---|---|
| | Experiment | ATHENA 2D | Experiment | ATHENA 2D | Experiment | ATHENA 2D |
| 1 | 3.7 | 2.2 | 8.0 | 6.1 | 11.5 | 11.2 |
| 2.25 | 0.2 | Not detected | 3.9 | 3.4 | 10.7 | 10.6 |
| 5 | Not detected | Not detected | Not detected | Not detected | 0.7 | 0.7 |

From table 2 and table 3, we can make the following comments:

- for one FBH : the signal-to-noise ratio (SNR) decreases as the frequency increases ;
- the SNR decreases as the grain diameter increases;

Those two results indicate that the scattering phenomenon is increasing with the mean grain diameter and with the frequency. These results are consistent with those obtained on the attenuation coefficient in the Rayleigh and stochastic scattering domains [15]. It is important to note that a mean grain size of 850 μm and a 5 MHz frequency conduct to a $\lambda/D$ ratio equal to 1.4. This value is characteristic of the stochastic domain and is associated to a maximal value of attenuation. Higher grain sizes or higher frequencies should be studied to analyze the variation of SNR in the geometric domain in which the attenuation is only dependent on the grain size and decreases when D increases. Finally, modelling results fit well the experimental data, with differences inferior to 2 dB over all the configurations tested.

This demonstrates the relevance of the modelling approach consisting in describing the microstructure at the twin and grain scale and confirms that the Voronoï diagrams are well adapted to represent a polycrystalline material. Consequently this study presents a validation of the finite elements code ATHENA 2D to simulate the scattering of a longitudinal wave propagating in a polycrystalline material.

## 5. Conclusion

In conclusion, the present study gives an experimental validation of ATHENA 2D code to model the scattering phenomenon of an ultrasonic compression wave propagating at 0° incidence in a polycrystalline material. The description of the microstructure at the grain scale is necessary to simulate the scattering phenomena and the level of refinement must integrate the crystallographic defects that are the grain boundaries and twins. It has furthermore been demonstrated that the Voronoï diagrams are appropriate to describe the structure. With this approach, it has been demonstrated that the FE model gives quantitative accurate results. Furthermore, it is important to notice that the multi-scattering is intrinsically integrated in the model. This is a very important difference and advantage of FE code compared to analytical models or semi-analytical models which systematically make the assumption of single scattering.

In perspectives, this study will be carried on in different directions. The first one will be dedicated to the validation of the code for shear waves and angular incidences. Finite elements modelling could also be an useful tool to develop methods of material characterization (grain size) from information given by the backscattered signals. For example, preliminary experimental works highlight differences between microstructures from the observation of a retro-scattering pick [18]. The last effort would be to develop the 3D code and associated 3D grain scale modelling. Indeed, thanks to the high computer performances and the joined parallelization of the code, this target is ambitious but realistic.

## 6. References

[1] Hirsekorn S 1982 Scattering of ultrasonic waves by polycrystals *Journal of the Acoustical Society of America* **72** pp 1021-1031
[2] Papadakis E P 1965 Revised grain-scattering formulas and tables Journal of the Acoustical Society of America **37** pp 703-710
[3] J. H. Rose J H 1991 Ultrasonic backscattering from polycrystalline aggregates using time-domain linear response theory *Review of progress in Quantitative Nondestructive Evaluation* pp 1677-1684
[4] Saniee J, Wang T and Bilgutay N M 1988 Statistical evaluation of backscattered ultrasonic grain noise *Journal of the Acoustical Society of America* **84** pp 400-408
[5] Reynolds W N and Smith R L 1984 Ultrasonic wave attenuation spectra in steels *Journal of Applied Physics* **17** pp 109-116
[6] Stanke F E and Kino G S 1984 A unified theory for elastic waves propagation in polycrystalline materials *Journal of the Acoustical Society of America* **75** pp 665-670
[7] Yalda-Mooshabad I, Margetan F J and Thompson R B 1996 Predicting Ultrasonic grain noise in polycrystals: a Monte-Carlo Model *Journal of the Acoustical Society of America*
[8] Turner J A and Ghoshal G 2004 Numerical models of longitudinal wave scattering in polycrystals

[9] Tsogka C, Becache E and Joly P 2000 An analysis of a new mixed finite elements for the approximation of wave propagations problems *SIAM Journal of Numerical Analysis* **37** pp 1053-1084

[10] Diaz J, Chassignole B, Duwig V, Fouquet T and Schumm A 2006 Structural noise in modelisation *Proceedings of the 9$^{th}$ European Conference on NDT (Berlin)*

[11] Chassignole B, Duwig V, Ploix M A, Guy P and El Guerjouma R 2009 Modelling the attenuation in the ATHENA finite elements code for the ultrasonic testing of austenitic stainless steel welds *Ultrasonics* **49** pp 653-658

[12] Garriga-Majo D 1993 Etude métallurgique de la corrosion sous contrainte en milieux aqueux de l'alliage 600 *In PhD Thesis, Université d'Orsay*

[13] Dupond O, Chassignole B, Doudet L and Birac C 2007 Methodology for modelling ultrasonic inspection of an austenitic stainless steel weld *Proceedings of 6$^{th}$ ICNDE in Relation to Structural Integrity for Nuclear and Pressurised Components (Budapest)*

[14] Becache E, Joly P and Tsogka C 2001 Application of the fictitious domain method to 2D linear elastodynamic problems *Journal of Computational Acoustic* **9** pp 1175-1202

[15] Nicolas F 2009 Etude de l'influence de la microstructure sur la diffusion ultrasonore en vue de l'amélioration du contrôle non destructif des matériaux polycristallins *In PhD Thesis, Université de la Méditerranée Aix-Marseille II*

[16] Feuilly N, Dupond O, Chassignole B, Moysan J and Corneloup G. 2008 Relation between ultrasonic backscattering and microstructure for polycrystalline materials *Review of progress in Quantitative Nondestructive Evaluation* pp 1216-1223

[17] Holden T M, Holt R A and Tomé C N 1999 Intergranular strains in Inconel-600 *Materials Science and Engineering* **A282** pp 131-136

[18] Aubry A and Derode A 2007 Ultrasonic imaging of highly scattering media from local measurements of the diffusion constant : separation of coherent and incoherent intensities *Phys Rev E* **75**

Journal of Physics: Conference Series **269** (2011) 012011 doi:10.1088/1742-6596/269/1/012011

# Temperature dependence of bulk viscosity in water using acoustic spectroscopy

**M J Holmes**[1], **N G Parker and M J W Povey**
School of Food Science and Nutrition. University of Leeds, LS2 9JT, UK

E-mail: prcmjh@leeds.ac.uk

**Abstract.** Despite its fundamental role in the dynamics of compressible fluids, bulk viscosity has received little experimental attention and there remains a paucity of measured data. Acoustic spectroscopy provides a robust and accurate approach to measuring this parameter. Working from the Navier-Stokes model of a compressible fluid one can show that the bulk viscosity makes a significant and measurable contribution to the frequency-squared acoustic attenuation. Here we employ this methodology to determine the bulk viscosity of Millipore water over a temperature range of 7 to 50°C. The measured attenuation spectra are consistent with the theoretical predictions, while the bulk viscosity of water is found to be approximately three times larger than its shear counterpart, reinforcing its significance in acoustic propagation. Moreover, our results demonstrate that this technique can be readily and generally applied to fluids to accurately determine their temperature dependent bulk viscosities.

## 1. Introduction

The parameter of bulk or volume viscosity $\mu$ receives little attention, despite being defined in the Navier-Stokes equation for a compressible liquid. It is of fundamental importance in the field of acoustic propagation where it plays the equivalent role in viscosity to the bulk modulus $K$ in solids. Specifically, the total viscosity for fluids $\eta_{total} = \mu + 4\eta/3$ is related to the longitudinal modulus $M = K + 4G/3$ in solids, where $G$ is the shear modulus and $\eta$ is shear viscosity, through the time dependence relation $i\omega\eta_{total} \rightarrow M$. In the absence of scattering, the classical theory of acoustic attenuation predicts a frequency-squared dependence on the viscosity. Shear viscosity is routinely measured using commercial rheological instrumentation, and bulk viscosity may be determined from a measurement of the frequency dependent attenuation, the shear viscosity and the ratio of specific heats for the fluid. However, very few measurements of this rheological parameter have been made despite its fundamental importance in the dynamics of compressible fluids.

The general description of the equations of motion for sound propagation in a compressible fluid is supplied by the Navier-Stokes and continuity equations [1, 2] in which the parameter of bulk viscosity is a natural consequence. Additionally, energy and thermodynamic equations are introduced [3-5] which describe the energy in the system using the state variables temperature, pressure and density. Considering only first order terms in time and assuming periodic state solutions of the form $\exp(i\omega t)$ enables the equations to be simplified by replacing partial time derivatives with the factor $-i\omega$ (where $\omega = 2\pi f$ is the angular frequency, $i$ is the imaginary number and $f$ is the frequency of the acoustic wave) to yield the following equations (1) and (2),

$$\omega^2 \mathbf{v} + \left(\frac{v^2}{\gamma} - i\omega\frac{N\eta}{\rho}\right)\nabla(\nabla\cdot\mathbf{v}) + i\omega\frac{\eta}{\rho}\nabla\times\nabla\times\mathbf{v} = -\frac{i\omega\beta v^2}{\gamma}\nabla T, \qquad N = \left(\frac{4}{3} + \frac{\mu}{\eta}\right) \quad (1)$$

$$\gamma\frac{\tau}{\rho C_p}\nabla^2 T + i\omega T = \frac{(\gamma - 1)}{\beta}\nabla\cdot\mathbf{v} \quad . \quad (2)$$

Here $\mathbf{v}$ is the velocity vector of the fluid, $T$ is the temperature, $\gamma$ is the ratio of specific heats, $\tau$ is the thermal conductivity, $\rho$ is the density, $C_p$ is the specific heat at constant pressure, $\beta$ is the bulk compressibility, $\eta$ is the shear viscosity, $\mu$ is the bulk viscosity and $v$ is the velocity of sound. In equation (1) the physical origins of the bulk viscosity is clear: it acts as a dissipative coefficient to dilatational-compressional motion of the fluid (described by the $\nabla\cdot\mathbf{v}$ term). Thus one can expect the

[1] To whom any correspondence should be addressed.

bulk viscosity to play a role in fluid dynamics wherever such motions exist, notably in acoustic propagation and shock waves.

For one dimensional plane wave propagation through fluids that are thermally non-conductive and viscous, solutions of the Navier-Stokes equations yield the following expression for the attenuation coefficient α [10]

$$2\left(\frac{\alpha v}{\omega}\right)^2 = \frac{1}{\sqrt{1+t^2\omega^2}} - \frac{1}{1+t^2\omega^2} \qquad (3)$$

where $t$ is the viscous relaxation time given by,

$$t = \frac{\eta}{\rho v^2}\left(\frac{4}{3}+\frac{\mu}{\eta}\right). \qquad (4)$$

Plotting α as a function of frequency reveals a Gaussian-like distribution profile having a maximum value at a critical frequency which corresponds to a timescale approximately equal to the viscous relaxation time. For Newtonian fluids this frequency is in the region of 1000GHz. In practice this is difficult to achieve, with current ultrasound instrumentation typically limited to the long wavelength limit given by $\omega t \ll 1$.

Equations (1) and (2) are further simplified by representing the velocity vector **v** in terms of scalar potentials which describe the longitudinal compressional field $\varphi$ and the thermal field $\psi$ together with a transverse shear vector potential **A** in the form, $\mathbf{v} = -\nabla(\varphi + \psi) + \nabla \times \mathbf{A}$. This allows equations (1) and (2) to be combined into a biharmonic type equation which may be then decoupled into independent Helmholtz equations for each given field as shown in equations (5) with respective acoustic, thermal and shear wavenumbers $K$, $L$ and $M$ (see [6-9] for full details),

$$(\nabla^2 + K^2)\,\varphi = 0 \qquad (\nabla^2 + L^2)\,\psi = 0 \qquad (\nabla^2 + M^2)\mathbf{A} = 0. \qquad (5)$$

In particular, the wavenumbers $K$, $L$ and $M$ are given by,

$$K = \frac{\omega}{v} + i\alpha = \frac{\omega}{v} + i\frac{\eta\omega^2}{2\rho v^3}\left[\frac{4}{3}+\frac{\mu}{\eta}+\frac{(\gamma-1)\tau}{\eta C_p}\right], \qquad (6)$$

and,

$$L = \left(\frac{\omega\rho C_p}{2\tau}\right)^{1/2}(1+i), \qquad M = \left(\frac{\omega\rho}{2\eta}\right)^{1/2}(1+i)\,. \qquad (7)$$

The dissipative terms in the wavenumbers are characterised by imaginary components, with the imaginary term in equation (6) parameterising the attenuation of the longitudinal acoustic wave in the long wavelength limit. The low frequency asymptotic expression for equation (3) may be derived and is given as,

$$\alpha_{long} = \frac{\eta\omega^2}{2\rho v^3}\left[\frac{4}{3}+\frac{\mu}{\eta}+\frac{(\gamma-1)\tau}{\eta C_p}\right]. \qquad (8)$$

Importantly, we see from equation (8) that this acoustic attenuation scales as the square of the frequency. Since this prediction is based upon the Navier-Stokes equation (1), which by definition describes a Newtonian fluid, a frequency-squared attenuation spectrum is synonymous with Newtonian fluidity. Vice versa, deviations from a frequency-squared spectrum provide a signature of non-Newtonian fluidity, offering an independent test to complement the well-established techniques based on shear rheometry.

From equation (8) we see that the total acoustic attenuation features a significant contribution from the bulk viscosity, highlighting the importance of considering the bulk viscosity when dealing in acoustic propagation. Equation (8) provides the theoretical and experimental route in calculating the bulk viscosity: by measuring the attenuation and sound speed, and assuming other parameters are either known or independently measurable, then one may calculate the bulk viscosity through the equation,

$$\mu = \left[\frac{2\alpha\rho v^3}{\omega^2}+\frac{4\eta}{3}+\frac{(\gamma-1)\tau}{C_p}\right]. \qquad (9)$$

Dukhin and Goetz [10] compared three techniques for bulk viscosity measurement: Brillouin spectroscopy, laser transient grating spectroscopy and acoustic spectroscopy. They concluded that acoustic spectroscopy is best placed and has the advantage of theoretical verification. They then proceeded to determine the bulk viscosity of water and various other fluids at a fixed temperature of 25°C. In this paper, we build upon the work conducted by Dukhin and Goetz [10] by extending the experimental measurements of bulk viscosity to the temperature domain. Specifically, we measure the bulk viscosity for Millipore water and reveal its temperature dependence over the range 7-50°C. In the next section we shall describe the measurement procedure for accurately determining the attenuation spectrum, and in Section 3 we will present and discuss the validation and results of our measurements, notably the temperature-dependence of the bulk viscosity of water. This work is novel in that a single study presents the bulk viscosity over several temperatures and, additionally, determines a functional fit to the temperature dependence. This approach may then readily be pursued in a systematic way to determine temperature dependent bulk viscosities of other fluids.

## 2. Measurement Techniques and Instruments

Acoustic attenuation measurements were carried out using an Ultrasizer MSV by Malvern Ltd [11] (see Figure 1), an instrument which performs acoustic spectroscopy of liquids and emulsions in the frequency range 1 – 100MHz. In brief, the instrument operates in transmission mode where one transducer emits sound into the sample while an opposing transducer detects the sound waves and generates an associated voltage signal. Two such pairs, one operating at low frequency and one at high frequency, are employed. Due to their broad-band response, the acoustic frequency of each pair can be swept over a wide range. The attenuation is measured from the decay of the received voltage with propagation distance through the sample. The diffraction of the beam, which also causes variations in the received voltage, is calibrated automatically and its effects factored out. Furthermore, by comparing the received signal at two propagation distances the non-trivial response of the transducer and electronic circuitry and any other systematic errors can also be factored out. Hence, an absolute measure of the acoustic attenuation is determined. Finally, the transducer separation is calculated automatically for a given fluid so as to exploit the full dynamic range of the measurement system and maximise the signal-to-noise ratio.

The Ultrasizer requires 500ml of sample and so, although offering an effective measurement system, is not suited for limited sample volumes. The Ultrasizer is connected to an external Huber Ministat temperature control unit which enables the sample temperature to be varied across a range of 0 – 50°C. During the measurement process the sample under test is agitated by a stirrer whose angular speed can be varied in the range 0-500 rpm. The agitator is employed to reduce thermal fluctuations in the sample (for emulsions it also suppresses sedimentation and creaming). This combination maintains the sample at the selected temperature to within ± 0.5°C, as measured by the internal thermometer of the Ultrasizer, and so provides a stable temperature to determine the associated acoustic attenuation. Note that the acoustic power levels are of the order of milliwatts and so the associated heating of the sample is negligible.

In this work we consider 0.22$\mu$m Millipore water as the sample. We restrict ourselves to the temperature range of 7 - 50°C. The lower limit is set to avoid microbubble formation at low temperature which can significantly affect ultrasonic measurements and potentially introduce erroneous attenuation measurements (see [12] for example). The upper limit is set by the operating conditions of the instrument. The instrument implements automated processes to conduct measurements at 50 different frequencies selected within the range 10 – 100 MHz. NB we set the lower bound to be 10 MHz rather than the instrument limit of 1 MHz since attenuations at low frequencies can be minimal and difficult to measure.

At each selected frequency of the acoustic spectra, 30 repeat attenuation measurements were made and, from these, an average attenuation value was calculated. We have verified that these 30 data values were symmetric about the mean value and could be fitted to a normal distribution. Finally, using the mean attenuation values across the chosen temperatures we calculated the mean bulk viscosity according to equation (9). Both average attenuation values and their standard error can then be quoted (as performed in Table 1).

**Figure 1.** The Malvern Ultrasizer™ [14] with its associated temperature control unit. The sample vessel is shown foremost in the picture. This requires 500ml of the sample under test. During the measurement process, the sample is agitated with a stirrer to reduce thermal fluctuations. The Ultrasizer can perform attenuation measurements in the frequency range 1-100MHz.

## 3. Results and Discussion

Our first task was to verify that the measured attenuation was not affected by the agitation speed. In Figure 2 we present (on a logarithmic scale) the attenuation spectrum of Millipore water at 25°C for agitator speeds of 150, 300 and 450 rpm. As can be seen, no discernable difference exists and so we subsequently measured all other samples using the intermediate speed of 300 rpm.

For the 300rpm results in Figure 2 we have fit them with a single polynomial in frequency of the form $f^{\delta}$. We find that an excellent fit ($R^2$=1) is provided by an exponent of $\delta$=2.0012. Indeed, we observe that all of our spectra give an exponent within a similar proximity to 2. We are thus well justified in fitting to the frequency-squared prediction of equation (8).

In Table 1 we show the calculated values for the bulk viscosity (derived from equation (9) using the experimentally measured values for the attenuation) together with the parameters required to make use of equation (9).

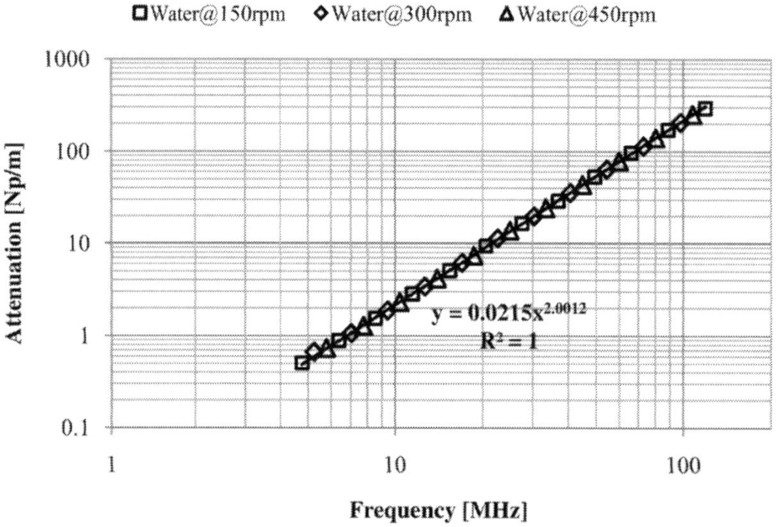


**Figure 2.** Attenuation-frequency plot for 0.22$\mu$m Millipore water at 25°C with agitation speeds 150, 300 and 450 rpm. As can be seen no appreciable difference was detected. Consequently all

measurements were made using the intermediate agitation of 300rpm. A polynomial best fit (for the 300 rpm case) is illustrated.

| Temperature | Density | Shear Viscosity | Thermal Conductivity | Specific Heat | Velocity | Bulk Viscosity | Standard error | Attenuation |
|---|---|---|---|---|---|---|---|---|
| $T$ | $\rho$ | $\eta$ | $\tau$ | $C_P$ | $v$ | $\mu$ | | $\alpha/f^2$ |
| °C | kg m$^{-3}$ | Pa s | W m$^{-1}$ K$^{-1}$ | J kg$^{-1}$ m$^{-3}$ | m s$^{-1}$ | Pa s | Pa s | Np m$^{-1}$ MHz$^{-2}$ |
| 7 | 999.81 | 1.43E-03 | 0.5747 | 4198.1 | 1434.92 | 4.50E-03 | 1.20E-05 | 4.28E-02 |
| 10 | 999.70 | 1.31E-03 | 0.5800 | 4192.1 | 1447.29 | 4.03E-03 | 1.17E-05 | 3.76E-02 |
| 15 | 998.97 | 1.14E-03 | 0.5900 | 4187.0 | 1465.96 | 3.38E-03 | 1.10E-05 | 3.07E-02 |
| 25 | 997.00 | 8.88E-04 | 0.6075 | 4180.1 | 1496.73 | 2.47E-03 | 1.08E-05 | 2.16E-02 |
| 40 | 992.22 | 6.53E-04 | 0.6305 | 4178.5 | 1528.89 | 1.84E-03 | 2.70E-05 | 1.51E-02 |
| 50 | 988.03 | 5.47E-04 | 0.6435 | 4180.6 | 1542.57 | 1.48E-03 | 2.76E-05 | 1.21E-02 |

**Table 1.** Parameter values for water for 6 selected temperatures. Bulk viscosity values are computed by averaging 30 measured attenuation values at each frequency. Each average value is then used to calculate the bulk viscosity using equation (6) at each frequency. A total mean bulk viscosity and its standard error are then computed by averaging over all frequencies. Velocity values taken from [12] and all remaining parameter values extracted from [13]. Parameter values at the temperature of 7°C were calculated using polynomial fits of the data in [13].

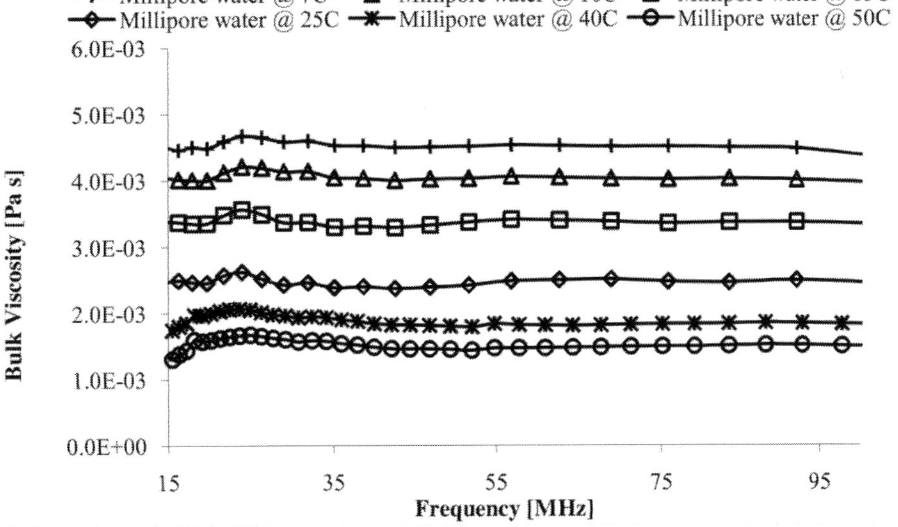


**Figure 3.** Bulk viscosity [Pa.s] for 0.22$\mu$m Millipore water against frequency [MHz] for the temperatures 7, 10, 15, 25, 40 and 50°C. We observe the bulk viscosity to be approximately constant with frequency, as expected for a Newtonian fluid.

In Figure 3 we have plotted the bulk viscosity as calculated using the measured attenuation $\alpha$ and equation (9). The approximately constant value is to be expected for a Newtonian fluid. In [10] Dukhin and Goetz reported a bulk viscosity value of 2.43E-03 Pa.s at 25°C for distilled water, which is in very good agreement with our value of 2.469E-03 [2.458E-03, 2.479E-03]. Similarly, they reported a ratio of bulk to shear viscosity as 2.73 whereas we obtained a value of 2.78. Another example cited in [10] is the result obtained by Litovitz and Davis [4] who reported a value of 3.09E-03 for the bulk viscosity of water at 15°C and viscosity ratio of 2.81, whereas in our case we record a value of 3.38E-03, with ratio 2.96. We additionally note the appearance of some anomalous features in the bulk viscosity in the vicinity of 25 MHz. Similar qualitative features are apparent in the results of Dukhin and Goetz [10] suggesting a common physical origin. This may be related to resonances in the molecular states and requires further study.

In Figure 4 we have plotted the frequency-averaged bulk viscosity as a function of temperature. For comparison, we plot the shear viscosity of pure water obtained from the literature [13]. The bulk viscosity is approximately 3 times larger than its shear counterpart throughout the temperature range, and both decrease with temperature in a similar monotonic form. We also plot the bulk viscosities determined from several other studies [4, 10, 12, 15]. Notably, Xu *et al* [15] used Brillioun scattering to determine the bulk viscosity of water over a similar temperature range. As can be seen, our results concur well with those obtained elsewhere.

A common empirical model for the temperature-dependence of shear viscosity is that of an exponential decay,

$$\mu = A_1 \exp(-A_2 T), \qquad (7)$$

($A_1$ and $A_2$ empirical constants) as first observed by Reynolds in 1886 [14]. Our bulk viscosity data closely follows such a model well with coefficients $A_1$=5.091712E-03 and $A_2$=2.545425E-02 ($R^2$=0.986). However, our data is most closely fitted by a cubic expression of the form,

$$\mu = B_0+B_1T+B_2T^2+B_3T^3 \qquad (8)$$

with coefficients $B_0$ = 5.94068E-03, $B_1$ = - 2.37073E-04, $B_2$ = 4.94789E-06 and $B_3$ = -3.97502E-08 ($R^2$ =0. 99991) .

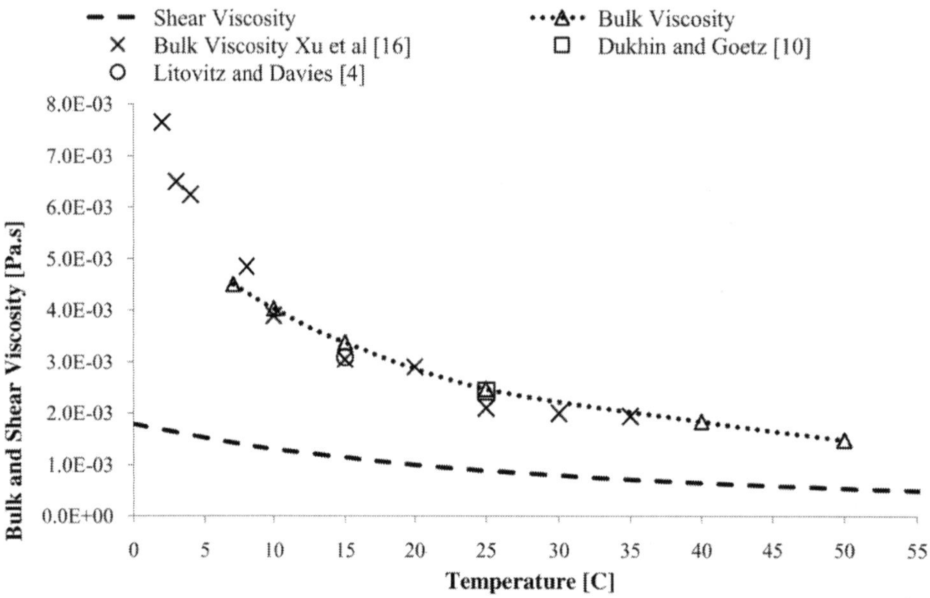


**Figure 4**. Bulk viscosity (··Δ··) [Pa s] for 0.22$\mu$m Millipore water against temperature [°C] using acoustic spectroscopy. Shear viscosity (- - -) taken from [12].Bulk viscosity data (×) by Xu *et al* [15] using Brillioun scattering, (o) Litovitz and Davies [4] and (□) Dukhin and Goetz [10]. Bulk viscosity is a factor a 3 times higher than the shear viscosity at low temperature.

## 4. Conclusions

We have successfully employed acoustic spectroscopy to determine the bulk viscosity of 0.22$\mu$m Millipore water across the temperature range 7-50°C. We have shown results consistent with the classical theory of acoustic attenuation exhibits a frequency squared dependence on the viscosity. The data obtained has shown that the bulk viscosity of water decays with temperature, and we specify exponential and polynomial fits of high $R^2$ value. Our results are in good agreement with other studies that have measured bulk viscosity using both the same and different techniques. Our work further demonstrates that acoustic spectroscopy provides a robust, repeatable and accurate approach to determining the bulk viscosity. The ratio of the bulk viscosity to shear viscosity is equal to approximately three throughout this temperature range, indicating that it cannot be ignored when dealing with thermal and fluid systems. The ratio decreases slightly with temperature, suggesting a convergence at higher temperatures, although further study is required to probe this. Our experimental measurements extend the limited bank of fluid bulk viscosity measurements that have been made to date. Such continued exploration of this fundamental fluid parameter is important to develop new fluid characterisation methods and provide theoretical studies with much needed data [16].

**Acknowledgement**
The authors would like to thank Dahe Liu for supplying the bulk viscosity data given in [15].

**References**

[1] Landau L D and Lifshitz E M 1959 *Fluid Mechanics* (London: Pergamon)
[2] Tritton D J 1988 *Physical Fluid Dynamics* (2nd edition) (Oxford: Clarendon Press), pp 52-60
[3] Morse P M and Ingard K U 1986 *Theoretical Acoustics* (New Jersey, Princeton University Press)
[4] Litovitz T A and Davis C M 1964 *Physical Acoustics* Vol 2 ed Mason W P (New York: Academic) chapter 5
[5] Zemansky M and Dittman R 1997 *Heat and Thermodynamics* (London: McGraw-Hill)
[6] Epstein P S and Carhart R R 1953 *J. Acoust. Soc. Am.* **25** 553-565
[7] Allegra J R and Hawley S A 1972 *J. Acoust. Soc. Am* **51** 1545-1564
[8] Harlen O G, Holmes M J, Povey M J W, Qiu Y and Sleeman B D 2001 *SIAM J. App. Math.* **61** 1906-1931
[9] Pinfield V J, Harlen O G, Povey M J W and Sleeman B D 2005 *SIAM J. Appl. Math.* **66** 489-509
[10] Dukhin A S and Goetz P J 2009 *J. Chem. Phys.* **130** 124519
[11] http://www malvern com/
[12] Povey M J W 1997 *Ultrasonic Techniques For Fluids Characterization* (San Diego: Academic Press)
[13] Lide D R (ed) 2002 *CRC Handbook of Chemistry and Physics* 83rd Edition (Boca Raton, FL.: CRC Press/Taylor)
[14] Reynolds O 1886 *Phil. Trans. R. Soc. Lond.* **177** 157.
[15] Xu J, Ren X, Gong W, Dai R and Liu D 2003 *Appl. Opt.* **42** 6704-6709.
[16] Guo G J and Zhang Y G 2001 *Mol. Phys.* **99** 283-289.

Journal of Physics: Conference Series **269** (2011) 012012 doi:10.1088/1742-6596/269/1/012012

# Evanescent CHOTs for the optical generation and detection of ultrahigh frequency SAWs

**A. Arca, T. Stratoudaki, R. Smith, M. Clark and M. G. Somekh**

Division of Electrical Systems and Optics , Faculty of Engineering, University of Nottingham, Nottingham, NG7 2RD, UK

E-mail: eexaa5@nottingham.ac.uk

**Abstract.** The use of optical gratings for the generation and detection of ultrahigh frequency narrowband surface acoustic waves is well known. Detection of surface acoustic waves (SAWs) using an optical grating, (without an interferometric set-up) relies on diffraction of light from the grating. Cheap Optical Transducers (CHOTs) for SAWs are optical ultrasonic transducers which are designed to optimise the diffraction from such gratings to yield a local interferometer. They have been demonstrated for typical NonDestructive evaluation (NDE) frequencies (0.5 and 100 MHz). In this paper, we discuss how the CHOTs operate and how their detection mechanism breaks down, when the optical diffraction orders from the grating become evanescent upon moving to multi GHz frequency range. We show that it is possible to design devices with enhanced sensitivity, in this range. This is done by adding a thin "background film" underneath grating fingers, in order to optimise and exploit the interaction between the deformation of the grating and the evanescent fields/ resonances involved in such gratings. The result is a novel operating mechanism, based on an energy balance between the 0-order reflection and resistive heating. These novel transducers are called evanescent wave CHOT (eCHOT). We have used Finite Element Method (FEM) modelling to link the physical displacements caused by the elastic waves with the optical behaviour. We demonstrated that the devices show enhanced sensitivity over a wider range of parameters, than possible with a device designed with conventional CHOT specifications, making the eCHOT an ideal candidate for GHz / nano scale ultrasonics. The eCHOT involves highly resistive thin films and yields thin structures advantageous for ultrahigh frequency ultrasonics. The operating mechanism of the eCHOT suggests that more exotic structures could be built to enhance the sensitivity of these devices.

## 1. Introduction

The optical detection of ultrahigh frequency (GHz) range surface acoustic waves (SAW) is known to be difficult, since the period of the surface modulation caused by such waves, the SAW wavelength ($\lambda_a$) , is small compared to the optical wavelength ($\lambda_o$) to cause diffraction of the detection beam. Generation and detection of such waves is important in NonDestructive Testing and Evaluation (NDT&E), when surface breaking cracks are to be analysed.

A key point in producing SAWs of desired frequency and direction is to create periodic disturbances. When utilising a laser, a periodic contrast in the absorption of optical energy, causes the generation of narrowband surface elastic waves. The amplitude of these waves is related to the amount of contrast in the absorption [1]. Usually, generation is achieved by writing a disturbance pattern on a nonabsorbing surface with an absorbing metal, or vice versa and illuminating with a laser pulse [2, 3, 4].

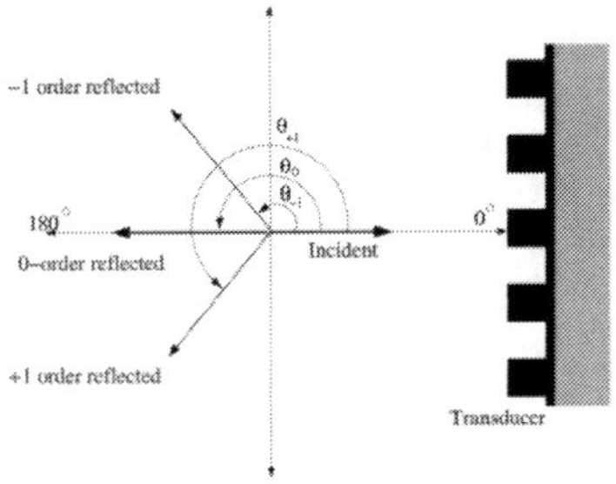


**Figure 1.** The explanation of the conventions about the farfield observation angles in CHOT and eCHOT simulations. The incident field is considered as a reference at 0° (normal to the surface). The transmitted field is also considered at 0° (on the other side of the sample). All the angles are measured anti-clockwise from this reference direction.

SAWs can be detected by a variety of techniques. These can be classified into interferometric and non-interferometric techniques. Non-interferometric ways include knife-edge, surface-grating, reflectivity methods [5]. Interferometric techniques include optical heterodyning, differential interferometry, velocity or time-delay interferometry [1, 5]. Interferometric systems can have high sensitivity, high bandwidth, and can work on rough surfaces and moving components. In practice, simple interferometry is hard to use and very sensitive, and it will have added complexity in order to deal with problems such as strong sample vibration, air currents and laser pointing instability [5].

Cheap Optical Transducers (CHOTs) for SAWs have been demonstrated for the generation and detection of elastic waves, up to 100 MHz [2], where the acoustic wavelength of interest was bigger than the optical wavelength used. CHOTs can be written on the sample to be analysed as periodically arranged metal stripes, i.e. a grating with period equal to $\lambda_a$, for generation and detection. The explanation of the general geometry of the transducers, together with the far field observation angle conventions are shown in figure 1, where the incident field travels from left to right. The generation CHOT (g-CHOT) system works using the above mentioned absorption contrast method. The detection mechanism of the CHOTs (d-CHOTs) does not suffer from the above mentioned sensitivity to outside effects [2]. The operation of the d-CHOTs is based on a local interferometric effect caused by the path difference between the field reflected from the background material and the grating fingers. The transducers are designed and optimised to facilitate the signal via an energy trade mechanism among the reflected optical diffraction orders.

The generation and detection of ultrahigh frequency SAWs have been looked at in the literature by the use of optical gratings [4, 6, 7]. However, the modulation of optical resonances inherent to such suboptical-wavelength transducers due to the interactions with the SAWs has not received much attention. The design of optical transducers by optimising and exploiting this interaction has not been done before.

In this work, we use FEM to investigate and optimise CHOTs for sub-optical wavelength sizes, when all the diffraction orders are evanescent. We find that, once the orders have become evanescent and the energy balance through the diffracted orders is abolished, another energy trade system through resistive heating can be obtained by adjusting the design parameters. This energy trade mechanism relies on the interaction between the optical resonances on the transducer (grating) and the change in transducer geometry due to the deformation caused by the SAW. This deformation causes optical evanescent fields to build up and disappear subsequently around the thin film, modulating the resistive heating in the transducer. This mechanism could be visualised as a thin film moving in and out of a region of concantrated optical field, and

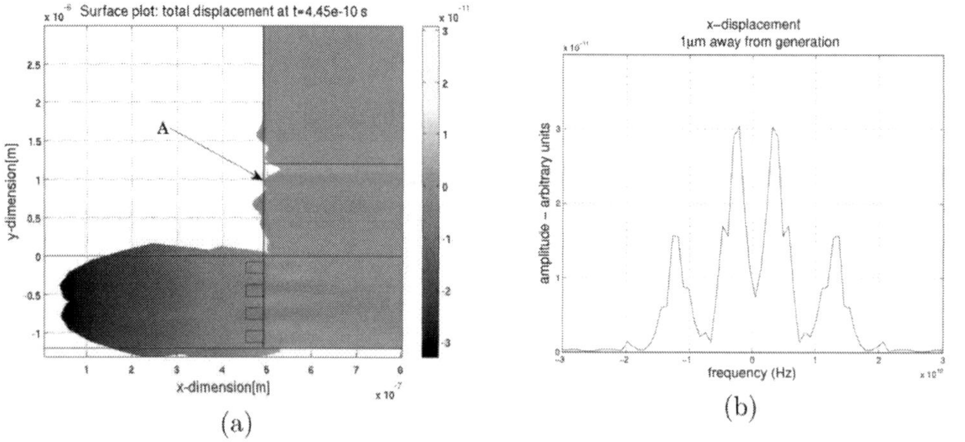


**Figure 2.** **(a)** The generation model of the g-eCHOT at time 445 ps (The optical pulse starts around t=0 s.). The surface plot is the total displacement. Additionally, for visualisation purposes, the displacement is also expressed as deformation. The 4 grating fingers are modelled in optical and heat transfer modules, which supply sufficient temperature information around the area. The big bulge in the bottom of the figure, where the transducer is located, represents the high displacement amplitudes due to the heating from the optical source. This model assumes free-space optical wavelength, $\lambda_o$ = 800 nm, and acoustic wavelength, $\lambda_a$ = 300 nm. Finger thickness is 40 nm and the background film thickness is 2 nm. **(b)** The fourier transform of the x-displacement in the thin film 1 $\mu$m away from the generation site (point A on figure 2(a) ). Two peaks are seen in the figure. The peak at $\approx$ 12 GHz is the designed frequency (corresponding to $\lambda_a$ = 300 nm ). The peak at $\approx$ 3.1 GHz is a modelling artifact, since 4 fingers are being modelled, it corresponds to $\lambda_a$ = 1200 nm.

consequent change in the amount of resistive heating on the thin film. We have optimised this mechanism to design transducers called eCHOTs (evanescent CHOTs) which can generate and detect ultrahigh frequency surface elastic waves with enhanced signal amplitudes, when compared to traditional CHOTs, for a wide range of parameters. The very thin nature of these transducers is also considered to be an advantage due to the small perturbation they pose to the acoustic waves.

## 2. Model Description

The modelling was done using Finite Element Method [1]. In order to model g-eCHOTs, we have coupled three types of models, optical, heat transfer by conduction, and plane strain. The optical model is used to calculate the resistive heating, which is coupled as a heat source to the "Heat transfer by conduction" model. The heating and the resultant temperature distribution are simulated in the heat transfer model. The temperature distribution in the structure is then coupled to a "Plane strain" model , where the thermal expansion and the resultant displacement are calculated. Hence the effect of optical heating is seen as mechanical displacement.

For d-CHOT and d-eCHOT, a purely optical model is used. The displacement due to the elastic wave is modeled as a sinusoidal deformation of the transducer. The phase of the imposed sinusoid is varied to simulate the passage of a surface acoustic wave. This is referred to as the "SAW phase". The reflected and transmitted optical fields are obtained in the far field as the detection metric.

[1] The implementation of FEM was done with a commercial software Comsol Multiphysics

Journal of Physics: Conference Series **269** (2011) 012012 doi:10.1088/1742-6596/269/1/012012

### 2.1. *Farfield Spectra*

The reflected/transmitted farfield spectra are obtained by doing near to farfield transformation using Stratton-Chu Formulations. The Stratton-Chu equation (equation 1) is applied on a circular area on the reflected and/or transmitted side. An appropriate apodisation technique is used to deal with the plane waves which enter and leave the circular boundary undisturbed, so as to simulate a net source enclosed in the circular area.

$$E_p = \frac{jk_n}{4\pi} r_0 \times \int [\mathbf{n} \times \mathbf{E_{ap}} - \eta_n \mathbf{r}_0 \times (\mathbf{n} \times \mathbf{H_{ap}})] \exp(jk_n \mathbf{r}.\mathbf{r}_0) dS \qquad (1)$$

Where,
$\mathbf{E_{ap}}$ : Electric Field on the aperture(apodised)
$\mathbf{H_{ap}}$ : Magnetic Field on the aperture(apodised)
$\mathbf{r}_0$ : unit vector pointing from origin to the field point p
$\mathbf{n}$ : unit normal to surface S
$\eta_n$ : impedance of the ambient medium
$k_n$ : wave number in the ambient medium
$\mathbf{r}$ : radius vector of the surface S.(not a unit vector)
$\mathbf{E_p}$ : calculated far field at point p

## 3. Results

### 3.1. *g-CHOTs and g-eCHOTs*

The heating in the g-CHOT and the g-eCHOT is provided by a laser pulse with pulse length less than the time it takes for an acoustic wave to travel through a grating period. Absorption contrast comes from the thickness dependent absorption constant of Aluminium. In the g-eCHOTs, the pulse length used was approximately 100 fs. The generation mechanism of the g-eCHOT is similar to g-CHOTs, i.e. spatial contrast in resistive heating, thermoelastic generation with a laser pulse. The model used for the generation is shown in figure 2(a). In order to determine the generated acoustic frequency, we have taken the fourier transform of the displacement at point 1 $\mu$m away from the generation area (point A in figure 2(a)), which is shown in figure 2(b). In figure 2(b) the fourier transform of the x-displacement(normal component) is shown. There are two distinct peaks corresponding to $\approx$ 12 GHz and $\approx$ 3 GHz. While the former is the frequency of the desired acoustic waves, the latter is a modelling artifact and occurs simply because there are 4 grating fingers in the model. In this model, there is an assumption that the optical source is infinitely long along the periodicity of the grating. However, only four periods are used to represent the above periodicity relationship. Therefore, the periodicity effect due to one finger and four fingers are seen in the models simultaneously. This effect is not physically relevant, hence cannot be seen in an experimental setting.

### 3.2. *d-CHOT*

d-CHOTs are reflective gratings which look like metal fingers deposited on a background film. The finger thickness is arranged in order to provide an in situ interferometric effect between the light reflected off the fingers and the background film. When the path difference between these components is arranged to be $\lambda_o/4$, then the modulation of this height due to the passage of an acoustic wave, can cause the maximum change in the 0-order reflected light. For this to occur, the finger height must be around $\lambda_o/8$. As the SAW interacts with the d-CHOTs, the diffracting efficiency of the transducer changes and the energy is redistributed among the diffracted orders. A difference in the intensity of the 0-order reflected light is compensated by the $\pm 1$ diffracted orders. This is shown in figure 3(a). The reflectivity of the transducer(CHOT) is recorded as the signal, while the SAW interacts with the CHOT. The difference between the maximum and

Journal of Physics: Conference Series **269** (2011) 012012 doi:10.1088/1742-6596/269/1/012012

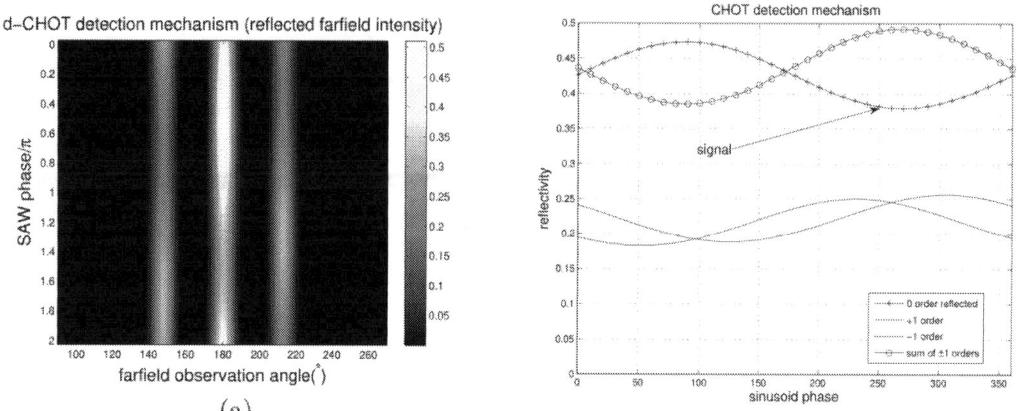


(a) (b)

**Figure 3.** **(a)** The operation characteristics of the d-CHOT. The reflected farfield intensity spectrum of the d-CHOT mechanism. The 0-order reflected intensity is noted to be the signal (corresponding to farfield angle 180°). The ±1 reflected diffraction orders are seen at farfield observation angles, 147° and 213°, according to the conventions in figure 1. **(b)** The intensity of different propagating components vs the SAW phase (normalised with incident intensity). This plot shows what happens to the intensity of the different field components as the SAW interacts with the d-CHOT, i.e. as the SAW phase changes. It should be noted that none of the diffracted orders are individually 180° out of phase with the 0-order reflected, however, their sum is completely out of phase with the 0-order reflected field, which suggests that the energy difference in the 0-order reflection is compensated equally by the diffracted orders.

the minimum of the recorded reflectivity , i.e. signal, is known to be twice the maximum optical signal amplitude.

When the acoustic wavelength of interest approaches to the optical wavelength used ($\lambda_o$), the orders open towards ±90° to eventually become evanescent. When using a substrate with refractive index, $n_g > 1$, some more decrease in the acoustic wavelength can be tolerated, if the background film is thin enough to allow transmission, through the transmitted orders. This is because, transmitted orders can still exist for acoustic wavelengths $\lambda_o/n_g < \lambda_a < \lambda_0$. However, for gratings whose periods are smaller than $\lambda_o/n_g$, the operation of this device must be reconsidered, since there will be no more diffracted orders on any side. Once all the diffracted orders become evanescent, only the transmission and absorption can carry away the energy from the signal.

d-CHOTs were designed for acoustic wavelength, $\lambda_a$ = 1500 nm and optical wavelength $\lambda_o$ = 800 nm, on glass substrates($v \approx 3500$ m/s). The corresponding SAW frequency is ≈ 2 GHz. Figure 3(a) shows intensity of the optical reflection components for this device in the farfield, as the SAW interacts with the transducer. The diffracted orders can be seen at farfield observation angles, -1-order : 147°, 0-order: 180°and +1-order: 213°(The farfield observation angle conventions are given in figure 1). The orders in figure 3(a) correspond to the curves in figure 3(b), as an explanation of the mechanism. As the phase of the sinusoid changes (SAW propagates), the intensity of these diffracted orders change out of phase with the 0-order reflection. The change in 0-order reflected intensity is accounted for by the change in intensity of the ±1 diffracted orders. The ±1-orders sum up to become exactly out of phase with the 0-order reflected wave.

Figure 4(a) shows the operating region for the d-CHOT at different grating finger and background film thickness values. The plotted value is the maximum optical signal amplitude

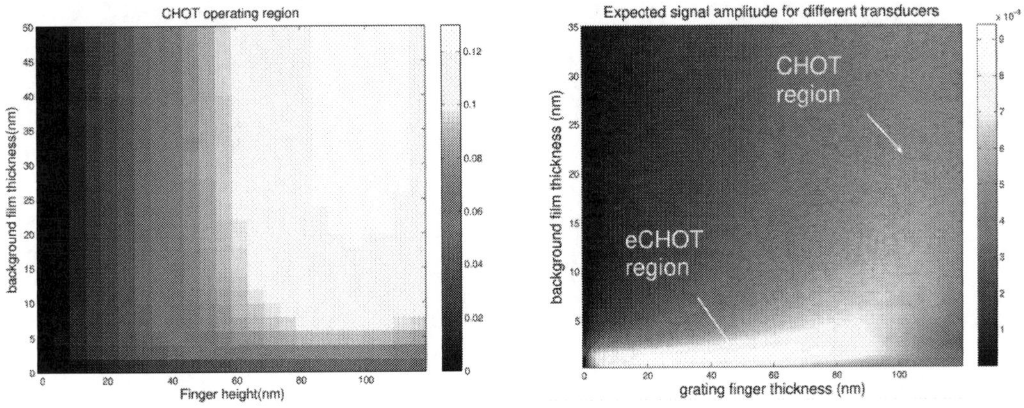


(a) (b)

**Figure 4.** **(a)** The effects of finger height and background film thickness on the optical signal amplitude of the CHOT. The plotted quantity is half the maximum change in the reflectivity as the transducer interacts with the SAW. This corresponds to the maximum expected signal amplitude. The specific CHOT device was built from Aluminum, for acoustic frequency 2 GHz (grating period 1.5 $\mu$m) and optical wavelength 800 nm. In this simulation the displacement amplitude used was 2 nm. **(b)** The effect of finger height and background film on the operation of the optical detection CHOT transducers. Specific image corresponds to a device designed for $\lambda_a$=300 nm. (The period of the grating is 300 nm, corresponding acoustic frequency 12 GHz ) The displacement amplitude used was 2 nm. As can be seen, the region that corresponds to a good operating point for (a), i.e. the CHOT region at the long wavelengths, is superceded by a region corresponding to very thin background films and relatively smaller finger heights for smaller wavelengths.

at the given finger height and background thickness value.

The optimum finger height is defined as the finger height at which the change in the observed reflected detection light is found to be the maximum, for a given amplitude of SAW. The optimum finger height is seen to be around 100 nm, where it corresponds to $\lambda_o/8$. The increase in the background film is seen to assist the d-CHOT operation, since the mechanism depends on reflected orders, rather than transmitted and absorbed.

### *3.3. eCHOT mechanism*

The detection mechanism of the eCHOTs is fundamentally different from that of CHOTs. When the light comes in with normal incidence, to the d-eCHOT, it is either reflected, transmitted or absorbed. It cannot be diffracted, since the period of the fingers in the eCHOT is smaller than the optical wavelength. This is seen in figure 5(a). In this figure the reflected farfield spectrum of the eCHOT (finger height 40 nm, background film thickness 2 nm) is plotted as the device interacts with the SAW. The reflected intensity is seen to change with the phase of the sinusoid. As can be seen there are no diffracted orders.

Figure 4(b) shows the effect of the finger height and the background film of a transducer made of Aluminium. As can be seen, there is a region corresponding to finger heights $\lambda_o/8$ which is similar to that of the CHOT criteria, labeled as "CHOT region". Also seen is the part of the image corresponding to low finger and background film thickness, which is labelled as the "eCHOT region". This is the region where the resistive heating is a major contribution to the device operation. As can be seen, the eCHOT area is wide in finger thickness.

The eCHOT operation relies on the interaction between the deformation due to the SAW and

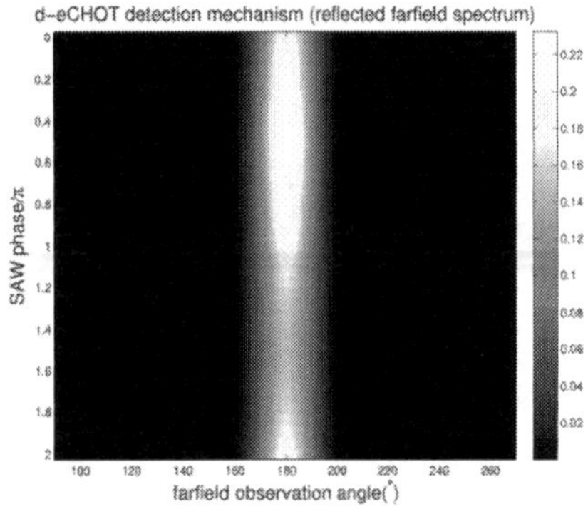


(a)

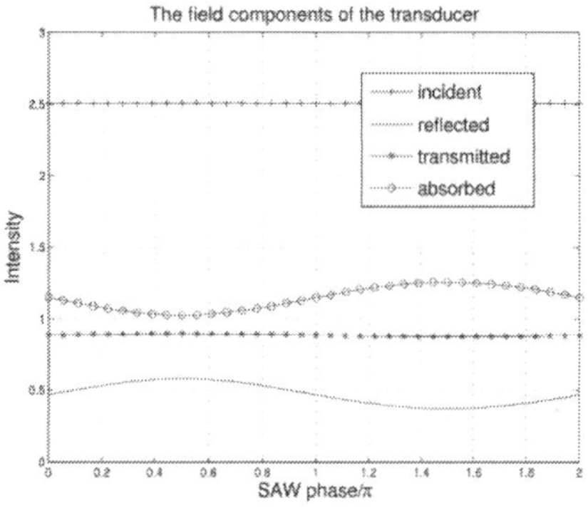


(b)

**Figure 5. (a)**The reflected farfield spectrum of the d-eCHOT as the phase of the sinusoid is varied. It can be seen that there is no diffracted orders. The only reflected component, namely the 0-order reflection, is seen to vary in intensity as the SAW interacts with the transducer. The considered transducer has 40 nm finger height and 2 nm background film. **(b)** The contribution of various optical components to the energy trade mechanism.These curves are obtained by extracting the field amplitudes from the farfield spectra, followed by the necessary normalisations. The absorption is calculated by subtracting the reflected and transmitted intensity from the incident. It can be seen that as the SAW propagates, the incident intensity is constant, whereas the reflected intensity (the signal) and the absorption are 180° out of phase with each other. The transmitted intensity also has a sinusoidal nature out of phase with the absorption, however, the amplitude of the sinusoid is too small to be visible on this axis. This plot shows that the energy is taken out of the specular reflection via absorption.

the optical evanescent fields/ resonances in the transducer. This is shown in figure 6, where the total field is being plotted at two different instances during the interaction of the SAW with the transducer. In figure 6 it is seen that at certain phases of the SAW, evanescent waves build up on the thin film and at some phases these waves clear off. This modulates the resistive heating in the transducer. This is shown in figure 7, where the resistive heating is being plotted for two distinct instances while the SAW interacts with the transducer. It is seen that the resistive heating in the middle of the thin film is different for the two instances, demonstrating that the transducer works by modulating the resistive heating.

## 4. Conclusion

We have used FEM modelling to extend the CHOTs concept, to the analysis of suboptical-wavelength-sized gratings in the generation and detection of ultrahigh frequency surface elastic waves. We have seen that by carefully adjusting the design parameters, an increase in the signal amplitude is possible for a wider range of finger thicknesses. We have introduced a new operating mechanism for optical ultrasonic transducers which is based on resistive heating. This mechanism can be used in the design of novel structures involving pointy geometries to confine evanescent fields or lossy surface plasmons.

## References

[1] C. B. Scruby and L. E. Drain, *Laser Ultrasonics: Techniques and Applications.* Adam Hilger, 1990.
[2] T. Stratoudaki, J. A. Hernandez, M. Clark, and M. G. Somekh, "Cheap optical transducers(CHOTs) for narrowband ultrasonic applications," *Measurement Science and Technology*, vol. 18, 2007.
[3] M. E. Siemens, Q. Li, M. M. Murnane, H. C. Kapteyn, R. Yang, E. H. Anderson, and K. A. Nelson,

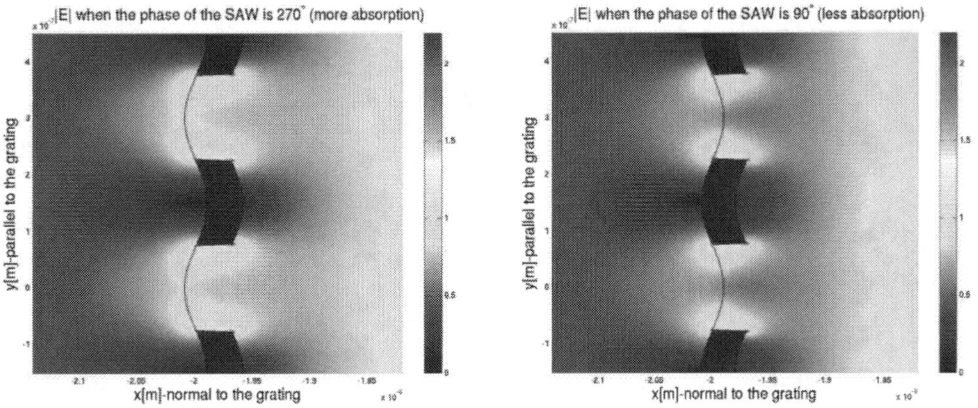


(a) (b)

**Figure 6.** The demonstration of the effect of evanescent fields in the eCHOT. **(a)**When the phase of the SAW is such that the evanescent field covers the thin film. The film is thus moved into the evanescent field, which increases the absorption. **(b)** When the phase of the SAW is such that the evanescent field does not cover the thin film. The thin film is thus moved out of the evanescent field, in which case the absorption is less.

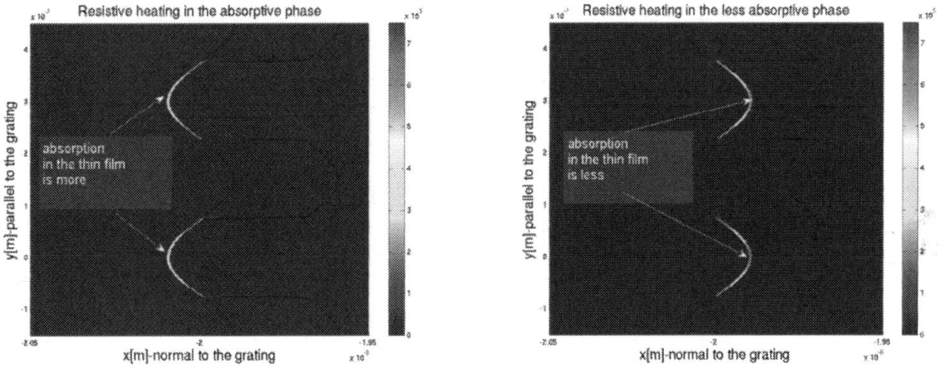


(a) (b)

**Figure 7.** The demonstration of the modulation of the absorption in the thin film in eCHOT. **(a)** Resistive heating in the thin film in the absorptive phase. (When the phase of the SAW is such that the thin film moves into an evanescent field increasing the absorption.) **(b)** Resistive heating in the thin film, in the less absorptive phase. (When the phase of the SAW is such that the thin film is moved out of the evanescent field yielding less absorption)

"High-frequency surface acoustic wave propagation in nanostructures characterized by coherent extreme ultraviolet beams," *Applied Physics Letters*, vol. 94, 2009.

[4] D. H. Hurley and K. L. Telschow, "Picosecond surface acoustic waves using a suboptical wavelength absorption grating," *Physical Review B*, vol. 66, 2002.

[5] J.-P. Monchalin, "Optical detection of ultrasound," *IEEE Transactions on ultrasonics, ferroelectrics and frequency control*, vol. UFFC-33, no. 5, 1986.

[6] B. Bonello, A. Ajinou, V. Richard, P. Djemia, and S. M. Cherif, "Surface acoustic waves in the GHz range generated by periodically patterned metallic stripes illuminated by an ultrashort laser pulse," *Journal of Acoustic Society of America*, vol. 110, 2001.

[7] D. H. Hurley, "Optical generation and spatially distinct interferometric detection of ultrahigh frequency surface acoustic waves," *Applied Physics Letters*, vol. 88, 2006.

Journal of Physics: Conference Series **269** (2011) 012013 doi:10.1088/1742-6596/269/1/012013

# Recent advances in semi-analytical scattering models for NDT simulation

**M. Darmon[1], S. Chatillon[1], S. Mahaut[1], P. Calmon[1], L.J. Fradkin[2] and V. Zernov[2]**

[1]CEA, LIST, Département Imagerie et Simulation pour le Contrôle, F-91191 Gif-sur-Yvette, France.
[2]Waves and Fields Research Group, London South Bank University, UK.

E-mail: michel.darmon@cea.fr

**Abstract.** For several years, CEA-LIST and partners have been developing ultrasonic simulation tools with the aim of modelling non-destructive evaluation. The existing ultrasonic modules allow us to simulate fully real ultrasonic inspection scenarios in a range of applications which requires the computation of the propagated beam, as well as its interaction with flaws. To fulfil requirements of an intensive use (for parametric studies), the choice has been made to adopt mainly analytical approximate or exact methods to model the scattering of ultrasound by flaws. The applied analytical theories (Kirchhoff and Born approximations, GTD, SOV...) were already described in previous GDR communication. Over the years, this "semi-analytical" approach has been enriched by adaptations and improvements of the existing models or by new models, in order to extend the applicability of the simulation tools. This paper is devoted to the following recent advances performed in the framework of this approach:

- The SOV method based on the exact analytical model for the scattering from a cylindrical cavity has been extended in 3D to account for field variations along the cylinder. This new 3D model leads to an improvement in simulation of small side-drilled holes.
- Concerning the geometrical theories of diffraction (GTD), subroutines for calculation of the 2D wedge diffraction coefficients (for bulk or Rayleigh incident waves) have been developed by the Waves and Fields Group and uniform corrections (UAT and UTD) are under investigation.
- Modelling of the contribution of the head wave and creeping wave to the echoes arising from a wedge.

Numerous experimental validations of the developed models are provided. New possibilities offered by these new developments are emphasized.

## 1. Introduction :

The CIVA software platform is developed at CEA-LIST and partners in the aim of simulating non-destructive evaluation [1]. Most of the developed models are based on semi-analytical methods since they are not heavy on computation time.

The ultrasonic simulation tools in CIVA allow one to fully predict the real ultrasonic inspection in a range of applications which requires the computation of the beam propagated, as well as its interaction with flaws. The transducer field is calculated using a pencil method derived from the Rayleigh integral for acoustic radiation [2]. Indeed, inside the coupling material, the Rayleigh integral is computed directly by summing the contributions to the field of each source point of the discretized transducer surface. In the specimen the pencil method which is an extension of ray theory [3] is applied.

The beam to flaw interaction is dealt with using different modelling approaches depending upon the defect and the inspection characteristics. Three kinds of models for the scattering of ultrasound by flaws have been integrated: approximate analytical solutions, exact analytical solutions and numerical modelling methods.

The developed approximate analytical solutions are respectively:

- the Kirchhoff approximation [4] to model specular reflexions from volumetric voids (spherical or hemispherical holes, SDHs) and cracks (rectangular, CAD or elliptical planar, FBHs, multifaceted). This approximation is mostly valid if the observation direction is close to reflexion and is particularly suitable to simulate specular reflexion, corner effects, etc. The corresponding integrated model requires the meshing of the defect surface.
- the geometrical theory of diffraction (GTD) to treat scattering from crack edges [5,6]. This approximation is valid away from specular angles and forward paths. The corresponding integrated model requires the meshing of the flaw contour.
- the modified Born approximation to deal with solid inclusions. It provides an analytical solution for some flaw geometries (spherical, cylindrical and ellipsoidal) without any meshing of the flaw [7].

An exact analytical solution for the scattering from a cylindrical cavity, based on the Separation Of Variables (SOV) method, has been used to simulate the response of a side drilled hole in a 2D configuration [4]. This model is discussed later.

The main general assumptions applied to deal with the application of the semi-analytical models were described in [4].

These previous methods have been experimentally validated in the most commonly used configurations [8]. A first example of experimental validation is now proposed. It deals with an application of the Kirchhoff approximation.

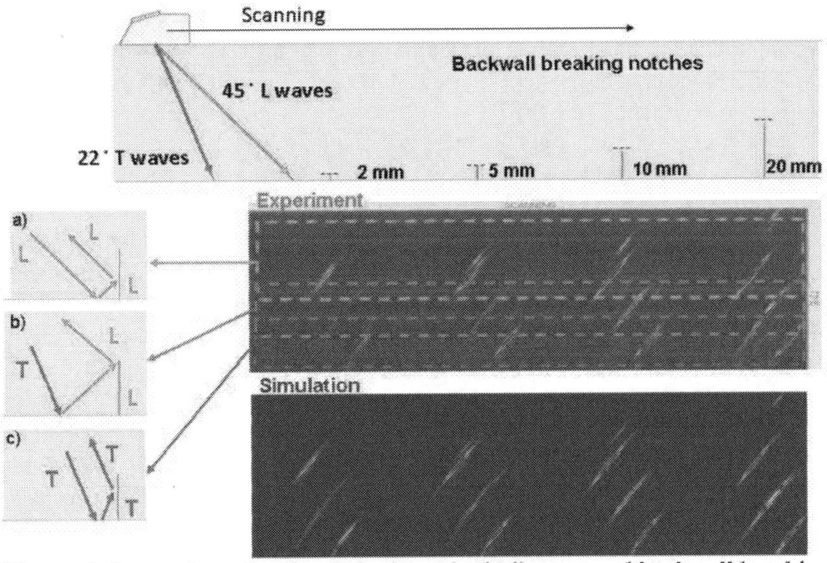


**Figure 1.** Inspection of a plane specimen including several backwall breaking flaws of various heights.

In the given example, a plane specimen including several backwall breaking flaws of various heights (from 2mm to 20mm) is inspected using a transducer generating both longitudinal and transverse waves. In figure 1 are shown the experimental and the simulated (using the Kirchhoff approximation) obtained B-Scans when scanning the specimen. In the experimental B-Scans are observed notably the following echoes:

- the P wave corner echo (a) including a backwall reflexion without mode conversion;
- In a similar way, the shear wave corner echo (c);
- A mixed corner echo (b) due to mode conversion at the backwall;

All these echoes are quantitatively and qualitatively reproduced in simulation. See [8] for details.

The following parts of the paper will be devoted to the advances performed at CEA-LIST in analytical diffraction models.

## 2. The Separation of Variables method (scattering from a cylindrical cavity).

### *2.1. Interest and principle of this method*

The semi-analytical Kirchhoff approximation, in a wide range of situations, leads to satisfactory quantitative predictions with a low computational overhead. Nonetheless, this model may cause some discrepancies in the prediction of small flaw responses, since it is a high frequency approximation. Particularly, experimental validations carried out in a pulse-echo configuration on small side drilled holes highlighted some differences between Kirchhoff simulation and the measured shear waves. These discrepancies could be explained by the theoretical principle of Kirchhoff approximation: it takes into account specular reflection on a cavity but not the creeping waves. These waves (called Franz waves [9]) propagate around the cavity circumference and may have a significant influence especially at low ka (product of the wave number k by the defect radius a). They are the analogues on curved surfaces of Stoneley waves (surface waves excited by grazing incidence) on flat surfaces. The creeping waves differ from Stoneley waves in two respects: they are attenuated because energy is radiated tangentially away from the cylinder; there are many circumferential modes n of different speeds but the damping increases with n.

That is why the separation of variables (SOV) method has been investigated. This method [10] is used to obtain an exact analytical model for the scattering from a cavity. The principle of the SOV method has been described in detail in [4]: it enables one to obtain series developments of the scattered waves. Terms of these series expressed are products of Hankel and complex exponential functions. A parametric study has been performed to optimize the number of series terms to compute so as to ensure the series convergence. The SOV diffraction coefficients have been theoretically validated by comparison with the literature [11].

Two models of echo simulation based on the SOV method have been developed. The first model developed (presented in GDR 2008) is a 2D one: it supposes that the fields radiated by both the emitter and the receiver are invariant along the cylinder length. To relax the previous hypothesis, the development of a "2.5D" SOV model has been carried out. This model takes into account the field variation along the flash line which is the first cylinder line insonified by the beam assumed to be a plane incident wave. It supposes an invariant diffraction coefficient along the flash line. This SOV model in CIVA is developed for a cylinder.

### *2.2. Experimental validations*

Numerous experimental validations of the SOV model have been performed [8,12] including comparisons with the Kirchhoff model. An example of experimental validation is presented in figure 2 below: it deals with inspection at 5MHz of two small side drilled holes (SDH) of respectively 0,5 and 1mm diameter (corresponding respectively to ka=4.9 and ka=2.4) with a SW45° contact transducer. The corner echo measured on a slot is taken as the amplitude reference. Given in the figure are comparisons of both maximum amplitudes and A-Scans corresponding to echoes obtained with several simulations (Kirchhoff, SOV and a finite element method FEM detailed in [4]) and with measure. The Kirchhoff model is unable to predict the creeping circumferential wave (chronologically the second and less intense observed echo) and leads to a discrepancy with the measurement because the creeping wave affects the maximum amplitude of the first echo (specular reflection on the cylinder). SOV and FEM give rise to a good agreement in amplitude with the measured result. The SOV model provides a more satisfactory prediction of the creeping wave behavior in term of both time of flight and relative amplitude compared to the specular echo.

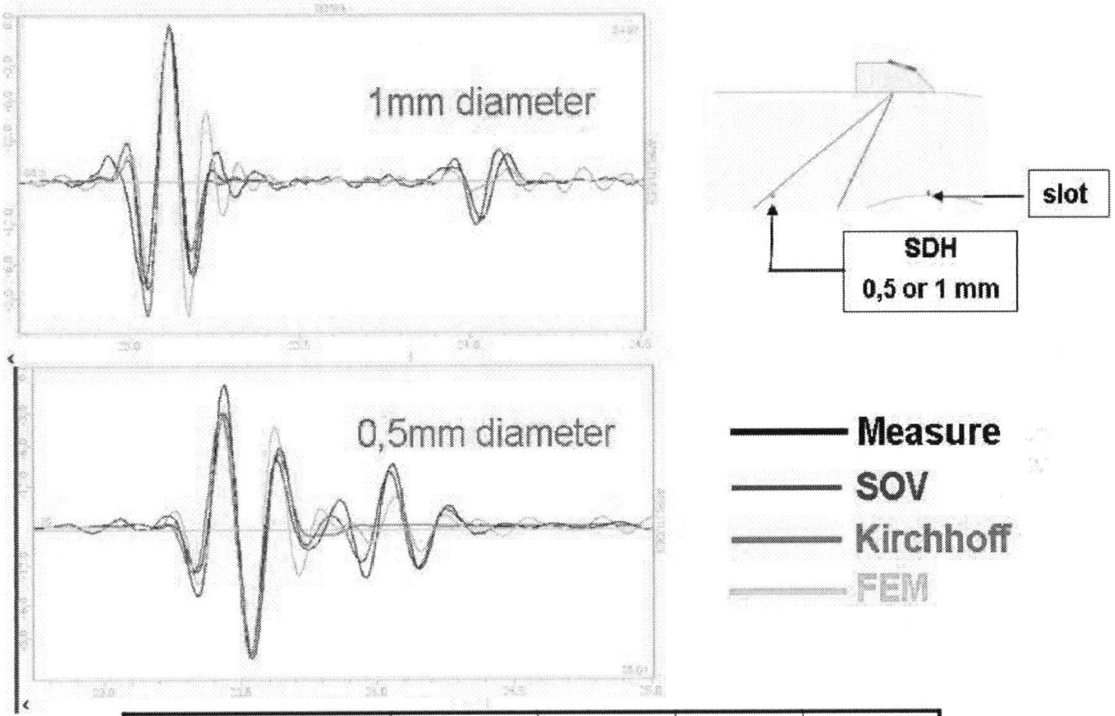


| Relative Amplitude (reference : slot) | Kirchhoff | SOV | FEM | Measure |
|---|---|---|---|---|
| SDH 1mm | - 9.5dB | - 11dB | - 13dB | - 13dB |
| SDH 0,5mm | - 10.9dB | - 16.3dB | - 15.7dB | - 16.5dB |

**Figure 2.** Inspection of a plane specimen including several backwall breaking flaws of various heights.

## 3. Models derived from the Geometrical Theory of Diffraction (GTD)

### *3.1. Introduction*

The Geometrical Theory of Diffraction (GTD) is commonly used to model scattering from flaw edges or wedges.

In the case of embedded cracks, two different strategies [6] have been investigated at CEA-LIST concerning the calculation of the GTD coefficients: a so-called "projected 2D" option, which is based on the projection of the incoming and scattered wave vectors over the plane normal to the flaw edge, and a "pure" 3D GTD code, using developments carried out by Larissa Fradkin and co-workers at London South Bank University (LSBU).

In addition to embedded cracks, the modelling of the scattering from wedges is under investigation at LSBU which is studying the performances of two different 2D codes for the wedge 2D GTD diffraction coefficient [13].

Journal of Physics: Conference Series **269** (2011) 012013 doi:10.1088/1742-6596/269/1/012013

*3.2. Simulation of edges diffraction from embedded cracks*

The Kirchhoff approximation deals with specular reflection contrary to GTD but is not always accurate for diffracted echoes. Indeed, the Kirchhoff approximation lies in the fact that the near field scattered by a crack is approximated by the reflected field as stated by geometrical acoustics.

On the other hand, the Geometrical Theory of Diffraction (GTD) is commonly used to deal with the prediction of diffraction phenomena. GTD is also an alternative model which relies on a ray theory [5]: contrary to geometrical acoustics, it is notably able to predict a non zero contribution to the diffracted field by a rectangular flaw in the shadow area.

Nevertheless GTD fails at predicting accurate diffracted amplitudes notably in the case the of direction of observation close to :

- the direction of specular reflection;
- the shadow boundary;

In the two previous directions, the GTD coefficient diverges and it is even invalid in a wide area in the vicinity of these directions (represented in the Figure below)

- the critical longitudinal angle for shear waves: in that case, the GTD coefficient is not continuous.

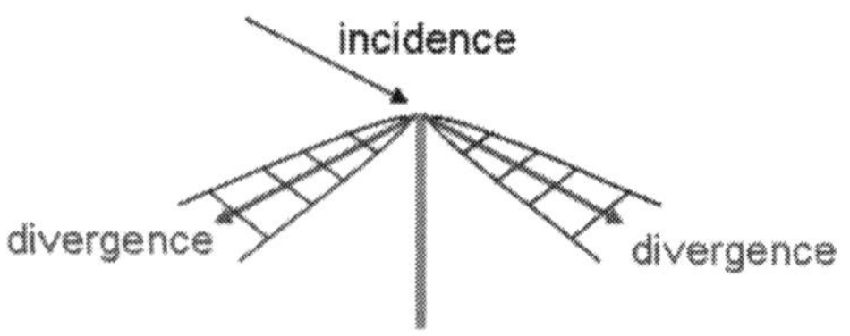


**Figure 3.** Description of the geometrical reflected and shadow areas of a rectangular flaw.

To improve the GTD predictions in these cases, our first step was to include a modification of GTD coefficients around these special observation directions.

*3.2.1. Modification of GTD coefficients around special observation directions.*

The developed GTD models are applied successfully in the majority of the inspection configurations and notably in TOFD inspections. But, they can provide problematic results in some complex cases notably for 3D configurations. Indeed the divergences or discontinuities of the GTD diffraction coefficients can lead to the presence of artefacts in the simulated results. Let us consider for instance a pulse-echo configuration of inspection of an elliptical crack (see Figure 5.a) which corresponds to an incidence with an angle quite critical to the normal to the flaw plane. Such a flaw can be detected thanks to the echoes scattered by the top and the bottom edges which can be distinguished in a true B-Scan (as those represented on Figure 5.b) superimposed on the flaw location.

In Figure 4.a and 4.b are respectively shown, versus the observation angle, the modulus and the phase of the 2D GTD coefficient for an incident angle corresponding to the previous configuration. In that pulse echo inspection, the observation angle is near the critical angle ($\theta = 57°$) and consequently the GTD coefficient suffers from a discontinuity in the neighbourhood (in term both of module and phase). Besides, we observe a divergence of the coefficient at the specular direction ($\theta = 120°$) and at the forward direction ($\theta = 240°$) which is accompanied by a sign change ($\pi$ phase shift).

In order to deal with such problematic configurations with the GTD models, we have decided to include simple treatments to avoid the presence of artefacts in the simulated results. In the following figure are thus presented the simulated C-Scans obtained without and with the treatment. The B-Scan with the treatment is represented too (superimposition of the true reconstructed B-Scan on the flaw location) and shows that the elliptical crack could be detected thanks to the echoes scattered respectively by the top and bottom edges. These edges are easily observed in the treated C-Scan with two distinguishable lobes whereas the initial C-Scan highlights an artefact which prevents a satisfactory detection of the flaw.

Journal of Physics: Conference Series **269** (2011) 012013 doi:10.1088/1742-6596/269/1/012013

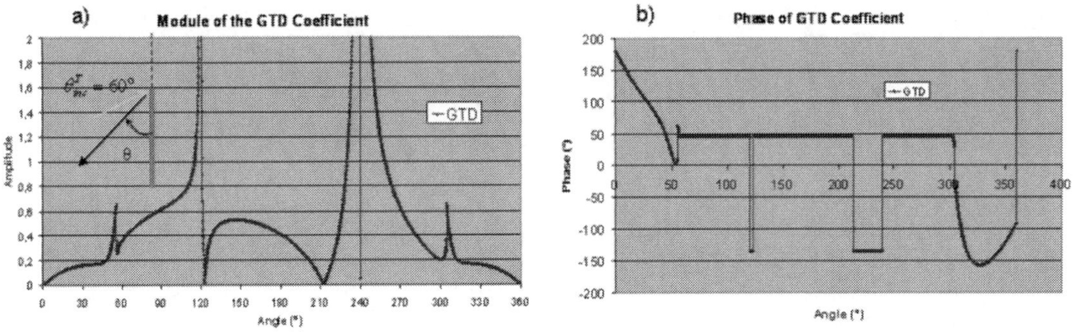


**Figure 4.** Module and Phase of the 2D GTD coefficients (initial and modified) for a 60° incident angle for P waves.

Consequently this treatment provides a qualitative improvement of GTD in some complex configurations. Nevertheless specific experimental validations have to be carried out in these problematic configurations.

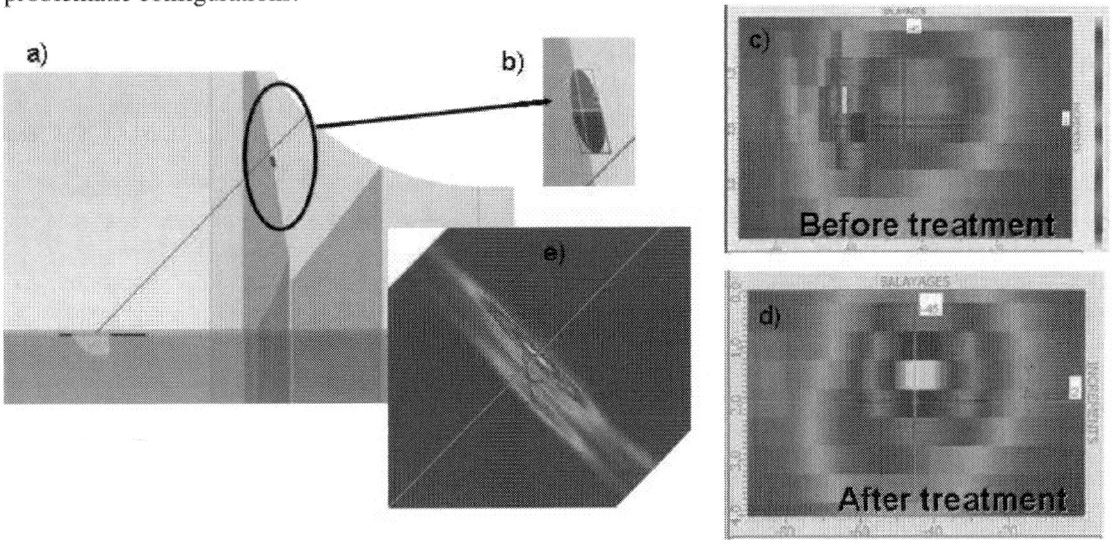


**Figure 5.** a) Configuration of SV wave pulse-echo inspection of an elliptical crack near the critical angle; b) elliptical crack; c) and d) C-Scans before and after treatment e) true B-Scan after treatment.

The limitations of the geometrical theory of diffraction show the interest of a more generic theory of diffraction. Our future step will be as a perspective to study and develop a model based on an uniform geometrical theory of diffraction [14] (UGTD) which merges the complementary theories of Kirchhoff and GTD in order to eliminate their drawbacks.

### *3.2.2. Theoretical principle of UGTD in the acoustic scalar case*

The UGTD principle has already been studied for cracks embedded in a acoustic medium by LSBU. It consists, for a semi-infinite planar crack, in decomposing the Kirchhoff integral field $u^{KA}$ into a geometrical contribution $u^{KA\_GE}$ and into an edge diffraction contribution. Outside shadow and reflected areas, the geometrical contribution $u^{KA\_GE}$ models the sum of the reflected and incident fields and the edge diffraction contribution has the form in 2D of a cylindrical scattered wave weighted by a specific Kirchhoff diffraction coefficient $D^{Kirchoff}$. Inside shadow and reflected areas, the expressions

Journal of Physics: Conference Series **269** (2011) 012013 doi:10.1088/1742-6596/269/1/012013

of these two Kirchhoff contributions are modified thanks to a uniform asymptotic evaluation of the Kirchhoff integral [15]. Then the main hypothesis of the developed UGTD is to express everywhere the UGTD field as the sum of the geometrical Kirchhoff contribution and the GTD field:

$$u^{\mathrm{UGTD}} = u^{\mathrm{KA\_GE}} + u^{\mathrm{GTD}} = u^{\mathrm{KA}} + (D^{\mathrm{GTD}} - D^{\mathrm{Kirchhoff}})\sqrt{\frac{i}{2\pi}}\frac{e^{ikr}}{\sqrt{kr}} \tag{1}$$

where r is the distance between the edge and the observation point and k the wave number. As a consequence, the UGTD method consists in adding the total Kirchhoff field and a modified GTD field (a cylindrical scattered wave in 2D) characterized by a diffraction coefficient which is the difference $D^{\mathrm{UGTD}} = D^{\mathrm{GTD}} - D^{\mathrm{Kirchhoff}}$ .The diffraction coefficients can be found analytically and depend of the incident $\phi_0$ and observation $\phi$ angles versus the normal to the edge in the flaw plane:

$$D^{\mathrm{Kirchhoff}} = -\frac{1}{2}\left(\tan\frac{\phi-\phi_0}{2} + \tan\frac{\phi+\phi_0}{2}\right), \quad D^{\mathrm{GTD}} = -\frac{1}{2}\left(\frac{1}{\cos\frac{\phi-\phi_0}{2}} + \frac{1}{\cos\frac{\phi+\phi_0}{2}}\right) \tag{2}$$

The UGTD diffraction coefficient $D^{\mathrm{UGTD}} = D^{\mathrm{GTD}} - D^{\mathrm{Kirchoff}}$ has no singularities at shadow boundaries where the GTD coefficients diverge (for $\phi \pm \phi_0 = \pi$). The UGTD is able to model both reflection and diffraction phenomena and relaxes the numerical failures of the initial GTD at shadow boundaries. This approach could be applied in elastodynamics too.

An application of this theory will be studied with the aim of modelling the echoes due to immersed targets.

## 4. Modelling of creeping and head waves.

### *4.1. Modelling of the creeping wave reflection*

#### *4.1.1. Observation of discrepancies between simulated and experimental results for critical incidence angle at the backwall*

The contribution of a head wave is usually small compared to the compressional and shear bulk waves in corner echo inspection, as mostly geometrical (specular or quasi specular) echoes are concerned. However, in some configurations, especially near critical angles (at backwall reflection, for instance), its contribution to echo response may be significant. This point is illustrated here by an experimental validation of corner echo prediction. In this study a stainless steel component composed of two different bottom slopes of 0° and 10° is considered. For each slope, the specimen contains a 10 mm height by 20 mm length rectangular backwall breaking flaw. The pulse-echo response of these defects is measured by a 45° refracted SV-waves planar contact transducer. Simulation of interaction between incident wave and flaw is performed with CIVA software (therefore, only using bulk waves contribution modelled thanks to the Kirchhoff model).

Figure 6 presents the geometry of the problem and the ray path of SV-mode corner echo for 0° and 10° configuration and the simulated and experimental B-Scan image, as well as the superposition of both echodynamic (amplitude versus scanning) curves. Results were normalized using a side drilled hole used as a reference. A very good agreement is observed for the 0° configuration, while some important differences of amplitude and overall behaviours are observed between these two results for the 10° slope configuration. For instance, quite a strong amplitude response is experimentally observed for such a configuration, while the probe is still "far" from the flaw, which means that an interaction occurs before the shear wave beam intersects the flaw. As the presented simulation only takes into account interactions of bulk waves with the flaw, it becomes obvious that for this configuration it is necessary to simulate head and creeping waves interaction (the shear waves beam hits the backwall with a nominal incidence angle of 35°, close to the critical angle, about 33° in stainless steel).

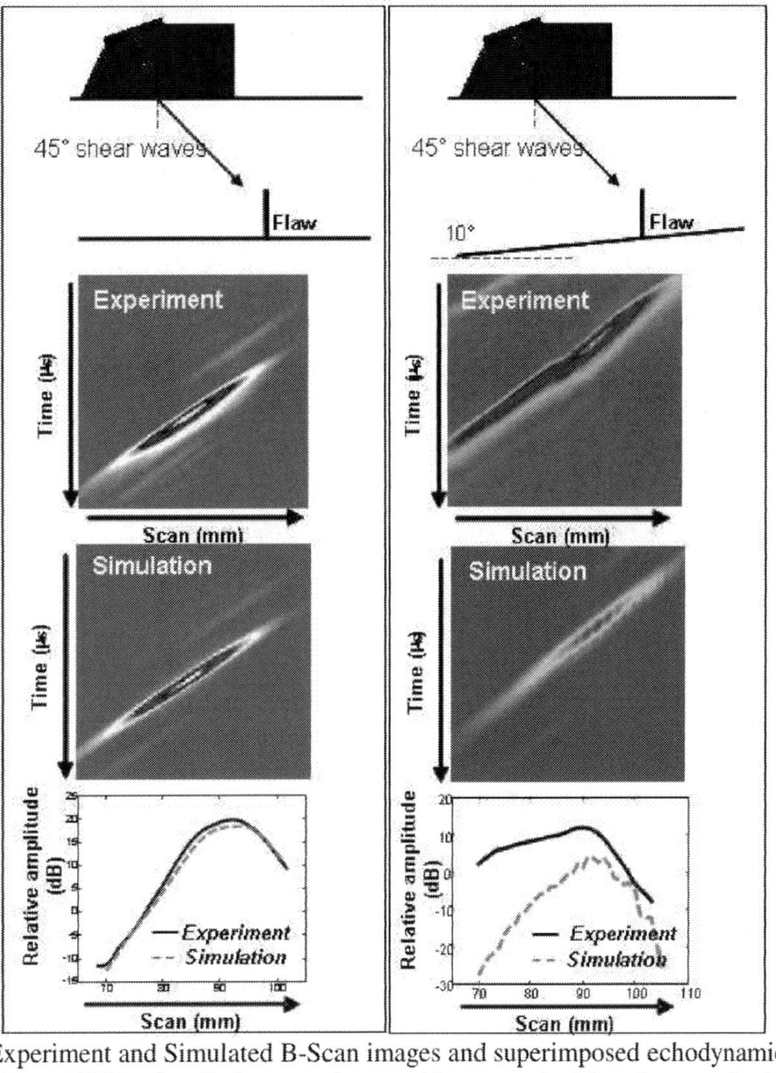


**Figure 6.** Experiment and Simulated B-Scan images and superimposed echodynamic curves of planar (left) and 10° backwall slope mock-ups with one vertical flaw inspected with a contact 45° shear waves probe.

*4.1.2. Simulation of the creeping wave reflection*

The contributions related to interactions of head and creeping waves with the flaw were not taken into account in the simulation until now. A time of flight analysis based on geometrical ray tracing has highlighted a potential interaction between a creeping wave and the flaw to explain the observed differences between simulated and experimental results in the 10° slope configuration. Geometrical paths related to this contribution are presented in figure 8.a). This interaction (a) corresponds to the reflection of the creeping wave on the flaw bottom. Indeed, at an incident critical angle, a creeping wave is generated at the backwall and is reflected on the flaw tip into another creeping wave. During its propagation along the backwall, this creeping wave will then generate a head wave propagating with an angle to the backwall normal equal to the critical angle.

In previous articles [16, 17], we have described the modelling of the creeping wave reflection (Figure 9.a) at a tip of a wedge made up of a backwall surface and a backwall breaking flaw. This model is based on a ray theory [3].

*4.1.3. Experimental validation of the creeping wave reflection simulation*
Modelling of the creeping wave reflection is now applied to simulate the echo response for the 10° slope configuration. To the initial simulation which models only interactions of the bulk wave with the flaw is added the contribution of the head wave radiated by a reflected creeping wave on the flaw bottom. The B-scan and the spatial repartition obtained with simulation is presented and compared to experiment in Figure 7. We can see by comparing figures (a) and (b) a good agreement between experimental and calculated B-scans. On figure (c) is shown first the initial calculation which models only bulk wave interactions. We observe then that the amplitude calculated with the addition of head wave improves modelling compared to experimental results (black line).

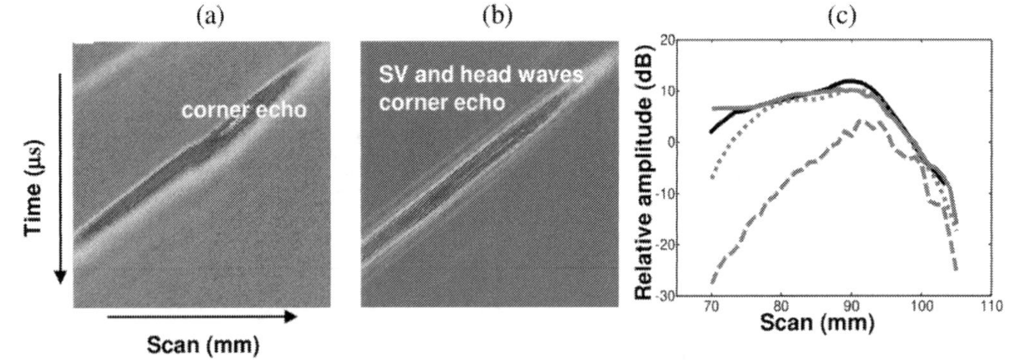


**Figure 7.** Measured (a) and calculated the addition of a head wave (b) B-scans of corner echo, and spatial distribution of corner echo (c) obtained with experiment (black line), finite element calculation (dotted grey line), CIVA calculation (dashed grey line) and CIVA calculation with add of head wave (grey line), according to the 10 degrees slope bottom of the specimen.

On this figure the result obtained with a finite element method FEM detailed in [4] is presented too (dotted grey line). We notice that it also provides a good prediction of experimental results.

*4.2. Other contributions of creeping waves and head waves*
Henceforth, the developed tools have been extended to simulate the reflection of a creeping wave which is generated along the flaw surface (Figure 8.b).Further, it is possible to take into account a backwall reflection before the generation of the creeping wave so that the contributions c) and d) (Figure 8) can be modelled.

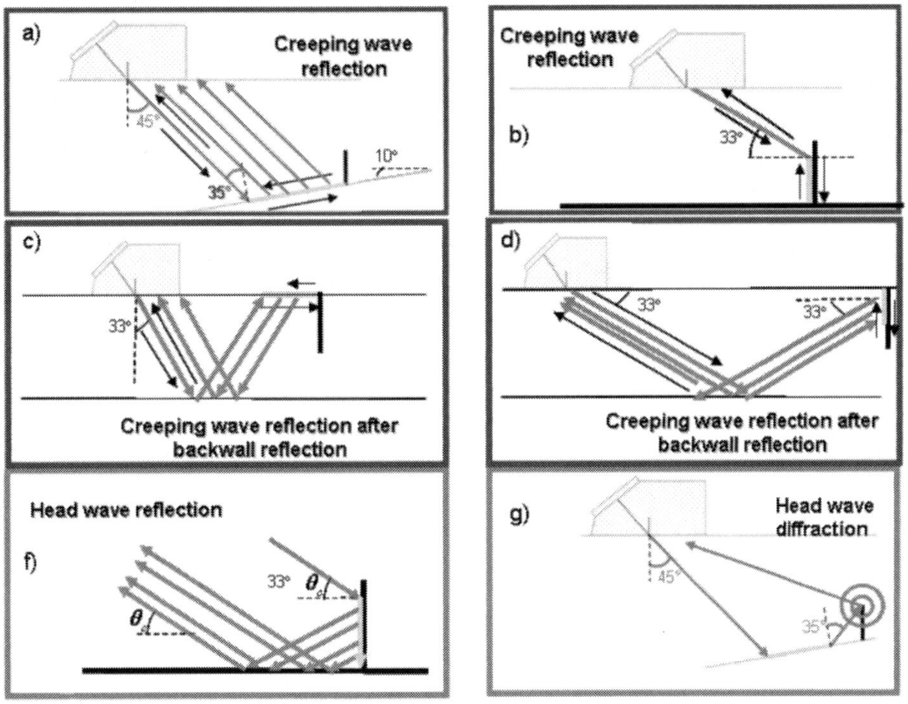


**Figure 8.** Contributions to the corner echo of creeping waves and head waves.

### *4.3. Account of a creeping wave after a backwall reflection*

To simulate the creeping wave reflection (Figure 9.a) at a wedge tip, we previously modelled [16, 17] the head wave amplitude. Indeed the displacement amplitude of the head wave received at a point $S_j$ on the probe and due to the reflection of the creeping pressure wave at the bottom of the flaw is:

$$\mathbf{u}_S\left(S_j,\omega\right) = -T_{C\to W} R_F T_{W\to C} \frac{v_0\left(\omega\right)e^{i\omega(t-\tau)}}{2\pi V_W} \frac{1}{\mathrm{i}\omega} \frac{R_{SL}^* R_{LS}^* V_L \tan\left(\theta^*\right)}{r^{1/2} l^{3/2}} \Delta S_{em} \mathbf{n}_W , \tag{3}$$

with $\Delta S_{em}$ the surface which surrounds the point source $S_i$, $v_0$ the particular velocity normal to probe surface, $T_{W\to C}$ the transmission coefficient from wedge to component (incoming longitudinal wave in the wedge to shear wave radiated inside the material), $T_{C\to W}$ the transmission coefficient from component to wedge. Further, $\mathbf{n}_W$ is the polarization direction of the wave propagating in the wedge with the velocity $V_W$, and $V_L$ is the longitudinal velocity in the component. The coefficients $R_{SL}^*$ and $R_{LS}^*$ are respectively normalized reflection coefficient [3] from shear wave to longitudinal wave and from longitudinal wave to shear wave at the critical angle $\theta^*$. Distances $r = r_i + r_j$ and $l = l_i + l_j$ are represented in the following figure 9.a.

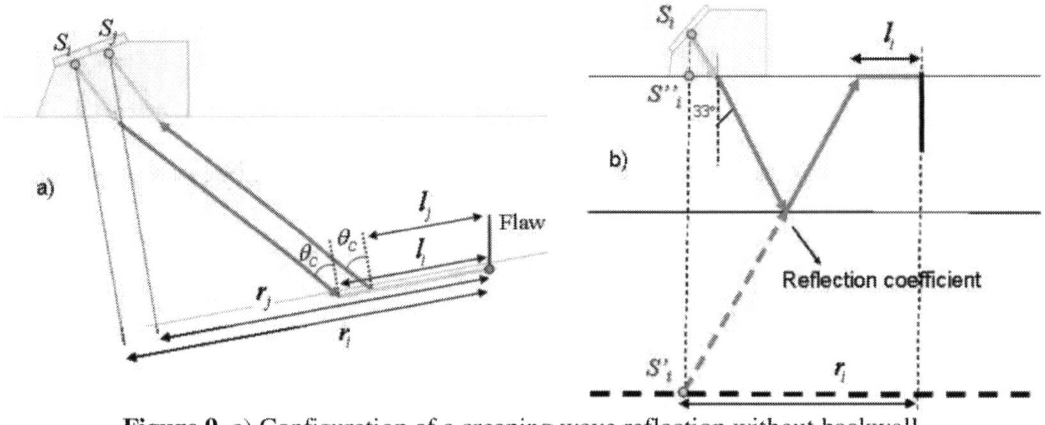


**Figure 9.** a) Configuration of a creeping wave reflection without backwall reflection; b) mirror strategy for a backwall reflection

To take into account a backwall reflection before the generation of the creeping wave, we extend the model developed to deal with the configuration without backwall reflection (figure 9.a).In that goal, we employ a mirror strategy to model the backwall reflection. The principle of this mirror effect is described on figure 9.b. for an incident path from the source to the wedge tip. A reception path from the wedge tip to a point $S_j$ on the crystal surface can be involved similarly. The point $S'_i$ is the image of the source point $S_i$ by symmetry with respect to the backwall surface and then $S''_i$ is the projection of $S'_i$ on the entry surface where the creeping wave propagates (in green on figure 9.b). In the case of a backwall reflection, the entry surface is the analogue of the backwall surface for the case without backwall reflection. Consequently, in case of no mode conversion at the backwall, the distance $r_i$ is calculated along the entry surface between $S''_i$ and the wedge tip. Finally, the amplitude of the head wave at a point $S_j$ on the probe is modelled using (3) by defining distances $r = r_i + r_j$ and $l = l_i + l_j$ thanks to the mirror strategy and by adding a plane wave reflection coefficient on the backwall.

Experimental measurements have to be carried out on both surface and backwall breaking flaws to validate the modelling of the creeping wave reflection along flaws with or without a backwall reflection.

## 5. Conclusion

This paper was devoted to the description of recent advances performed in the modelling of diffraction by cracks in the context of NDE. Three "semi-analytical" approaches, which have been enriched or extended in order to extend the applicability of the simulation tools, have been presented. Numerous experimental validations have been carried out for each developed model so as to highlight their accuracy.

The first invoked method is an exact analytical model based on Separation Of Variables (SOV): it has been developed for a cylindrical cavity without flaw disorientation (2,5D model). This SOV model provides a better prediction than the Kirchhoff approximation for SV waves and small flaws. It could be extended to a 3D configuration and to a spherical void.

We focused then on models derived from the Geometrical Theory of Diffraction (GTD) which is commonly used to deal with diffraction phenomena. A modification of the GTD coefficients has been included around special observation directions to improve the GTD simulation of the scattering from cracks edges. In future, progress towards uniform corrections of GTD in a view of a generic model (able to treat both reflection and diffraction effects) are envisaged. New GTD developments could enable also the treatment of flaws in anisotropic media (GTD coefficients for transversely-isotropic media) and the modelling of the scattering from wedges. The study of the interaction of a flaw with a Rayleigh surface wave is in progress.

As to the modelling of creeping and head waves, the simulation of the reflection of a creeping wave on a wedge tip is henceforth available if this wave propagates along a specimen or a flaw surface even after one backwall reflection. Experimental validations have confirmed that the account of the head wave reflection must be included in the simulation.

## 6. References

[1] CIVA software website, http://www-civa.cea.fr

[2] Gengembre N and Lhémery A 2000 Pencil method in elastodynamics. Application to ultrasonic field computation *Ultrasonics* **38** 495

[3] Červený V 2001 *Seismic Ray Theory* (Cambridge U. Press)

[4] Darmon M, Leymarie N, Chatillon S, Mahaut S 2009 Modelling of scattering of ultrasounds by flaws for NDT *Ultrasonic wave propagation in non homogeneous media* ed A Léger and M Deschamps (Springer Proc. Phys. **128**) 61

[5] Achenbach J D and Gautesen A K 1986 Edge Diffraction in Acoustics and Elastodynamics *Low and High Frequency Asymptotics - Vol 2* ed Varadan V K and V V (Elsevier) 335-401.

[6] Darmon M, Chatillon S, Mahaut S, Fradkin L J and Gautesen A 2008 Simulation of disoriented flaws in a TOFD technique configuration using GTD approach *Review of Progress in QNDE* **27**, ed D O Thompson and D E Chimenti (Melville: AIP Conf. Proc. **820**) 155

[7] Darmon M, Calmon P and Bele B 2004 An integrated model to simulate the scattering of ultrasounds by inclusions in steels *Ultrasonics* **42** 237

[8] Raillon R, Bey S, Dubois A, Mahaut S and Darmon M 2010 Results of the 2009 UT modelling benchmark obtained with CIVA: responses of notches, side-drilled holes and flat bottom holes of various sizes *Review of Progress in QNDE* **29**, ed D O Thompson and D E Chimenti (Melville: AIP Conf. Proc. **1211**) 2157

[9] Uberall H 1973 *Physical Acoustics - Vol. 10* ed W P Mason and R N (Thurston Academic Press)

[10] Achenbach J D, Brind R J, Gubernatis J E 1984 High-frequency scattering of elastic waves from cylindrical cavities *Wave Motion* **6** 41

[11] Schmerr L W Jr and Song J-S 2007 *Ultrasonic Nondestructive Evaluation Systems: Models and Measurements* (Springer)

[12] Mahaut S, Chatillon S, Darmon M, Leymarie N, Raillon R and Calmon P 2010 An overview of ultrasonic beam propagation and flaw scattering models in the Civa software *Review of Progress in QNDE* **29**, ed D O Thompson and D E Chimenti (Melville: AIP Conf. Proc. **1211**) 2133

[13] Fradkin L J, Gautesen A, Zernov V, Darmon M 2010 Elastic wave diffraction by infinite wedges *J. Phys.: Conf. Ser.* present volume

[14] McMaken, H 1984 A uniform theory of diffraction for elastic solids *J. Acoust. Soc. Am.* **75** 1352

[15] Borovikov V A 1994 *Uniform Stationary Phase Method* (The Institution of Electrical Engineers London)

[16] Huet G, Darmon M, Lhémery A and Mahaut S 2009 Modelling of corner echo ultrasonic inspection with bulk and creeping waves *Ultrasonic wave propagation in non homogeneous media* ed A Léger and M Deschamps (Springer Proc. Phys. **128**) 217

[17] Mahaut S, Huet G and Darmon M 2009 Modeling of corner echo in UT inspection combining bulk and head waves effect *Review of Progress in QNDE* **28** ed D O Thompson and D E Chimenti (AIP Conference Proceedings **1096** Melville 2009) 73

Journal of Physics: Conference Series **269** (2011) 012014 doi:10.1088/1742-6596/269/1/012014

# Ultrasound characterization of red blood cells distribution: a wave scattering simulation study

**Emilie Franceschini, Bruno Lombard, Joël Piraux**

Laboratoire de Mécanique et d'Acoustique LMA - CNRS UPR 7051, 13402 Marseille cedex 20, France

E-mail: `franceschini@lma.cnrs-mrs.fr`

**Abstract.** Ultrasonic backscattered signals from blood contain frequency-dependent information that can be used to obtain quantitative parameters describing the aggregation state of red blood cells (RBCs). However the relation between the parameters describing the aggregation level and the backscatterer coefficient needs to be better clarified. For that purpose, numerical wave simulations were performed to generate backscattered signals that mimic the response of two-dimensional (2D) RBC distributions to an ultrasound excitation. The simulated signals were computed with a time-domain method that has the advantages of requiring no physical approximations (within the framework of linear acoustics) and of limiting the numerical artefacts induced by the discretization of object interfaces. In the simple case of disaggregated RBCs, the relationship between the backscatter amplitude and scatterer concentration was studied. Backscatter coefficients (BSC) in the frequency range 10 to 20 MHz were calculated for weak scattering infinite cylinders (radius 2.8 $\mu$m) at concentrations ranging from 6 to 36%. At low concentration, the BSC increased with scatterer concentrations; at higher concentrations, the BSC reached a maximum and then decreased with increasing concentration, as it was noted by previous authors in *in vitro* blood experiments. In the case of aggregated RBCs, the relationship between the backscatter frequency dependence and level of aggregation at a concentration of 24% was studied for a larger frequency band (10 - 40 MHz). All these results were compared with a weak scattering model based on the analytical computing of the structure factor.

## 1. Introduction

The aggregation of red blood cells (RBCs) is a physiological reversible process. When RBCs are under low shear rates (<10 $s^{-1}$), they interact strongly with each other and form complex 3D rouleaux structures. When the shear rate increases, rouleaux structures desaggregate. This phenomenon is not pathological. However, RBC hyper-aggregation, an abnormal increase of RBCs aggregation, is a pathological state associated with several circulatory diseases such as deep venous thrombosis, atherosclerosis and diabetes mellitus. These pathologies inflict particular sites (inferior members for thrombosis, arterial bifurcations for atherosclerosis, the foot and eye for diabetes). Ultrasonic imaging holds the perspective of further elucidating the role of RBC aggregation on the etiology of these diseases *in vivo* and *in situ*.

Classical blood characterization techniques consist of measuring the magnitude and frequency dependence of the radio frequency backscatter spectrum. A large number of *in vitro* experiments on blood have been performed to study how these two indices vary with shear rate, flow turbulence, hematocrit (i.e. concentration of RBCs), and RBC aggregation level [1]-[4].

Theoretical efforts have been also made to develop ultrasound backscattering models for non-aggregating [5] and aggregating RBCs [6]. Twersky [5] proposed a packing factor to describe the backscattering coefficient of a distribution of hard particles. This model succeeds to explain the relationship between the backscatter amplitude and hematocrit for non-aggregating RBCs but fails to predict the frequency dependence observed in *in vitro* experiments when considering aggregating RBCs. Yu and Cloutier [6] recently parametrized the backscattering coefficient from blood: two indices describing RBC aggregation, the packing factor and mean aggregate diameter, were extracted from the Structure Factor Size Estimator. However this estimator still needs to be quantitavely evaluated.

The purpose of this paper is to perform numerical ultrasound wave simulations through 2D RBC distributions in order to validate a weak scattering theoretical model based on a particle approach [7]. This theoretical model is commonly employed for approximating the ultrasound backscattering from RBCs [8]-[12]. In order to simulate ultrasonic propagation through RBCs, we use a time-domain numerical method [13] [14] that has the advantages of requiring no physical approximation (within the framework of linear acoustics) and of limiting the numerical artefacts induced by the discretization of object interfaces. Our method therefore automatically accounts for multiple scattering, refraction, and reflection, and the accuracy of the numerical solution is independent of the cell number. In the simple case of disaggregated RBCs, the relationship between the backscatter amplitude and scatterer concentration is studied. In the aggregating case, the relationship between the backscatter frequency dependence and level of aggregation at a concentration of 24% is studied in the frequency range 10 to 40 MHz. All the results obtained from the direct numerical simulation are compared with those expected from theory.

## 2. Ultrasound backscattering theory

Ultrasonic scattering from blood is mainly caused by the RBCs. Indeed, blood can be mechanically described as a colloidal suspension of RBCs in plasma. RBCs constitute the vast majority (97%) of the cellular content of blood and occupy a large volume fraction (hematocrit) of 35-45% under normal conditions. In this 2D study, RBCs are modeled as parallel infinite cylinders of radius $a$, that have a small acoustical impedance contrast relatively to the host homogeneous medium (i.e. the plasma): $\gamma_z = (Z_{RBC} - Z_{plasma})/Z_{plasma} = 0.11$. RBCs and plasma are assumed to be fluid and non-dissipative media. In the following, we recall the scattering model commonly used for a single RBC and then for many RBCs.

### 2.1. *Single-RBC scattering model*

Consider a backscattering configuration that consists of the insonification of one fluid RBC by a monochromatic plane wave $p_i(\mathbf{r}) = p_0 e^{i\mathbf{kr}}$, where $\mathbf{r}$ is the spatial vector, $p_0$ is the incident pressure amplitude, and $\mathbf{k}$ represents the wave vector ($|\mathbf{k}| = k = 2\pi/\lambda$, $\lambda$ being the wavelength). In the far field of the particle, the field backscattered by the particle has the asymptotic form:

$$p_s(\mathbf{r}) = p_0 \frac{e^{i\mathbf{kr}}}{\sqrt{r}} \left[\sigma_b(k)\right]^{1/2}, \tag{1}$$

where $\sigma_b$ is the backscattering cross section per unit area of one particle that can be calculated using the theory of Faran [15]. For the simplest case where the particle size is small compared to wavelength ($ka << 1$), we use Rayleigh scatterer form [16]

$$\sigma_b(k) = \frac{1}{2\pi^2} k^3 A_p^2 \gamma_z^2 \left(\frac{J_1(2ka)}{ka}\right)^2, \tag{2}$$

where $A_p = \pi a^2$ is the particle area and $J_1$ is the first order Bessel function of the first kind.

### 2.2. *Many-RBC scattering model*

The theoretical model of ultrasound backscattering by blood is based on a particle approach[6] [7], which consists of summing contributions from individual RBCs and modeling the RBC interaction by a structure factor. By using the Born approximation (weak scattering) and considering a collection of $n$ identical and cylindrical RBCs in the backscattering configuration described in the previous section, the differential backscattering cross section per unit area (also called the backscattering coefficient BSC) can be written as [5]:

$$BSC_{theo}(k) = m\sigma_b(k)S(k), \tag{3}$$

where $m$ is the number density of RBCs in blood that is related with the hematocrit $H$ as $m = H/A_p$. The function $S$ is the structure factor representing the spatial positioning of particles and defined by:

$$S(\mathbf{k}) = E\left[\frac{1}{n}\left|\sum_{i=1}^{n} e^{-i2\mathbf{k}\mathbf{r}_i}\right|^2\right] \tag{4}$$

where $E$ represents the expected value of a random variable and $\mathbf{r}_i$ the position vectors defining the center of the $i$th scatterer in space. Since the medium is isotropic, $S(\mathbf{k})$ depends only on $k$; for the clarity of the notation, we confused $S(\mathbf{k})$ and $S(k)$ as done in Eq. (3). The aggregation phenomenon is thus supposed to only affect the structure factor since RBCs properties and hematocrit remain constant in the region of interest.

In the low frequency limit, the structure factor tends towards a constant value $S(k) \to S(0) = W$ called the packing factor that has been extensively studied for non-aggregating particles. The most often used packing factor expression is based on the Percus-Yevick pair-correlation function for identical, hard and radially symmetric particles. The Perkus-Yevick packing factor $W_{PY}$ was first applied to blood by Shung [17] and is related in the 2D space with the hematocrit as [5] [17] [7]

$$W_{PY} = \frac{(1-H)^3}{1+H}. \tag{5}$$

At very low $H$, $W_{PY}$ is equal to unity. As $H$ increases, $W_{PY}$ decays to zero so that the BSC also approaches zero at very high $H$ because in a densely packed medium one can always find an RBC that will destructively interfere with the contribution from another RBC.

## 3. Methods

### 3.1. *Computation of a RBCs distribution*

We describe here how random distributions for non-aggregating and aggregating RBCs were computed. These distributions were employed in the direct numerical simulation and in the computation of the theoretical BSC.

*Computation of a random distribution for non-aggregating RBCs.* First, we specify the RBC radius $a = 2.75\mu$m and the surface size $S$ to be investigated. For each studied hematocrit $H$, the number of RBCs $n$ is given by $n = HS/A_p$ and then a simple method allows the determination of the RBC spatial positions as described in the following. The center position of each cylinder is determined by a pair of random variables $(x_i, y_j)$. Each time a new cylinder is placed, a procedure is applied to check the distance between the newly generated cylinder and the cylinders generated before. If a newly generated scatterer overlaps with any other scatterers, the scatterer will be discarded, otherwise the scatterer's position is saved. This procedure is repeated until the required hematocrit is reached. Note that, for each hematocrit, two types of distribution were performed:

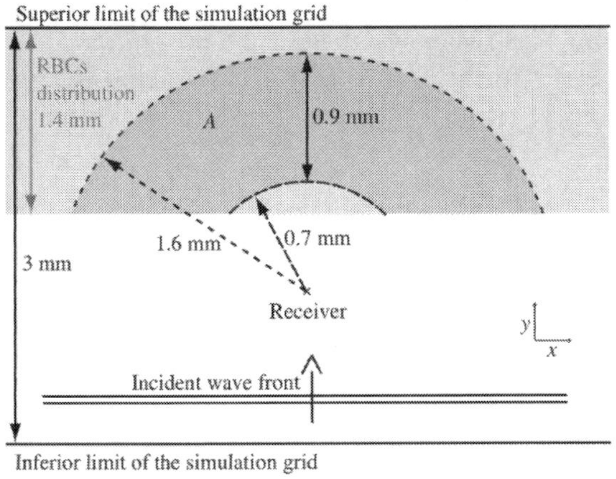


**Figure 1.** Sketch of the computational domain. The area of interest $A$ is the zone containing the RBCs distribution and located between the two arcs of circles. The center of these circles corresponds to the receiver position and the circle radii are 0.7 mm and 1.6 mm.

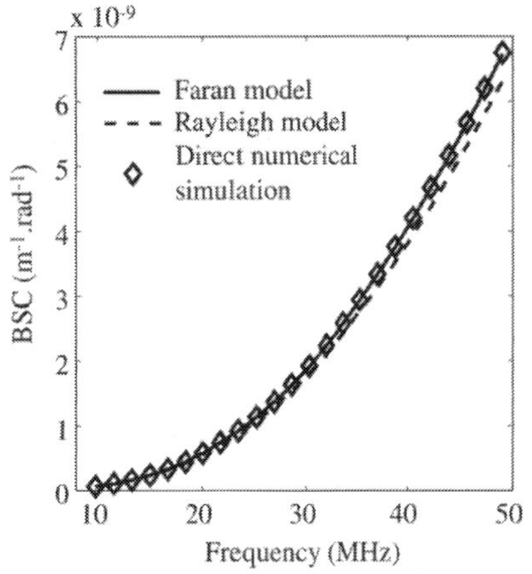


**Figure 2.** Backscattering cross sections per unit area $\sigma_b$ of a single RBC. The solid and dashed lines display $\sigma_b$ given by, respectively, the Faran and Rayleigh models. Simulations results are shown by the symbols.

- one with an exclusion length $l_e$=0.9$\mu$m between scatterer (i.e. a non-zero minimal length between scatterer interfaces) for computational purpose in the direct numerical simulation (see section 3.2),
- one without an exclusion length, i.e. scatterers could be in contact.

*Computation of a random distribution for aggregating RBCs.* First, discs of radius 6.60 $\mu$m (and respectively 9.08 $\mu$m) are randomly generated, these discs correspond to aggregates containing 3 RBCs (and respectively 6 RBCs). An exclusion length $l_e$ is ensured between aggregates. For a specific hematocrit, the locations of the scatterers inside each aggregate are generated randomly with an exclusion length $l_e$ between scatterers.

### *3.2. Direct numerical simulation*

The numerical method used to compute the solution of the wave equation was previously described [13] [14]. The time-domain numerical integration of acoustic equations is done by a second-order Lax-Wendroff scheme. The discretization of interfaces is done by an immersed interface method, ensuring accurate description of geometrical features and of the jump conditions. The computations involve grids of 2 mm×3 mm with mesh sizes $\Delta x = \Delta y = 0.195$ $\mu$m (i.e. grids of 10240×15360 nodes) requiring the parallelization of algorithms on a cluster of PCs. The RBCs are distributed on a 2 mm × 1.4 mm rectangular subdomain. As required by the immersed interface method, an exclusion length $l_e$ between each scatterer is taken equal to $4\Delta x$=0.9 $\mu$m.

The computational grid is excited by a plane wave centered at 25 MHz whose support is originally outside the region of interest. That is why in the following the studied frequency range will be from 10 to 40 MHz. The $x$ axis of the computational domain are periodized in order to simulate the wave propagation in a semi-infinite medium (see Fig. 1). At each time step, the simulated acoustic pressure is recorded on 256 receivers. These receivers are placed at every 8 $\mu$m and in parallel to the incident wave front in order to simulate a transducer moved laterally. Signals were selected with a rectangular window of length 0.9 mm centered at a distance $d = 1.15$ mm from the receivers. The power spectra of the 256 backscattered signals were then averaged to provide the mean backscattered power spectrum $\overline{P}$. The BSC obtained from the simulated data was computed as

$$BSC_{sim}(k) = \frac{d}{A}\frac{\overline{P}}{P_{inc}}, \tag{6}$$

where $P_{inc}$ is the power spectrum of the incident signal and $A$ is the area of interest determined by acoustic ray theory. Indeed, since signals are selected with a rectangular window corresponding to depths between 0.7 mm and 1.6 mm from the receivers, the RBCs contributing to the selected signals are localized between two circles of radii 0.7 mm and 1.6 mm and of center the receiver positions. The area of interest $A$ is thus the zone containing the RBCs distribution and located between these two circles as shown in Fig. 1.

### *3.3. Computation of the theoretical BSC*

The computation of the theoretical BSC requires the knowledge of the structure factor as described in Eq. (3). Since $S$ is by definition a statistical quantity, an average of all structure factors obtained from several random particle distributions can give an estimated value of $S$. For each particle distribution, a matrix $D$ was computed by dividing the square simulation plane $L^2$ in $N_p^2$ pixels and by counting the number of particles falling into each pixel, $N_p$ being sufficiently high in order that there is a maximum of one particle by pixel. We chose to take $L$=300 $\mu$m and $N_p$=256 pixels as a tradeoff between the computational domain size and the computational time. This matrix $D$ represents the microscopic density function defined by $D(\mathbf{r}) = \sum_{i=1}^{n} \delta(\mathbf{r} - \mathbf{r}_i)$, where $\delta$ is the Dirac distribution. According to Eq.(4), the structure factor can thus be described by:

$$S(\mathbf{k}) = E\left[\frac{1}{n}\left|\int D(\mathbf{r})e^{-i2\mathbf{kr}}d\mathbf{r}\right|^2\right]. \tag{7}$$

The structure factor was thus computed by averaging 2-D fast Fourier transforms (2D FFT) of 200 density matrix $D$. This FFT gave the structure factor values $S(\mathbf{k})$ on a centered grid of wavevectors between $\pm\pi N_p/2L$ with a step of $\Delta k = \pi/L$. The theoretical BSC is then obtained using Eq. (3).

## 4. Results and discussion

### *4.1. Non-aggregating case*

Figure 2 shows the simulated $\sigma_b$ as a function of frequency in a linear scale. Also represented are two theoretical $\sigma_b$ obtained from the Rayleigh and Faran models. One can notice that the Faran model matches well the $\sigma_b$ obtained with the direct numerical simulation at all frequencies, whereas the Rayleigh model is only valid within 5% accuracy until a frequency of 22 MHz. Note that this result was expected since the Rayleigh model is accurate within 5% for $ka$ <0.25. In the following, all the theoretical BSCs presented will be calculated using the Faran model.

For non-aggregating RBCs at low frequencies, the relationship between the BSC amplitude and hematocrit was investigated. Figure 3a presents four averaged BSCs obtained by averaging BSCs between frequencies in the validity domain of Rayleigh behavior (i.e. between 10 and 22 MHz). The diamond markers represent the simulation results. The solid line shows the prediction by the hard sphere Percus-Yevick packing factor. The dashed lines indicate the particle theory approach by considering RBC distributions with and without an exclusion length.

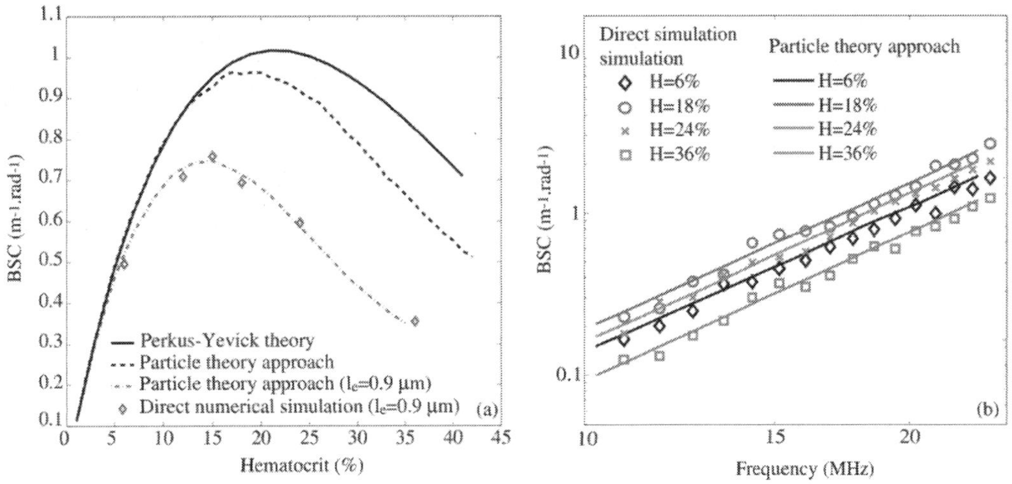


**Figure 3.** (a) Plots of backscatter coefficients versus the hematocrit for non-aggregated RBCs. The solid curve represents the $BSC_{theo}$ calculated with the Percus-Yevick packing factor, the dashed lines the particle theory approach $BSC_{theo}$ obtained by computing the structure factor with and without the exclusion length. The diamond markers express simulation results $BSC_{sim}$. (b) Frequency-dependent backscattering coefficients for an hematocrit of 6, 18, 24 and 36% in the case of non-aggregated RBCs. The solid line represents the curve for the particle theory approach. Simulations results are shown by the symbols.

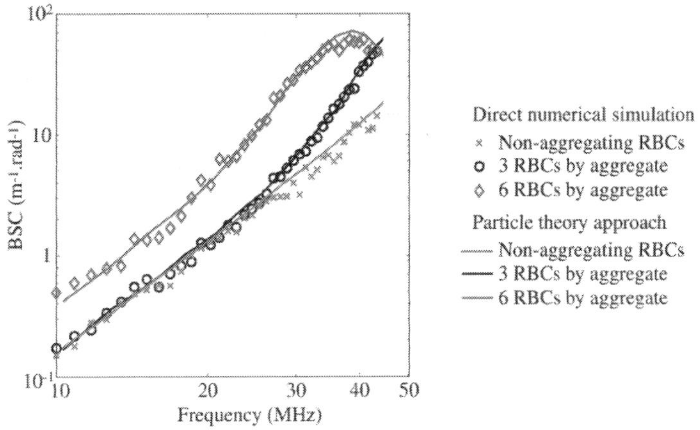


**Figure 4.** Frequency-dependent backscattering coefficients for a constant hematocrit of 24 % in the case of non-aggregated and aggregated RBCs. The solid line represents the curve for the particle theory approach. Simulations results are shown by the symbols.

This figure demonstrates that both simulation results and particle theory approach with the same exclusion length are in good agreement at low frequencies. For the results without an exclusion length, the peak of the BSC obtained by the particle theory approach is seen at about 19%, in close agreement with the Percus-Yevick packing theory at about 22% as observed by Zhang and co-workers [8] with a Gaussian pulse incident wave (see Fig. 10 in Ref. 19). However, from a quantitative point of view, the Percus-Yevick packing theory does not match the particle theory approach for hematocrits superior to 14 %.

Typical results of the frequency-dependent $BSC_{sim}$ and $BSC_{theo}$ for an hematocrit of 6, 18, 24 and 36% are given in Fig. 3b in a log compressed scale. For both BSCs, the distributions were realized with the exclusion length $l_e$. Standard deviations are not shown for clarity. The weak scattering model based on the structure factor matches quite well the simulation results. For all hematocrits, the BSCs are Rayleigh at low frequencies (up to 22 MHz) i.e. the BSCs have third-power frequency dependence.

To conclude the non-aggregation case study, for the RBC distributions having an exclusion length, the direct numerical simulation validates the structure factor theory for frequencies between 10 and 22 MHz. We can reasonably extrapolate this result to RBC distributions without an exclusion length. As shown in Fig. 3a, it will be thus more accurate to calculate the theoretical BSC via the particle theory approach rather than the Percus-Yevick packing theory for dense medium.

### *4.2. Aggregating case*

Figure 4 shows the simulated and theoretical BSCs as a function of frequency in a log compressed scale in the case of aggregating RBCs. A constant hematocrit H=24% and two aggregate sizes (i.e. 3 and 6 RBCs by aggregate) were used for these computations. The markers represent the simulation results and the solid line shows the prediction by the particle theory approach. As observed in the case of the non-aggregating case in Fig. 3b, the particle theory approach matches quite well the simulation results.

Also represented in Fig. 4 are the results in the case of non-aggregated RBCs at an hematocrit of $H$=24% in order to compare non-aggregating and aggregating cases. For non-aggregated RBCs, the slope of the BSC as a function of frequency is linear in our log-log scale (i.e. spectral slope equal to 3). When considering aggregated RBCs, the BSC becomes non-Rayleigh: the slope becomes not linear in our log-log scale and the BSC amplitude increases as the level of aggregation becomes larger. Note that the behaviors (frequency-dependence and amplitude) of the BSC in both non-aggregated and aggregated RBCs were the same in *in vitro* experiments performed by Yu and Cloutier [6] in a controlled Couette flow. Besides no effect of aggregation is seen for the case of 3 RBCs by aggregate at low frequencies, whereas the increase of the BSC amplitude is important for the case of 6 RBCs by aggregate. It is due to the hematocrit of 24% used in our simulations since for the 3-RBC case at lower hematocrit ($H < 10\%$), a steady increase in backscatter at low frequencies is observed (data not shown here).

The impact of aggregation on the spectral slope was also evaluated in the whole frequency range of 10 - 40 MHz as performed by Van Der Heiden and co-workers [3] in an experimental study in a controlled Couette flow. In our numerical study, the spectral slope decreases when the level of aggregation increases: the spectral slopes are equal to 3, 2.75 and 2.05 respectively for 1, 3 and 6 RBCs by aggregate. A decrease was also observed in experiments by Van Der Heiden and co-workers [3]: the spectral slope of aggregated RBCs (slope = 1.3) was much lower that the spectral slope of disaggregated RBCs (slope = 3).

To conclude, in our numerical simulations, we can make a clear distinction between disaggregated and aggregated RBCs by observing the spectral slope, frequency-dependence and amplitude of the BSC. From a qualitative point of view, these parameters match quite well the experiments of controlled Couette flow found in the literature [3] [6]. Nevertheless, a quantitative comparison between our simulations and the experiments is difficult. Indeed, our simulations are in 2D and size aggregates are generally unknown in experiments of blood flow at high hematocrit ($H > 6\%$) since blood is opaque to light.

## 5. Conclusion

This study shows that the particle theory approach approximates accurately the backscattering field of ultrasonic waves when the wavelength of the incident field is much greater than the scatterer size, for hematocrits inferior to 36% and for frequencies between 10 and 40 MHz. The weak scattering model is thus sufficient for computing the backscatter coefficient from aggregated RBCs. Nevertheless, our work needs to be completed by the study of larger size aggregates and a larger frequency bandwidth (10 - 70 MHz) to reach the model limit.

[1] Yuan W and Shung K K 1988 *J.Acoust. Soc. Am.* **84**, 1195-1200
[2] Foster F S, Obara H, Bloomfield T, Ryan L K and Lockwood G R 1994 *IEEE Ultrasonics Symposium Proceedings* 1599-1602
[3] Van Der Heiden M S, De Kroon M G, Bom N and Borst C 1995 *Ultrasound Med. Biol.* **21** 817-26
[4] Cloutier G, Qin Z, Durand L G and Teh B G 1996 *IEEE Trans. Biomed. Eng.* **43** 441-50
[5] Twersky V 1987 *J. Acoust. Soc. Am.* **81** 1609-18
[6] Yu F T H and Cloutier G 2007 *J. Acoust. Soc. Am.* **122**, 645-56
[7] Mo L Y L and Cobbold R S C 1993 *Ultrasonic Scattering in Biological Tissues* (CRC, Boca Raton, FL) chapter 5 pp 125-170
[8] Zhang J, Rose J L and Shung K K 1994 *Ultrasound in Med. Biol.***20** 903-913
[9] Fontaine I, Bertrand M and Cloutier G 1999 *Biophysical Journal* **77** 2387-99
[10] Savery D and Cloutier G 2005 *IEEE Trans. Ultrason. Ferroelectr. Freq. Control* **52** 94-103
[11] Franceschini E, Yu F T H, Fenech M and Cloutier G 2007 *IEEE Ultrasonics Symposium Proceedings* 2503-06
[12] Saha R K and Cloutier G *Physical review* E**78** 061919
[13] Piraux J and Lombard B 2001 *Journal of Computational Physics* **168** 227-48
[14] Lombard B and Piraux J 2004 *Journal of Computational Physics* **195** 90-116
[15] Faran J J 1951 *J. Acoust. Soc. Am.* **23** 405-418
[16] Insana M F and Brown D G 1993 *Ultrasonic Scattering in Biological Tissues* (CRC, Boca Raton, FL) chapter 5 pp 75-124
[17] Shung K K 1982 *IEEE Trans. Ultras., Ferroelect., Freq. Contr.* **SU-26**, 327-31

Journal of Physics: Conference Series **269** (2011) 012015 doi:10.1088/1742-6596/269/1/012015

# Multiple scattering by cylinders randomly located in a fluid: effective properties

**Francine Luppé[a], Jean-Marc Conoir[b]**

[a] Laboratoire Ondes et Milieux Complexes, FRE CNRS 3102, Groupe Ondes Acoustiques, Université du Havre, place R. Schuman, 76610 Le Havre, France

[b] UPMC Université Paris 06, UMR 7190, Institut Jean Le Rond d'Alembert, F-75005 Paris, France.

Francine.luppe@univ-lehavre.fr

**Abstract**. We consider the interface between a homogeneous fluid and that same fluid with a random distribution of $n_0$ cylinders *per* square meter inside. A harmonic plane wave, frequency $\nu$ and wavenumber $k$, is incident upon that boundary under incidence angle $\alpha$. The reflection coefficient obtained with the Fikioris and Waterman approach is expanded into powers of $n_0/k^2$ up to order 2, using Linton and Martin's expansion of the wavenumber of the coherent wave. This coefficient is then compared to that obtained when a homogeneous viscous fluid replaces the random medium. When the two reflection coefficients are equal, the random fluid is acoustically equivalent to the viscous one, which is called in that case the effective fluid. The coherent wave in the random medium is thus described as the acoustic mode in the effective fluid. Equating the two reflection coefficients provides expressions for the effective properties of the random medium: mass density $\rho_{eff}$, and coefficient of dilatation viscosity $\eta_{eff}$, as the shear viscosity is set to zero. Both depend on $\alpha$ and $\nu$, unless low frequencies only are considered, in which case the dependence on $\alpha$ vanishes.

## 1. Introduction

It is well known that a harmonic plane wave that propagates through a dilute random collection of identical elastic cylinders in an ideal fluid undergoes multiple scattering and that the average field may be described as a plane heterogeneous wave propagating with a complex wavenumber $K$. The most recent analytic expression of the effective wavenumber $K$ was given in 2005 by Linton and Martin [1],

$$\frac{K^2}{k^2} = 1 - 4if(0)\frac{n_0}{k^2} + \frac{8}{\pi}\frac{n_0^2}{k^4}\int_0^\pi \cot\frac{\theta}{2}\frac{d}{d\theta}\left[f(\theta)f(-\theta)\right]d\theta \qquad (1)$$

with $k=\omega/c$ the real wavenumber in the absence of scatterers, $n_0$ the number of cylinders per unit surface, and $f(\theta)$ the far-field amplitude of the wave scattered by a single cylinder in direction $\theta$. Letting $a$ denote the radius of the cylinders, the concentration of scatterers is $\varphi=n_0\pi a^2$.

Linton and Martin's equation (Eq.(1)) may be recovered [2] from the Fikioris and Waterman's equation of dispersion [3], under the assumption that $n_0/k^2$ and $kb$, with $b$ the radius of the hole correction in Ref.[3], are small. As $b$ represents the minimum distance between centres of scatterers, $b > 2a$, and thus Eq.(1) is a low concentration ($n_0/k^2 \ll 1$) ***and*** low frequency ($ka \ll 1$) approximation. Its range of validity, however, may be larger in frequency than expected, as will be shown in the next section, and the question arises whether it is possible to define an effective medium (acoustic equivalent of the random one) with that same range of validity.

The effective medium in the following is supposed to be a viscous fluid with no shear viscosity, so that the coherent wave of wavenumber $K$ given by Eq.(1) is supposed to be the only propagating mode, *i.e.* the acoustic mode [4],

$$\frac{K^2}{k^2} = \left(1 - i\frac{\omega}{\rho_{eff}c^2}\eta_{eff}\right)^{-1}. \tag{2}$$

In Eq.(2), $\eta_{eff}$ is the dilatation viscosity, $\rho_{eff}$ the effective mass density, and the adiabatic sound speed of the viscous fluid is supposed to be that of the ideal fluid in the absence of scatterers. Both $\eta_{eff}$ and $\rho_{eff}$ are sought, from the equality of Eq.(1) and Eq.(2), and from the equality of the reflection coefficient $R$ at the interface between an ideal fluid and a viscous one [5],

$$R = \frac{-\rho K\cos\alpha_r + \rho_{eff}k\cos\alpha}{\rho K\cos\alpha_r + \rho_{eff}k\cos\alpha}, \tag{3}$$

with that at the interface between the same ideal fluid and the random medium [6],

$$R = -\frac{1 + \sum_{n\in\mathbb{Z}^*}(-1)^n D_n T_n e^{-in(\alpha_r+\alpha)}}{1 + \sum_{n\in\mathbb{Z}^*}^{+\infty} D_n T_n e^{in(\alpha_r-\alpha)}}\,\frac{K\cos\alpha_r - k\cos\alpha}{K\cos\alpha_r + k\cos\alpha}. \tag{4}$$

In Eq.(3), $K$ is given by Eq.(2), while in Eq.(4), it is given by Eq.(1). $\alpha$ is the incidence angle in the ideal fluid, $\alpha_r$ the refraction angle, $K\sin\alpha_r = k\sin\alpha$, $\rho$ the mass density of the ideal fluid, and the $T_n$ are the scattering coefficients of the cylinders. The $D_n$ coefficients are given from Eqs.(5,6),

$$\forall n \neq 0,\ X_n = D_n T_0 X_0,\quad \sum_{m\in\mathbb{Z}^*}\left(\delta_{nm} - \frac{2\pi b n_0}{\left(K^2 - k^2\right)a}A_{nm}T_m\right)X_m = \frac{2\pi b n_0}{\left(K^2 - k^2\right)a}A_{n0}T_0 X_0, \tag{5}$$

$$A_{nm} = KaJ_{n-m}'(Kb)H_{n-m}^{(1)}(kb) - kaJ_{n-m}(Kb)H_{n-m}^{(1)}{}'(kb). \tag{6}$$

In Ref.[6], the Fikioris-Waterman's approach was used to calculate $R$, under the assumption that the average acoustic field in the random medium was given by the coherent wave at arbitrary distance from the boundary. According to Refs.[2,3], however, the situation is much more complicated in the vicinity of the interface, namely at distances smaller or the order of $b$, and neglecting these complications should be relevant only at low frequency ($ka \ll 1$), as long as $b > 2a$ is assumed. Eqs.(5,6) will thus be simplified in section 3 by assuming low concentration ($n_0/k^2 \ll 1$) and letting $b$ tend to zero, and the effective parameters will be derived from here in section 4. The next section, as for it, is devoted to the discussion of the validity of Eq.(1) at intermediate frequencies.

All numerical studies are performed for steel cylinders in water. In water, $c$=1480m/s, $\rho$=1000kg/m$^3$. Steel has a mass density $\rho_S$=7816kg/m$^3$, a longitudinal velocity $c_L$=6000m/s, and a shear velocity $c_S$=3100m/s. Unless otherwise specified, the concentration, with $a$=1mm, is set to 0.1, so that $n_0a^2$=0.03, $n_0/k^2$ =0.98 at $ka$=0.18 and $n_0/k^2$ =0.10 at $ka$=0.56.

## 2. Validity range of Linton-Martin's expression of the effective wavenumber

As stated in the introduction section, Eq.(1) is the approximation of the solution $K_{FW}$ of Fikioris and Waterman's dispersion equation for $kb \ll 1$ and $n_0/k^2 \ll 1$. $K_{FW}$ is found by setting the determinant of the following homogeneous system to zero:

$$\sum_{m\in\mathbb{Z}}\left(\delta_{nm}-\frac{2\pi b n_0}{\left(K^2-k^2\right)a}A_{nm}T_m\right)X_m=0. \tag{7}$$

Fig.1 show the comparison between $K$, as given in Eq.(1), and $K_{FW}$, for different values of $b$, over the frequency range $ka\in[0,5]$. The $K_{FW}$ values for $b$=0 are obtained from Eqs.(7,8) (see section 3).

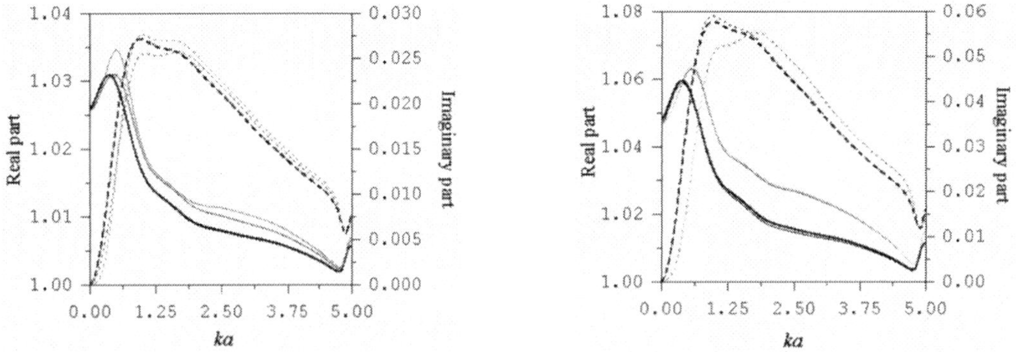


a) $\varphi$=0.1 b) $\varphi$=0.2

**Figs.1** Comparison between $K_{FW}/k$ (black, thin, and colours) and $K/k$ (black, thick) for different values of $b$.
Black, thin : $b$=0 - Blue : $b$=2.05$a$ - Green : $b$=2.1$a$ - Red : $b$=3.0$a$
Solid lines : real part. Dotted lines : imaginary parts.

At a concentration $\varphi$=0.1, the difference between $K_{FW}/k$ and $K/k$ is very small (at most 10% on the imaginary part for $b$=3$a$, otherwise less than 6% on the imaginary part and less than 1% on the real part), so that Eq.(1) still provides a reasonable approximation to $K_{FW}$ at intermediate frequencies and $b$ not too far from its minimum value 2$a$. The situation is completely different at concentration $\varphi$=0.2, for which $n_0/k^2$ is less than 0.1 for $ka$ values larger than 0.77. The difference between the imaginary parts of $K_{FW}$ and $K$ exceeds 10% practically on the whole range of frequency even for $b$=2.05$a$. This is the reason why the concentration $\varphi$ in the following is kept constant at $\varphi$=0.1. For this concentration, and even at intermediate frequency (1<$ka$<5) letting $b$ tend to zero independently of $a$ thus provides a reasonable analytic approximation of the "actual" effective wavenumber (supposed here to be $K_{FW}$, which can be found only numerically from Eq.(7)) *via* Eq.(1), which, in that case, is a low concentration approximation only.

## 3. The reflection coefficient of the random medium at low concentration

In order to get an analytic expression of $\rho_{eff}$ and $\eta_{eff}$, we need a more explicit expression of the reflection coefficient given in Eq.(4). This is the objective of this section. Following Linton and Martin, we look for an expansion of $R$ in powers of $n_0/k^2$ at low concentration. Letting $b$ tend to zero, Eq.(6) turns into

$$A_{nm}\simeq\frac{2a}{i\pi b}\left(\frac{K}{k}\right)^{|n-m|}. \tag{8}$$

The following integrals are first introduced, as $J(0)$ appears in Eq.(1) and both $I(\alpha)$ and $J(\alpha)$ will be useful later,

$$I(\alpha)=\sum_{n\in\mathbb{Z}}|n|T_n e^{in\alpha}=-\frac{1}{2\pi}\int_0^{\pi}\cot\frac{\theta}{2}\frac{d}{d\theta}\left[f(\alpha+\theta)+f(\alpha-\theta)\right]d\theta, \tag{9a}$$

$$J(\alpha)=\sum_{n\in\mathbb{Z}}\sum_{m\in\mathbb{Z}}|m-n|T_mT_ne^{in\alpha}=-\frac{1}{2\pi}\int_0^{\pi}\cot\frac{\theta}{2}\frac{d}{d\theta}\left[f(\theta)f(\alpha-\theta)+f(-\theta)f(\alpha+\theta)\right]d\theta. \quad (9b)$$

Now Eqs.(6,8) are written in a matrix form, $X=T_{00}M^{-1}SX_0$, after letting $X=(\ldots X_{m\neq 0}\ldots)^t$ and $S=(\ldots S_{m\neq 0}\ldots)^t$, with

$$S_m=\frac{1}{f(0)}\left[1+2i\frac{n_0}{k^2}\left[\frac{J(0)}{f(0)}-|m|f(0)\right]\right], M=I-\left[\frac{1}{f(0)}B+2i\frac{n_0}{k^2}\left[\frac{1}{f(0)}\frac{J(0)}{f(0)}B-C\right]\right], \quad (10)$$

$I$ the identity matrix, and matrices $B$ and $C$ defined from $B_{nm}=T_m, C_{nm}=|n-m|T_m, (n,m)\in\mathbb{Z}^{*2}$. The inverse of matrix $M$ is expanded as

$$M^{-1}=I+\sum_{q=1}^{+\infty}\left(\frac{B}{f(0)}\right)^q+2i\frac{n_0}{k^2}\left[\frac{J(0)}{f(0)}\sum_{q=1}^{+\infty}q\left(\frac{B}{f(0)}\right)^q-\sum_{q=1}^{+\infty}q\left(\frac{B}{f(0)}\right)^{q-1}C\right], \quad (11)$$

so that finally the approximate $D_n$ coefficients are obtained as

$$D_n=\frac{1}{T_0}-2i\frac{n_0}{k^2}\left[+\frac{1}{T_0}\frac{J(0)}{f(0)}+|n|+\frac{1}{f(0)}\sum_{m\in\mathbb{Z}^*}|n-m|T_m-\frac{1}{T_0}\frac{f(0)+2T_0}{f(0)}I(0)\right]. \quad (12)$$

Use of the Snell-Descartes laws $K\sin\alpha_r=k\sin\alpha$ along with Eq.(1) and Eq.(9b) leads to, for any incidence other than grazing,

$$\frac{K\cos\alpha_r}{k\cos\alpha}=1-2i\frac{n_0}{k^2}\frac{f(0)}{\cos^2\alpha}+2\frac{n_0^2}{k^4}\left[\left(\frac{f(0)}{\cos^2\alpha}\right)^2-2\frac{J(0)}{\cos^2\alpha}\right], \quad (13)$$

so that

$$R=i\frac{n_0}{k^2}\frac{f(\pi-2\alpha)}{\cos^2\alpha}+2\frac{n_0^2}{k^4}\frac{T_0}{\cos^2\alpha}\left[\begin{array}{l}-\frac{f(0)}{T_0}\left[\frac{f(\pi-2\alpha)}{\cos^2\alpha}-f'(\pi-2\alpha)tg\alpha\right]\\ -\frac{2f(\pi-2\alpha)-f(0)-T_0}{f(0)}I(0)+\frac{f(\pi-2\alpha)-T_0}{T_0f(0)}J(0)\\ +\frac{f(0)-T_0}{f(0)}I(\pi-2\alpha)+\frac{1}{f(0)}J(\pi-2\alpha)\end{array}\right]. \quad (14)$$

Fig.2 shows the comparison at normal incidence between $R$, as given in Eq.(14), and $R_{FW}$, obtained from Eq.(4), with $K=K_{FW}$. For $b$=0, Eq.(8) is used instead of Eq.(6).

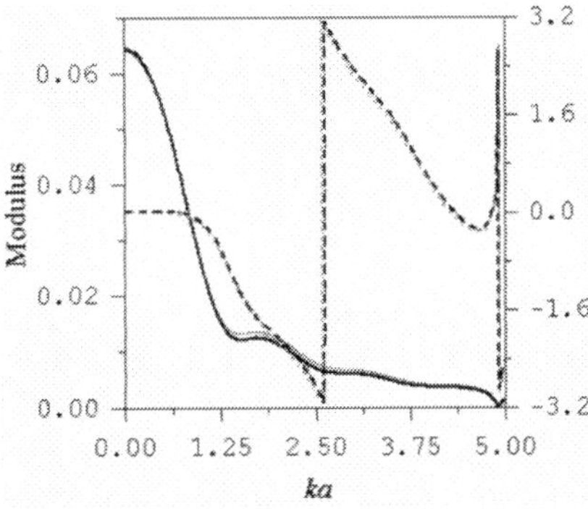


**Fig.2** Normal incidence. Comparison between $R_{FW}$ (black, thin and colours) and $R$ (black, thick) for different values of $b$.

Black, thin : $b$=0 - Blue : $b$=2.05$a$ - Green : $b$=2.1$a$ - Red : $b$=3.0$a$

Solid lines : Modulus. Dotted lines : phase.

Eq.(14) is clearly quite a good approximation to $R_{FW}$, even for values of $b$ as high as 3, whatever the frequency. Figs.3a,b show the same comparison, *versus* the incidence angle, at low ($ka$=0.5) and intermediate ($ka$=1.5) frequency. The effect of the value of $b$ on the curve is more visible than at normal incidence. At both frequencies, however, Eq.(14) gives practically the same result as Fikioris and Waterman with $b$=0. Eq.(14), then, is assumed correct in the following for $b$ tending to zero, for low and intermediate frequencies.

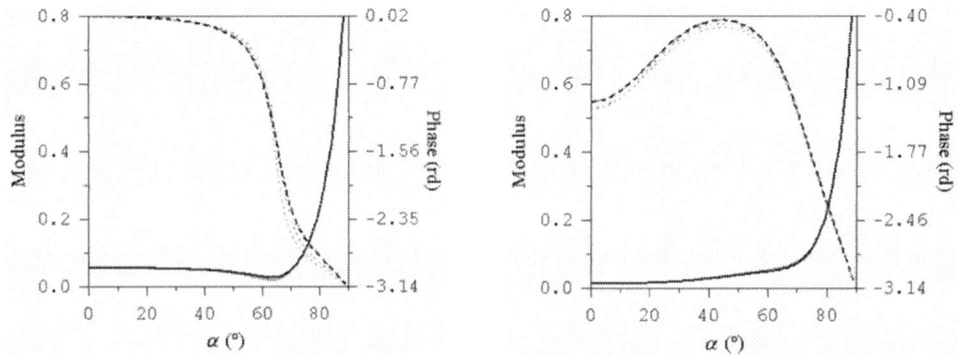


a) $ka$=0.5 ($n_0/k^2$=0.12) b) $ka$=1.5 ($n_0/k^2$=0.013)

**Figs.3** Comparison between $R_{FW}$ (black, thin and colours) and $R$ (black, thick) for different values of $b$.

Black, thin : $b$=0 - Blue : $b$=2.05$a$ - Green : $b$=2.1$a$ - Red : $b$=3.0$a$

Solid lines : Modulus. Dotted lines : phase.

## 4. The effective mass density and viscosity

Equating the reflection coefficient $R$ of the viscous fluid (Eq.(3)) to that of the random medium given by Eq.(14) provides the expression of the effective mass density as a function of the incidence angle,

$$\frac{\rho_{eff}}{\rho}=1-2i\frac{n_0}{k^2}\frac{f(0)-f(\pi-2\alpha)}{\cos^2\alpha}+4\frac{n_0^2}{k^4}\frac{1}{\cos^2\alpha}\left[\begin{array}{l}\frac{1}{2}\frac{(f(0))^2}{\cos^2\alpha}-\frac{1}{2}\left(\frac{f(\pi-2\alpha)}{\cos\alpha}\right)^2\\ +f'(\pi-2\alpha)f(0)tg\alpha\\ +\frac{f(\pi-2\alpha)-T_0-f(0)}{f(0)}J(0)\\ +T_0\frac{f(0)-T_0}{f(0)}I(\pi-2\alpha)+\frac{T_0}{f(0)}J(\pi-2\alpha)\\ -T_0\frac{2f(\pi-2\alpha)-f(0)-T_0}{f(0)}I(0)\end{array}\right].(15)$$

The angle dependence of $\rho_{eff}$ is exhibited in Figs.4a-b at low and intermediate frequencies.

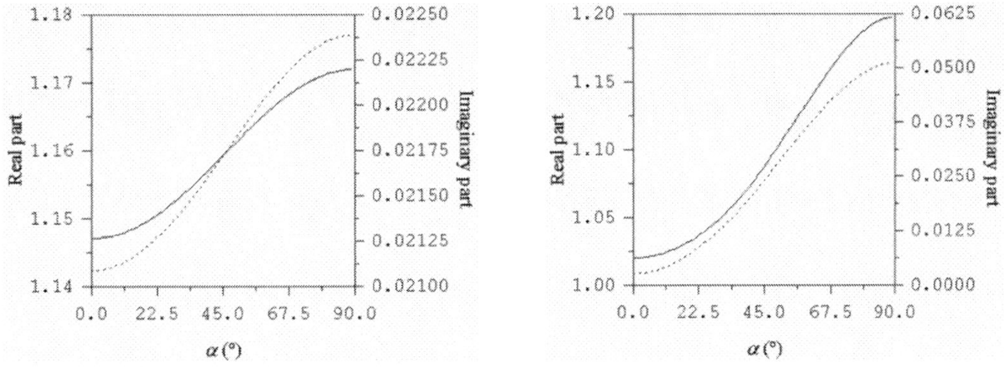


a) $ka$=0.5 ($n_0/k^2$=0.12) b) $ka$=1.5 ($n_0/k^2$=0.013)

**Figs.4** The dependence of $\rho_{eff}$ on the incidence angle.
Solid lines : Real part of $\rho_{eff}/\rho$. Dotted lines : Imaginary part of $\rho_{eff}/\rho$.

At low frequency ($ka$=0.5), the real part of $\rho_{eff}$ varies only by about 2% from normal to (quasi-) grazing incidence, and the imaginary part varies by about 5%. Neglecting this variation provides the following approximation to $\rho_{eff}$, obtained from Eq.(15) at normal incidence,

$$\frac{\rho_{eff}}{\rho} = 1 - 2i\frac{n_0}{k^2}\left(f(0)-f(\pi)\right) + 4\frac{n_0^2}{k^4}\begin{bmatrix} \frac{1}{2}\left(f(0)\right)^2 - \frac{1}{2}\left(f(\pi)\right)^2 + \frac{f(\pi)-T_0-f(0)}{f(0)}J(0) \\ +T_0\frac{f(0)-T_0}{f(0)}I(\pi) + \frac{T_0}{f(0)}J(\pi) - T_0\frac{2f(\pi)-f(0)-T_0}{f(0)}I(0) \end{bmatrix}, \quad (16)$$

which, at first order in $n_0/k^2$, is identical to that given in Ref.[7].

At $ka$=1.5, however, the dependence of $\rho_{eff}$ on the incidence angle cannot be neglected any more, as the variation of its real part is about 17% from normal to grazing incidence, and its imaginary part at normal incidence is multiplied roughly by 20 at grazing incidence.

Once $K$ and $\rho_{eff}$ are known, $\eta_{eff}$ is given by Eq.(2). Using Eq.(1) for $K$ and Eq.(16) for $\rho_{eff}$ provides

$$\eta_{eff} = -4\frac{n_0}{k^2}\frac{\omega\rho}{k^2}f(0) + 8i\frac{n_0^2}{k^4}\frac{\omega\rho}{k^2}\left[-2\left(f(0)\right)^2 + J(0) + f(0)\frac{f(0)-f(\pi-2\alpha)}{\cos^2\alpha}\right]. \quad (17)$$

The dependence of the nondimensional effective viscosity $\frac{k^2}{\omega\rho}\eta_{eff}$ on the incidence angle is shown in Figs.5. While rather negligible at $ka$=0.5, it increases again at $ka$=1.5 (20% variation of the real part).

## 5. Conclusion

At low frequency, the random medium may be considered as a viscous fluid, the parameters of which being complex, frequency and angle dependent. The dependence on the angle, however, is negligible enough to provide some demonstration of the consistency of the approach followed in the previous sections. At intermediate frequencies, the problem is still open, and the validity of our approach is subject to the possibility of neglecting the boundary layer mentioned in section 1 (see also Refs.[8-10]

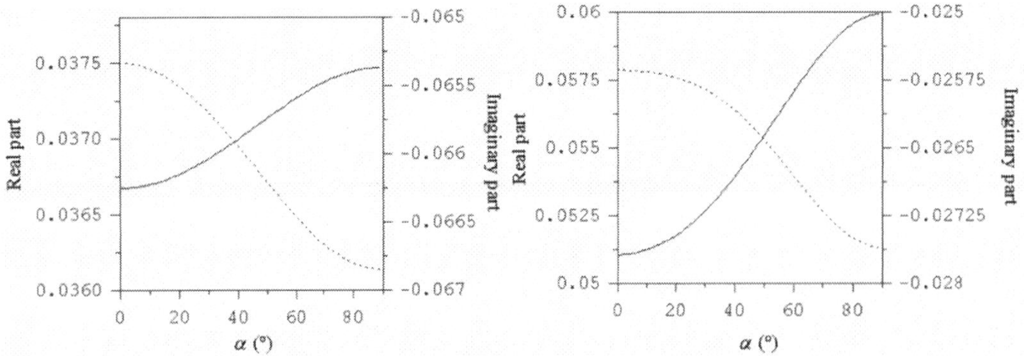


a) $ka$=0.5 ($n_0/k^2$=0.12) b) $ka$=1.5 ($n_0/k^2$=0.013)

**Figs.5** Effective nondimensional viscosity *versus* incidence angle
Solid line : real part – Dotted line : imaginary part

**References**

[1]C.M. Linton, P.A. Martin, "Multiple scattering by random configurations of circular cylinders : second-order corrections for the effective wavenumber", J. Acoust. Soc. Am. 117(6), 3413-3423 (2005)

[2]J.M. Conoir, A.N. Norris, "Effective wavenumbers and reflection coefficients for an elastic medium containing random configurations of cylindrical scatterers", Wave Motion, in Press (2009)

[3]J.G. Fikioris and P.C. Waterman, "Multiple scattering of waves. II 'hole corrections' in the scalar case", J. Math. Phys. 5(10), 1413-1420 (1964)

[4]F. Coulouvrat "On the equations of nonlinear acoustics", J. Acoustique 5, 321-359 (1992)

[5]M. Deschamps, C.L. Cheng, "Ondes hétérogènes et réflexion-réfraction à l'interface liquide-non-absorbant/liquide-thermovisqueux", ("Heterogeneous waves and reflection-refraction at non-absorbing and thermoviscous liquids interface"), Acustica 68, 96-103 (1989)

[6]P.Y. Le Bas, F. Luppé, J.M. Conoir, " Reflection and transmission by randomly spaced elastic cylinders in a fluid slab-like region", J. Acoust. Soc. Am. 117(3), 1088-1097 (2005)

[7]C. Aristégui, Y.C. Angel, " Effective mass density and stiffness derived from P-wave multiple scattering", Wave Motion 44, 153-164 (2007)

[8] M. Chekroun, L. Le Marec, B. Lombard, J. Piraux, O. Abraham, « Comparison between a multiple scattering method and direct numerical simulations for elastic wave propagation in concrete », in « Ultrasonic Wave Propgation in Non Homogeneous Media », Springer Proc. In phys.128, ISBN 978-3-540-89104-8, 317-327 (2009)

[9] D. Abrahams, W. Parnell, "Wave reflection and transmission at the boundary of a composite elastic material; an exact Wiener-Hopf solution", GDR-AFPAC meeting, 18-22 january 2010, Lake District, UK

[10] J. Dubois, C. Aristégui, O. Poncelet, A.L. Shuvalov, "Coherent responses of a screen containing random distributions of scatterers: comparison between different approaches", GDR-AFPAC meeting, 18-22 january 2010, Lake District, UK

# Modelling the backscatter from spherical cavities in a solid matrix: can an effective medium layer model mimic the scattering response?

**Valerie J. Pinfield and Richard E. Challis**

Electrical Systems and Applied Optics Division, Faculty of Engineering, University of Nottingham, University Park, Nottingham NG7 2RD

valerie.pinfield@nottingham.ac.uk

**Abstract**. Industrial applications are increasingly turning to modern composite layered materials to satisfy strength requirements whilst reducing component weight. An important group of such materials are fibre/resin composites in which long fibres are laid down in layers in a resin matrix. Whilst delamination flaws, where layers separate from each other, are detectable using traditional ultrasonic techniques, the presence of porosity in any particular layer is harder to detect. The reflected signal from a layered material can already be modelled successfully by using the acoustic impedance of the layers and summing reflections from layer boundaries. However, it is not yet known how to incorporate porosity into such a model. The aim of the work reported here was to model the backscatter from randomly distributed spherical cavities within one layer, and to establish whether an effective medium, with a derived acoustic impedance, could reproduce the characteristics of that scattering. Since effective medium models are much more readily implemented in simulations of multi-layer structures than scattering per se, it was felt desirable to simplify the scattering response into an effective medium representation.

A model was constructed in which spherical cavities were placed randomly in a solid continuous matrix and the system backscattering response was calculated. The scattering from the cavities was determined by using the Rayleigh partial-wave method, and taking the received signal at the transducer to be equivalent to the far field limit. It was concluded that even at relatively low porosity levels, the received signal was still "layer-like" and an effective medium model was a good approximation for the scattering behaviour.

## 1. Introduction

Industrial applications are increasingly turning to modern composite layered materials to satisfy strength requirements whilst reducing component weight. An important group of such materials are fibre/resin composites in which long fibres are laid down in layers in a resin matrix. Amongst the significant faults which need to be detected by inspection methods, are delamination flaws, which can be detected by standard ultrasonic techniques, and porosity, which cannot. Currently, models can be used to simulate the reflected signal from a typical multi-layer composite system by summing the reflections from each interface using estimates for the acoustic impedance of each layer. However, it is not yet fully understood how to incorporate porosity into such a multi-layer model. A number of workers have investigated the problem by using modified elastic properties [1-2], for the mixed fibre/resin/porosity composite whilst others have considered the frequency-dependent attenuation resulting from the cavities [3-4]. Multi-layer models which incorporate porosity through such property mixture-rules and porosity-induced attenuation have been recently reported [5]. Other workers have considered the porous layer as an equivalent homogeneous medium having effective properties derived from the ensemble-averaged reflection and transmission responses [6-10].

However, since a real sample has fixed locations for the cavities, any measurement made during inspection is not an ensemble-averaged response, but the actual scattered signal from cavities at specific locations. The aim of the current work was to investigate the conditions under which an ensemble-averaged or effective-medium description of the material is a good approximation for the signal received from distributed cavities in a single snapshot measurement. We are exploring the emergence of the effective-medium behavior, rather than studying the emerged behavior itself. The question was addressed by formulating a model for the backscattered signal from distributed porosity, and comparing with a simple impedance-based effective-medium model. Whilst our longer-term aim is to establish properties for porous fibre/resin composite, the work reported here investigates the properties of a material which has porosity in a homogeneous matrix. Scattering from fibres themselves will be considered at a later stage.

The two models are presented in the next section: one an "effective-medium" model which treats the reflections from the layer interfaces using the acoustic impedance of the material, and the second a scattering model which considers the signal scattered by individual cavities in the layer. Results from each model are shown and compared in section 4. and final conclusions are summarized in section 5.

## 2. Scattering from a Single Spherical Cavity

Before considering the system as a whole, it is first necessary to establish the scattering characteristics of the individual pores, which are modelled as spheres of vacuum for simplicity. The Rayleigh partial-wave method can be used to determine the scattering coefficients for a single spherical cavity in a solid medium using a planar incident wave [11] which in the notation of later workers (see Challis et al. [12] for definitions) can be written in the limit of low frequency as

$$A_0 \approx \frac{2}{3} i \,(ka)^3 \quad A_1 = -\frac{1}{9} i \,(ka)^3 \quad A_2 \approx \frac{1}{24} i \,(ka)^3 \tag{1}$$

These results assume a ratio of shear to longitudinal wave speeds of 0.5 as found in many materials; within this assumption the coefficients at low frequency depend only on the wavenumber of the medium, $k$ and the radius of the cavity, $a$, and not on other physical properties of the medium. In the model, the scattering coefficients are calculated numerically from the boundary condition matrix equation which extends the validity up to $ka$~0.8; these analytical results, valid at low frequency, permit the frequency and radius dependence to be identified.

The displacement potential of the scattered longitudinal wave in a solid material in the far-field is

$$\varphi \to f(\theta,\phi)\frac{e^{ikr}}{r} \text{ as } r \to \infty \tag{2}$$

with $f(\theta,\varphi)$ the scattering amplitude, given in the forward and backward directions respectively by

$$f_0 = \frac{1}{ik}\sum_{n=0}^{\infty}(2n+1)A_n \approx \frac{13}{24}k^2a^3, \qquad f_\pi = \frac{1}{ik}\sum_{n=0}^{\infty}(2n+1)A_n(-1)^n \approx \frac{29}{24}k^2a^3 \tag{3}$$

Hence, both forward and backward scattered amplitudes are dependent on the square of frequency and the cube of cavity radius at low frequency. A time-dependence of $e^{-i\omega t}$ has been assumed throughout.

## 3. The Two Models

The system configurations for the two models are shown in Figure 1. In both cases, only a single region of modified material is considered for simplicity – that is the space between $z_{min}$ and $z_{max}$ from the transducer face. The solid material of layers 1 and 3 and the matrix surrounding the cavities have the same acoustic impedance and density as a 70% carbon fibre-resin composite material. The transducer is directly coupled to the solid material.

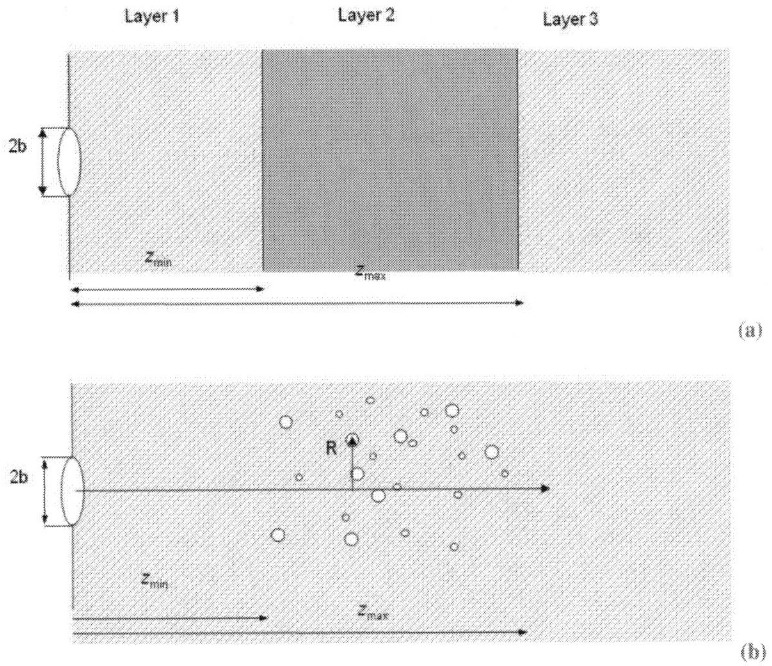


**Figure 1** Configuration of the effective medium (a) and discrete scatterer (b) models.

### 3.1. Effective medium model

Following other workers, the porous layer is treated as a homogeneous material with effective properties, which determine the reflection and transmission responses of the layer [7-8]. For the effective medium model (Figure 1a), layer 2 represents a fibre/resin composite layer with porosity, and layers 1 and 3 non-porous composite material with density, $\rho$ and sound speed, $c$. Each material is defined by its acoustic impedance, and the signal received at the transducer is obtained by summing the reflections from the front and back interfaces of layer 2. The density of layer 2 can be obtained from the cavity volume fraction, $\phi$, thus $\rho_2 = \rho(1-\phi)$. The sound speed in the porous layer, $c_2$, can be estimated using the effective wavenumber for a medium of randomly distributed scatterers. We have used the single-scattering approximation derived by Foldy [13] extended to non-isotropic scatterers, later confirmed by the multiple-scattering models of workers such as Lloyd and Berry [14]. Thus, the wavespeed can be estimated from the equation

$$\frac{c^2}{c_2^2} \approx 1 + \frac{3\varphi}{k^2 a^3} \operatorname{Re} f(0) \approx 1 + 3 \cdot \frac{13}{24}\varphi \tag{4}$$

using the scattering amplitude of the cavities (1) at the very low frequency limit (*ka*~0.2) to illustrate the frequency dependence. In the model, the calculations are carried out with the full boundary matrix solution, so are valid over a wider range up to around *ka*~0.8. Unsubscripted quantities refer to the non-porous composite. The sound speed in the low frequency, long wavelength limit (wavelength much longer than the cavity radius) is approximately independent of frequency, despite the strong frequency-dependence of the scattering amplitude. Thus the acoustic impedance of the porous layer $Z_2 = \rho_2 c_2$, is also frequency-independent at low frequency, and less than the impedance of the non-porous material. Other studies which have adopted an effective acoustic impedance to model the response of a layer of scatterers have extended the definition by using the form $Z = \rho\omega/k$ [15-16]. We can express the ratio of the acoustic impedance in the porous and non-porous composite as $\hat{Z} = Z_2/Z$. The signal received at the transducer is the sum of reflections from the interfaces and can be readily expressed as a system frequency response [17]

$$F_{\text{eff}}(\omega) = F_0(\omega)H(\omega) \tag{5}$$

where $F_0(\omega)$ is the frequency-domain transducer response and

$$H(\omega) = e^{2(i\omega/c-\alpha)z_{\min}} \cdot r_{12}\left[1 - t_{12}t_{21}\, e^{2(i\omega/c-\alpha)d} R(\omega)\right] \tag{6}$$

where the wavenumber $k = \omega/c + i\alpha$ with attenuation $\alpha$ ; $t_{12}$ and $t_{21}$ are the pressure transmission coefficients in each direction at the boundary between layers 1 and 2, and $r_{12}$ is the pressure reflection coefficient at that boundary for waves emanating from the transducer, given by

$$r_{12} = (\hat{Z}-1)/(\hat{Z}+1), \quad t_{12} = 2\hat{Z}/(\hat{Z}+1) \text{ and } t_{21} = 2/(\hat{Z}+1) \tag{7}$$

The multiple reflection term sums to $R(\omega) = \left[1 - r_{12}^2\, e^{2(i\omega/c-\alpha)d}\right]^{-1}$ and $d = z_{\max} - z_{\min}$ is the thickness of the layer. We now have a model for the signal received at the transducer from a single layer which has different properties from those around it. We have derived acoustic impedance for a porous layer based on its sound speed and density, but using the layer reflection model any impedance can be supplied for layer 2. By comparing the reflection response with that of the discrete model we will derive the acoustic impedance of the material. Scattering from individual cavities is not included in this model, and diffraction due to finite transducer size has been neglected in both models.

### 3.2. Discrete scatterer model

Whereas the effective medium model treats only the reflections of a plane wave from the interfaces between regions of different properties, the discrete scatterer model considers each cavity as a secondary source of spherical sound waves which can be detected at the transducer. As shown in Figure 1b, the discrete scatterer model simulates the scattering by spherical cavities located in the region $z_{\min}$ to $z_{\max}$ from the transducer. There are no planar interfaces between materials of different types in this model: the fibre-resin composite (taken as homogeneous) modeled in regions 1 and 3 of the effective medium configuration (Figure 1a) is the same as the matrix surrounding the cavities in the scatterer model. Scatterers are placed randomly in the specified region (overlap is ignored), within the transducer cylindrical "beam" volume, and the incident wave at each scatterer is assumed to be planar. Only single-scattering is modeled.

The signal received at the transducer is taken to be the sum of the far-field scattered waves produced by each of the cavities. For a single scatterer, located at a distance $z_j$ from the transducer, the far-field frequency-domain signal received at the transducer is given by

$$F(\omega) = F_0(\omega)\frac{f_\pi}{z_j} e^{-2\alpha z_j} e^{2i\omega z_j/c} \tag{8}$$

For a number of scatterers, $N_{Sc}$, distributed in the space, the received signal can be obtained by summation

$$F(\omega) = F_0(\omega)\sum_{j=1}^{N_{SC}} \frac{f_\pi}{z_j} e^{-2\alpha z_j} e^{2i\omega z_j/c} \tag{9}$$

The simulation predicts the signal received from a single configuration of discrete scatterer locations. As such it represents a typical measurement in a real sample, in which the locations of the cavities are fixed, in contrast to an ensemble-averaged model which is by definition an average over all possible scatterer locations.

### 3.3. System Properties

The properties for the modelled systems are given in Table 1. The fibre/resin composite material is taken to be a homogeneous solid, with physical properties similar to that of a 70% fibre-resin composite mixture, derived from previous models [18]. The longitudinal sound speed was 3035 $ms^{-1}$, shear wave speed 1517 $ms^{-1}$ and density 1564 $kgm^{-3}$. The attenuation of the composite was taken to be zero for the calculations, in order to investigate the effects of scattering only, in the absence of attenuation effects. Scattering from the fibres themselves will be addressed in a later study. A small radius transducer was chosen in order to minimize the effects of diffraction which were considered in an additional study. The transducer response was obtained experimentally using a pair of 10 MHz centre-frequency transducers in water and the signal was sampled using an oscilloscope and then sub-sampled to the required sampling frequency of 50MHz. This process gave an experimentally valid signal with which to perform the calculations. The long wavelength approximation is only valid up to a frequency of ~10MHz for the pore size used in the calculations; however, the full solution for the scattering coefficients (valid at any *ka*) and sound speed was used in the calculations at all frequencies. Only the analytical solutions to the scattering coefficients assume a low frequency limit; all other parts of the model are valid at any frequency range. Calculations were carried out using MATLAB [19].

| | | | |
|---|---|---|---|
| Distance $z_{min}$ | 10 mm | Transducer centre frequency | 10 MHz |
| Layer thickness | 1.2 mm | Sampling frequency | 50 MHz |
| Transducer radius | 1 mm | Number of samples in signal | 1024 |
| Cavity radius | 10 μm | | |

Table 1 Properties of the systems modelled.

## 4. Results

Calculations were carried out to model the signal received at the transducer according to the two models: the effective medium model and the discrete scatterer model. The time-domain signal predicted by the effective medium model is shown in Figure 2, alongside the transducer signal. All signals have been time-shifted so that the zero occurs at the first signal expected to be received from the front interface of layer 2. The received signal predicted by the effective medium model includes a reflection from the front interface of the layer, and a second reflection from the back interface, with the appropriate time-delay. A comparison with the transducer signal shows that the effective medium model predicts that the signal is inverted at the front interface, but not at the back interface, due to the lower acoustic impedance in the layer compared with the material around it. The shape of the waveform is almost identical to the transducer signal, consistent with the flat frequency response in the effective medium model.

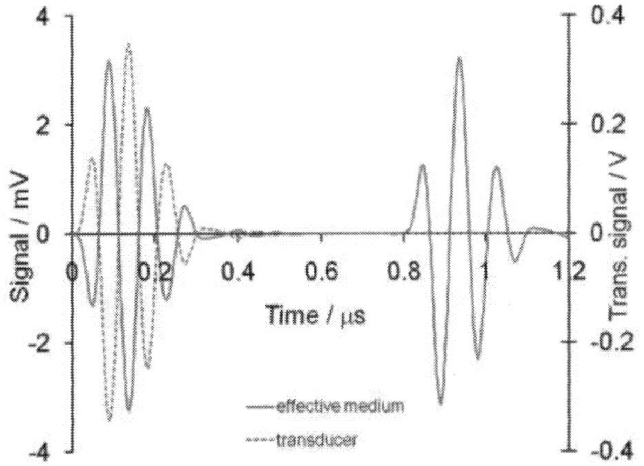


Figure 2 Time-domain signal from the effective medium model (red, solid line) compared with the transducer signal (blue, dashed line). Signals are time-shifted by the propagation time from transducer to the front interface and back.

To investigate the results of the discrete scatterer model, we first observe the signal received from a single isolated cavity embedded in the composite matrix (Figure 3). The shape of the waveform is somewhat modified, due to the frequency-dependence of the scattering amplitude (see section 2. ), but the signal is largely in-phase with the transducer signal. This is consistent with the analytical result (8), with $f_\pi$ real and positive.

Next we consider the signal received when there are a number of cavities located in the defined region. The predicted signals, given by (9), are shown in Figure 4a-c, scaled by volume fraction in order to permit plotting of all results on the same axes. Each plot is the result of a single configuration of the scatterer positions. At the lowest volume fraction (1%) there is a significant scattered signal for the whole duration of the time-window, as the transducer receives scattered sound waves from each individual cavity placed anywhere in the region. However, there is a stronger signal corresponding to the times at which a front-face and back-face reflection would appear. Destructive interference between scattered waves from nearby cavities in the middle of the region results in a stronger signal corresponding to the front and back-face reflections. As the concentration of cavities is increased, the interference effect becomes stronger, and the apparent "front- and back-face" signals are more clearly demarked from the much smaller signal in the intervening period. At the highest concentration, the signal looks very much like that expected from a homogeneous layer in the same region.

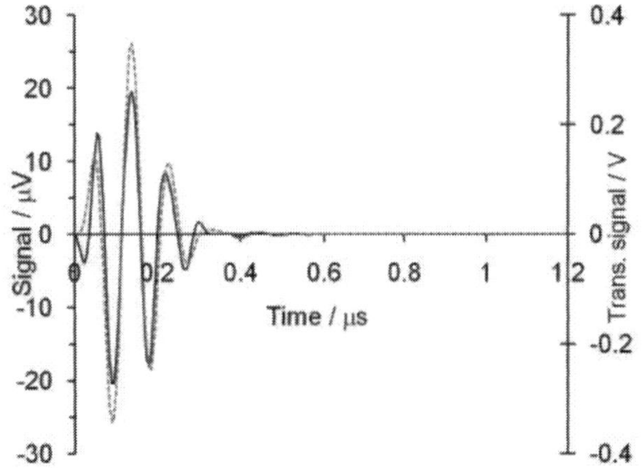


Figure 3 Time-domain signal received from a single spherical cavity (black, solid line) located at a distance of 10mm from the transducer, time-shifted to remove the time-delay, compared with the transducer signal (blue, dashed line).

A measure of the layer-like nature of the signal can be obtained by calculating the ratio of the largest peak height in middle region to the largest peak height in the first group. In Figure 5 this parameter is plotted as a function of concentration, clearly showing that even at 1% porosity the middle interference region is significantly smaller than the apparent "front-interface" reflection. It should be noted, that there are no interfaces between different materials in this model, only cavities in an otherwise homogeneous medium. The response has similarities to reflections from interfaces due to destructive interference between signals scattered by individual cavities within the body of the porous layer.

Here we have demonstrated a typical signal which would be seen from a measurement of a real system i.e. a single configuration of scatterer (cavity) positions. Increasing the number of scatterers in the discrete scatterer model leads to a system response which is closer to an effective medium response, with apparent layer boundary reflections. Thus, as the number density of cavities increases, the system response becomes akin to an expected ensemble-averaged response. Although at 20% porosity, multiple scattering effects are likely to become significant, the results of the single-scattering model are presented to illustrate the tendency towards homogeneous layer characteristics at higher concentrations. Our results illustrate the degree of appropriateness of applying effective, ensemble-averaged properties at various concentrations. For the small, spherical cavities studied here, a composite containing even a relatively low concentration (1-2%) of cavities could feasibly be treated as an effective medium as long as they are randomly distributed. Only a single snapshot of scatterer positions has been studied, as would be observed in an experimental measurement on a real composite, but clear trends towards effective medium characteristics have been observed. These effects are explored further in a later paper. Although effective-medium behavior does emerge from the discrete-scatterer model, there are differences between the emerging response and the effective medium model. These will be investigated in a later paper, but in summary a difference in both phase shift and frequency response has been identified between the simple effective medium chosen here, and the discrete scatterer model.

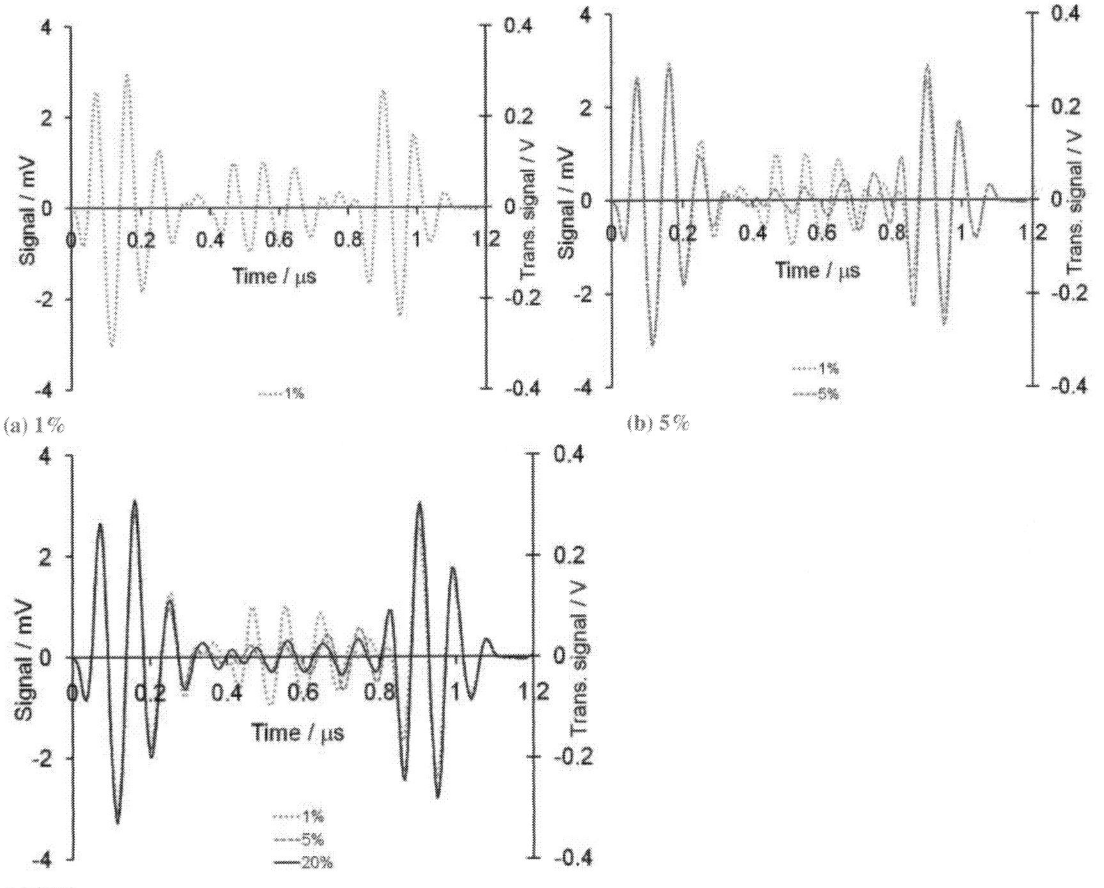


(a) 1%

(b) 5%

(c) 20%

**Figure 4 Time-domain signal received from randomly distributed spherical cavities at various volume fractions, scaled by volume fraction and time-shifted (a) 1% (b) 5% (c) 20% (earlier results greyed out for comparison).**

## 5. Conclusions

The aim of the work was to establish whether a simple "layer" model would be able to reproduce the characteristics of scattering from cavities in fibre/resin composites. We have established that

- the backscattered signal from randomly distributed spherical cavities does have strong similarities to the signal reflected by a layer, including an apparent "front-" and "back-interface" reflection.
- as the number of scatterers (cavities) is increased, the system response becomes closer to that of an effective medium model.
- the simple effective medium model reproduced many of the features of the discrete scatterer model, but with some differences in the time domain, and in the frequency domain response.

Hence, we conclude that an acoustic-impedance-based multi-layer model will be able to incorporate the effects of porosity (even at relatively low concentrations) in fibre/resin composite materials by use of modified effective properties in an effective-medium model. Further study of the time and frequency domain response is in progress and will be presented in a future publication. It is anticipated that this development will improve techniques for porosity detection in such materials.

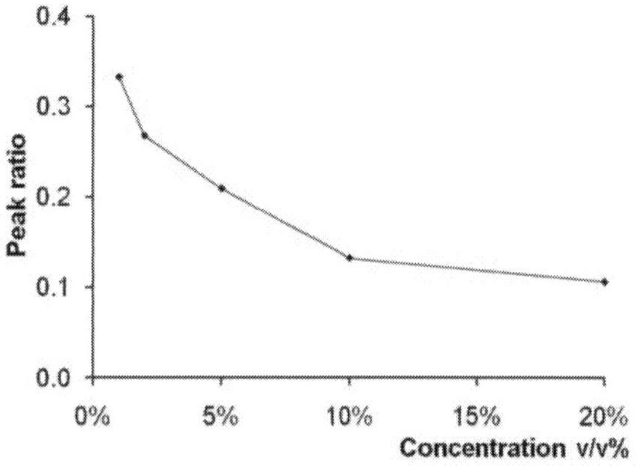


Figure 5 The ratio of the largest peak in the middle region to the largest peak in the first group as a function of concentration. Each point is from a single realisation of scatterer positions.

## 6. Acknowledgements

Thanks are due to Dr. Andrew Holmes and Mr. Robert Smith (QinetiQ Ltd., Farnborough, UK) for helpful discussions.

## References

[1] Martin B G 1977 Ultrasonic attenuation due to voids in fibre-reinforced solids containing voids. *Journal of applied physics* **48** 3368-73

[2] Greszczuk L B 1971 Interfilamentary stresses in Filamentary Composites *American Institute of Aeronautics and Astronautics Journal* **9** 1274-80

[3] Nair S M, Hsu D K and Rose J H 1989 Porosity estimation using the frequency dependency of the ultrasonic attenuation *Journal of nondestructive evaluation* **8** 13-26

[4] Adler L, Rose J H and Mobley C 1986 Ultrasonic Method to Determine Gas Porosity in Aluminum-Alloy Castings - Theory and Experiment *Journal of Applied Physics* **59** 336-47

[5] Mienczakowski M J, Holmes A K and Challis R E 2008 *Review of Progress in Quantitative Nondestructive Evaluation, Vol 27a and 27b,* ed D O Thompson and D E Chimenti pp 995-1001

[6] Parnell W J, Abrahams I D and Brazier-Smith P R 2010 Effective properties of a composite half-space: exploring the relationship between homogenization and multiple scattering theories *Quarterly Journal of Mechanics and Applied Mathematics* **63** 1-32

[7] Angel Y C and Aristegui C 2005 Analysis of sound propagation in a fluid through a screen of scatterers *J. Acoust. Soc. Am.* **118** 72-82

[8] Robert S and Conoir J M 2007 Reflection and transmission process from a slab-like region containing a random distribution of cylindrical scatterers in an elastic matrix *Acta Acustica United with Acustica* **93** 1-12

[9] Aristegui C and Angel Y C 2002 New results for isotropic point scatterers: Foldy revisited *Wave Motion* **36** 383-99

[10] Le Bas P Y, Luppe F and Conoir J M 2005 Reflection and transmission by randomly spaced elastic cylinders in a fluid slab-like region *J. Acoust. Soc. Am.* **117** 1088-97

[11] Ying C F and Truell R 1956 Scattering of a plane longitudinal wave by a spherical obstacle in an isotropically elastic solid *Journal of applied physics* **27** 1086-97
[12] Challis R E, Povey M J W, Mather M L and Holmes A K 2005 Ultrasound techniques for characterizing colloidal dispersions *Rep. Prog. Phys.* **68** 1541-637
[13] Foldy L L 1945 The multiple scattering of waves *Physical review* **67** 107-19
[14] Lloyd P and Berry M V 1967 Wave propagation through an assembly of spheres IV Relations between different multiple scattering theories *Proceedings of the physical society, London* **91** 678-88
[15] Aristegui C and Angel Y C 2007 Effective mass density and stiffness derived from P-wave multiple scattering *Wave Motion* **44** 153-64
[16] Conoir J M, Robert S, El Mouhtadi A and Luppe F 2009 Reflection and transmission at low concentration by a depth-varying random distribution of cylinders in a fluid slab-like region *Wave Motion* **46** 522-38
[17] Challis R E, Freemantle R J, White J D H and Wilkinson G P 1995 Ultrasonic compression wave NDT of adhered metal lap joints of uncertain dimensions *INSIGHT* **37** 954-63
[18] Mienczakowski, M. 2010. Thesis submitted for PhD, University of Nottingham
[19] MATLAB® version R2008a (v. 7.6.0.324), The Mathworks Inc., Natick, U.S.A.

Journal of Physics: Conference Series **269** (2011) 012017 doi:10.1088/1742-6596/269/1/012017

# Thin layer Characterization by ZGV Lamb modes

**Maximin Cès, Dominique Clorennec, Daniel Royer and Claire Prada**

Laboratoire Ondes et Acoustique, ESPCI- Université Paris 7- CNRS UMR 7587, 10 rue Vauquelin, 75231 Paris Cedex 05- France

E-mail: maximin.ces@espci.fr

**Abstract.** Ultrasonic non-destructive testing of plates can be performed with Lamb modes guided by the structure. Non contact generation and detection of the elastic waves can be achieved with optical means such as a pulsed laser source and an interferometer. With this setup, we propose a method using zero group velocity (ZGV) Lamb modes rather than propagating modes. These ZGV modes have noteworthy properties, in particular their group velocity vanishes, whereas their phase velocity remains finite. Thus, a significant part of the energy deposited by the pulsed laser can be trapped in the source area. For example, in a homogeneous isotropic plate and at the minimum frequency of the $S_1$-Lamb mode a very sharp resonance can be observed, the frequency of which only depends on the plate thickness, for a given material. In fact, other ZGV modes exist and the set of ZGV resonance frequencies provide a local and absolute measurement of Poisson's ratio. These non-propagating modes can also be used to characterize multi-layered structures. Experimentally, we observed that a thin (500 nm) gold layer deposited on a thick (1.5 mm) Duralumin plate induces a sensitive down-shift of the set of ZGV resonance frequencies. This shift, which is typically 5 kHz for the $S_1$-Lamb mode at 1.924 MHz, can be approximated by a formula providing the layer thickness. Thickness down to 100 nm can be estimated by this method. Such a sensitivity with conventional ultrasound inspection by acoustic microscopy would require an operating frequency in the GHz range.

## 1. Introduction

Coatings are commonly used in industrial processes for protecting surfaces against aggressive environment such as corroding media, high temperatures, friction or simply wear. Over past decades, various ultrasonic techniques have been developed to characterize the deposited material. Using Laser Based Ultrasonics, the measurements can be achieved without any mechanical contact with the part under test. Recently, techniques using zero group velocity Lamb modes provided local measurements of material parameters and thickness of homogeneous isotropic plates [1].

Zero group velocity Lamb modes (ZGV) are plate modes whose group velocities vanish for non-zero wavenumbers. Such non-propagative modes have been experimentally observed when source and detection are superimposed: they give rise to local resonances. Those resonances, predicted by Tolstoy and Usdin [2], have recently been highlighted [1, 3, 4, 5]. Sharp peaks can be observed in the spectrum of the normal displacement. Prada et al. [1] have shown for isotropic plates that the bundle of these ZGV resonances is related to the thickness and the material parameters.

The objective of this study is to apply this method to inhomogeneous samples like isotropic plates covered with thin layers and to determine the film characteristics.

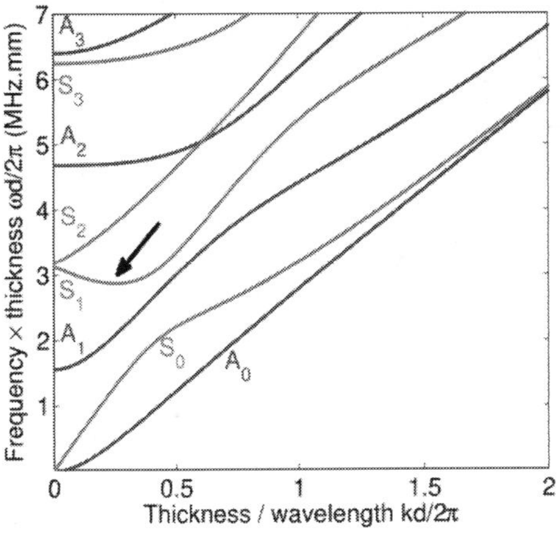


**Figure 1.** Lamb mode dispersion curves for a Duralumin plate of thickness *d*. Red: symmetric modes, Blue: antisymmetric modes, Arrow: ZGV point.

## 2. Zero group velocity Lamb modes

As shown in Fig. 1, plotted for a Duralumin plate ($V_L$=6.398 mm/$\mu$s; $V_T$=3.122 mm/$\mu$s), some branches of the dispersion curves $\omega(k)$ exhibit minima for non-zero wavenumbers. At the corresponding point, the group velocity vanishes, whereas the phase velocity remains finite, hence the name *Zero Goup Velocity* given to these modes. The existence and the relative values of these frequencies depend on the Poisson's ratio. Experimentally, this interesting phenomenon can be observed by detecting the normal displacement at the source point or less than half a wavelength apart from the source. The spectrum of the signal exhibits sharp peaks whose frequencies match ZGV modes.

These sharp resonances illustrate the non-propagative property of ZGV modes: the laser generation gives rise to numerous propagating and non-propagating Lamb modes and only Lamb modes whose energy velocity, i.e. group velocity, is zero and whose phase velocity is finite remain after a few microseconds. The resonance frequency gives an absolute measurement of bulk wave velocities, if the plate thickness is known [6].

In our experiment, elastic waves are generated in the sample using a 20-ns laser pulse and the local displacement in the neighbourhood of the source is measured by an optical interferometer. The normal displacement can oscillate during a very long time. The recorded signal is shown in Fig. 2, in the case of a 1.5-mm thick Duralumin plate. The Fourier transform of the measured signal reveals the expected resonances. Apart from the continuous background due to the $A_0$ mode, each peak corresponds to a ZGV mode. Their amplitude is decaying, because the laser impulse does not excite all the modes with the same efficiency: low frequency components are more excited than higher ones due to the frequency content of the laser pulse.

The existence of ZGV modes depends on Poisson's ratio. This phenomenon can be explained in terms of repulsion at $k \neq 0$ between modes of the same family and of different parity, such as $S_{2m+1}$ and $S_{2n}$ or $A_{2n}$ and $A_{2m+1}$ [1]. The smaller the difference between the cut-off frequencies of two modes, the stronger the repulsion between these modes. Shuvalov et al. [7] give a simple criterion to predict the ZGV modes existence from the curvature of the dispersion curves for $k = 0$. For the first ZGV mode, the ZGV frequency depends on the cut-off frequency ($k = 0$) through a multiplicative factor denoted $\beta$ in the *Impact Echo* (I. E.) method: $f_{ZGV} = \beta V_L/2d$ [4]. This coefficient depends only on Poisson's ratio $\nu$. It can be generalized to other ZGV modes. The value of the resonance frequency obtained for a ZGV mode provides the *local* plate thickness if the longitudinal and transverse velocities are known [6]. Theoretically, in a thickness resonance the whole surface is vibrating in phase, because of the vanishing wave number, whereas a ZGV resonance is a local resonance, with a finite wave number. Compared to the use of a

Journal of Physics: Conference Series **269** (2011) 012017 doi:10.1088/1742-6596/269/1/012017

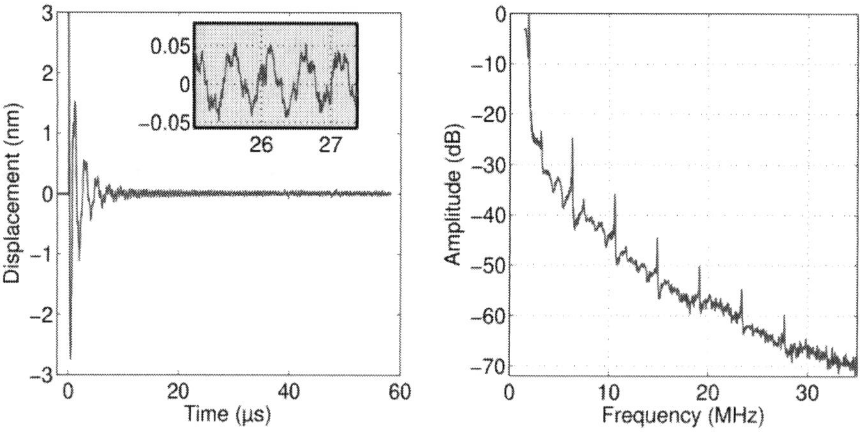


**Figure 2.** Normal displacement detected at the source point on a 1.5 mm-thick Duralumin plate, on the left, and its Fourier transform on the right. The large low frequency oscillation is due to the $A_0$ mode, which appears in the spectrum as the continuous background. The sharpest peaks correspond to ZGV modes, the thickness resonances occur at frequencies 3.2 MHz, 7.5 MHz, 11.7 MHz. The spectrum is represented above 1.9 MHz, because at lower frequencies $A_0$ mode dominates.

thickness resonance, the relevance of such a measurement is the locality of the result.

## 3. Experimental setup

The experimental setup is shown in Fig. 3. The thermoelastic source is a Q-switched Nd:YAG laser (optical wavelength 1064 nm) providing a pulse having 20 ns duration and 4 mJ of energy. The spot diameter of the unfocused beam is equal to 1 mm, in order to generate efficiently the first mode [1, 8]. Lamb waves were detected by a heterodyne interferometer equipped with a 100 mW, frequency doubled Nd:YAG laser (wavelength 532 nm) [9]. This interferometer is sensitive to any phase shift along the path of the optical probe beam reflected by the moving surface. In the experiments, the source and detection points are superimposed. The normal displacement is recorded using an oscilloscope linked to a computer, which permits to process the data.

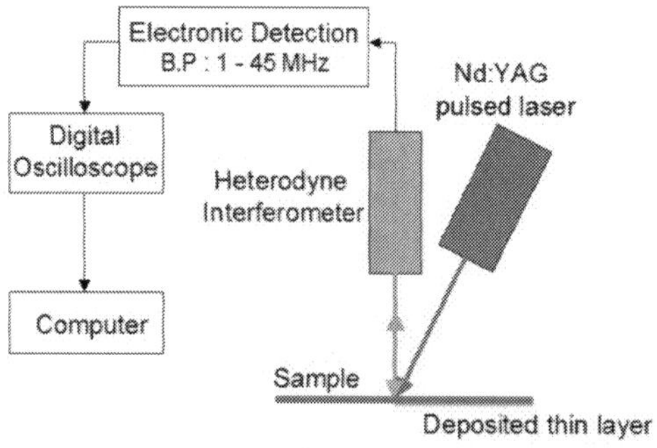


**Figure 3.** Experimental setup

## 4. Measurements

A set of measurements are performed on Duralumin samples. We assume that the plate is isotropic. The sample is placed on a motorized stage for the scanning, and three sets of measurements were carried out on the bare face using the measurement system described above. On the opposite side a thin gold layer is deposited. Sample #1 is a 150-mm square Duralumin plate of thickness 1.513 mm half-covered with a 500-nm thick gold film. Sample #2 is a Duralumin plate of the same dimensions half-covered with a 100-nm thick gold film. A 3-nm thick chromium layer was deposited before the gold layer. We assume that the values of Poisson's ratio are $\nu_{Duralumin} = 0.344$ and $\nu_{gold} = 0.421$ [10].

On sample # 1, the normal displacement was recorded at two points 3 cm apart, one on the bare region, the other on the gold-covered region.

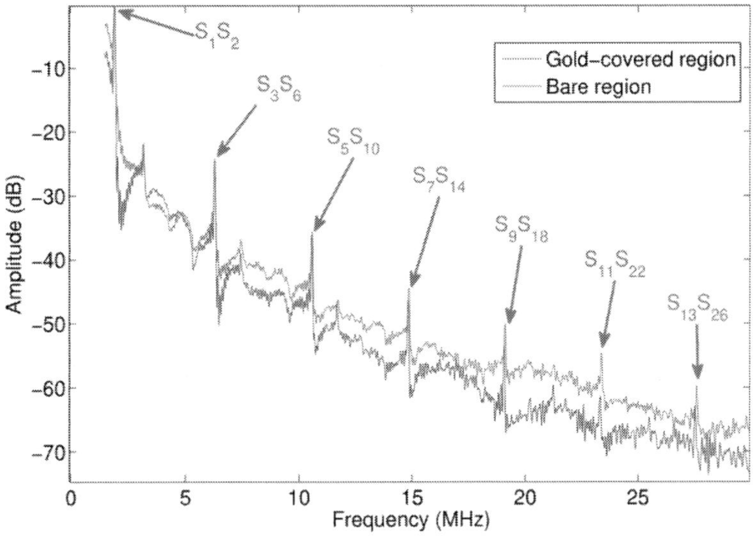


**Figure 4.** Spectrum of 10 $\mu$s signal detected on sample #1.

The spectrum shown in Fig. 4 was obtained after processing the first 10 $\mu$s of signal. Seven peaks can be distinguished, corresponding to the $S_1S_2$, $S_3S_6$, $S_5S_{10}$, $S_7S_{14}$, $S_9S_{18}$, $S_{11}S_{22}$ and $S_{13}S_{26}$ ZGV modes. On the enlargement in Fig. 5, it is clear that the deposit modifies the ZGV frequencies. The frequencies are down-shifted for the gold-covered region.

On sample #1, a 50 mm-scan has been performed over the two different regions, with a 0.25 mm spatial step. The resonance frequency shifts are shown in Fig. 6.

The frequency step between the two regions is obvious for the three first observable ZGV modes. In Table 1, the frequency differences between the two regions are compared to simulations performed with *Disperse* software [11], using the following material parameters [10]:

- 500-nm thick gold layer, mass-density $\rho = 19281\ kg/m^3$, $V_L = 3.240$ mm/$\mu$s, $V_T = 1.200$ mm/$\mu$s
- 1.5-mm thick Duralumin plate, mass-density $\rho = 2795\ kg/m^3$, $V_L = 6.398$ mm/$\mu$s, $V_T = 3.122$ mm/$\mu$s

On sample # 2, a 50 mm-scan was performed over the two different regions, with a 0.5 mm spatial step, at a 250 MHz sampling frequency. A "reference scan" was acquired before the deposit and a second scan after, so that the effect of the 100-nm deposit can clearly be observed

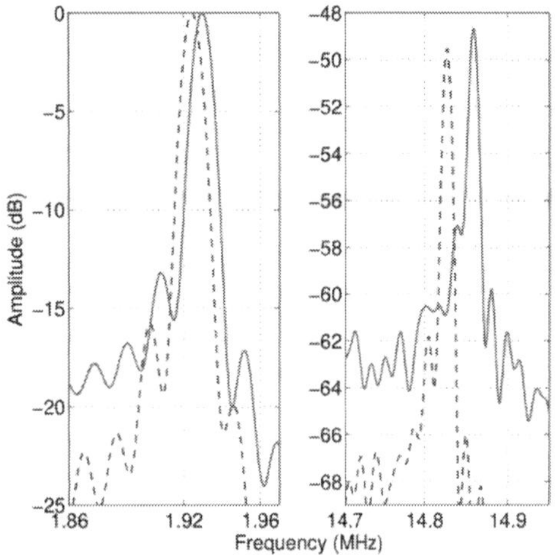


**Figure 5.** Enlargement of Fig. 4, for $S_1S_2$ ($S_7S_{14}$) resonance on the left (on the right). The solid (dashed) curve corresponds to the bare (gold-covered) region.

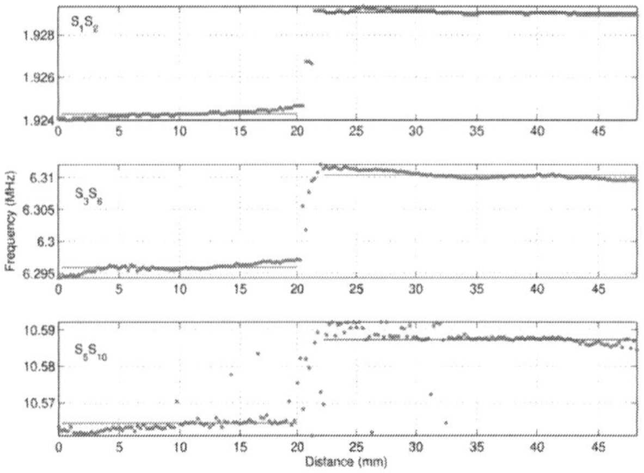


**Figure 6.** ZGV resonance frequencies obtained over 539 $\mu$s, for sample #1.

**Table 1.** Comparison between measurements and theoretical predictions on sample #1.

| Mean frequency difference (kHz) | $S_1S_2$ | $S_3S_6$ | $S_5S_{10}$ |
|---|---|---|---|
| measured | 4.8 | 14.4 | 22.6 |
| theoretical | 5.1 | 14.6 | 23.9 |
| Relative difference | 6 % | 1 % | 6 % |

independently of the local thickness variations of the plate. In Fig. 7, the three first ZGV resonances are plotted with respect to the distance, before the deposit, in green, and after the deposit, in blue. The curves should be superimposed on the bare region. Nevertheless, there is a tiny difference (260 Hz for $S_1S_2$, 980 Hz for $S_3S_6$ and 1.98 kHz for $S_5S_{10}$), which may be

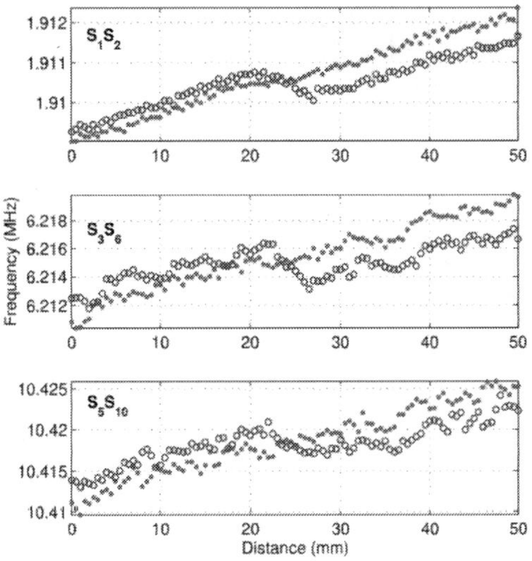


**Figure 7.** Sample #2: ZGV resonance frequencies with respect to distance obtained over 58 $\mu s$.

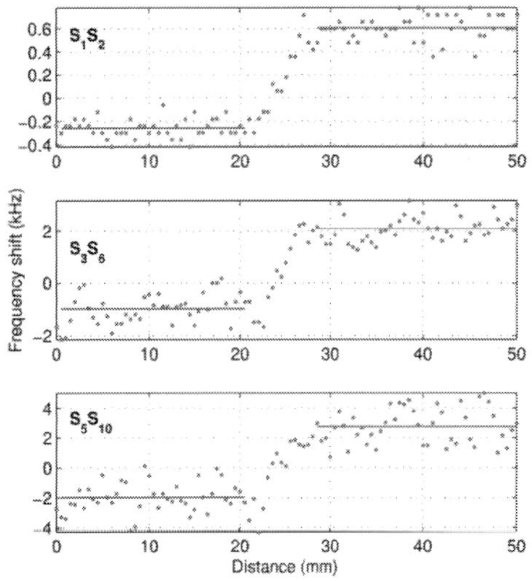


**Figure 8.** Sample #2: ZGV resonance frequency shifts with respect to distance obtained over 58 $\mu s$.

**Table 2.** Comparison between measurements and theoretical predictions on sample #2.

| Mean frequency difference (kHz) | $S_1S_2$ | $S_3S_6$ | $S_5S_{10}$ |
|---|---|---|---|
| measured | 0.86 | 3.07 | 4.73 |
| theoretical | 1.0 | 2.9 | 4.8 |
| Relative difference | 16 % | 6 % | 2 % |

due to a temperature variation of less than 1 ° C. Indeed ZGV resonance frequencies decrease when the temperature rises [8]. We can observe that the frequency is increasing as a function of distance, which means that the plate thickness is decreasing with respect to distance. It is thus more interesting to plot the difference between the three first ZGV resonance frequencies before and after the deposit, versus the distance (Fig. 8).

In Table 2, experimental results are compared to simulations performed with Disperse software [11]. The maximum standard deviation for the frequency difference is 100 Hz for $S_1S_2$, 580 Hz for $S_3S_6$ and 1.2 kHz for $S_5S_{10}$. Thus, it could be possible to detect layers three times thinner.

## 5. Discussion

The dispersion curves of a Duralumin plate of thickness 1.5 mm covered with thin gold layers of different thicknesses (0, 100, 200, 300, 500, 800 nm) have been calculated with Disperse software. Then the ZGV and thickness modes frequencies were extracted. In Fig.9, the frequency variations $\Delta f$ due to the deposit are drawn for $S_1S_2$, $S_3S_6$ and $S_5S_{10}$ ZGV Lamb modes, for both thickness resonances ($f_C$, with circles) and ZGV resonances ($f_{ZGV}$, with crosses), with respect to thickness deposit. Three important features can be noticed:

- Frequency variations in ZGV resonances and in thickness resonances are quite similar: $\Delta f_C \approx \Delta f_{ZGV}$, especially for higher order modes: the higher the order of the mode, the weaker the curvature of the dispersion curve $\omega$(k), so that ZGV and thickness resonance frequencies have closer and closer behaviour;
- for both ZGV and thickness resonances, the frequencies variations are proportional to the thickness of the deposit;
- for both ZGV and thickness resonances, the frequencies variations are proportional to the order of the mode.

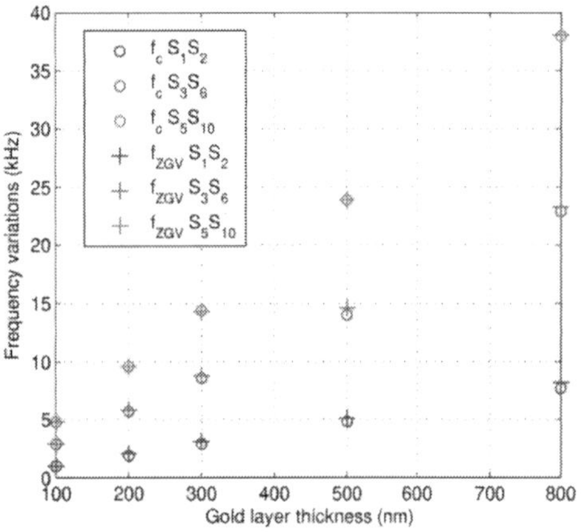


**Figure 9.** Frequency variations $\Delta f$ due to the deposit with respect to the gold layer thickness, for $S_1S_2$, $S_3S_6$ and $S_5S_{10}$ ZGV Lamb modes (theory).

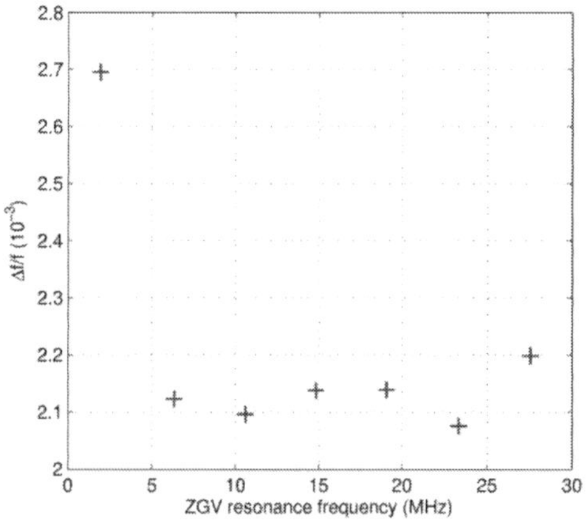


**Figure 10.** $\Delta f/f$ due to the 500 nm gold layer, with respect to the bare plate ZGV resonance frequency (experiment).

This behaviour can be experimentally observed. In Fig. 10, the $\Delta f_{ZGV}/f_{ZGV}$ ratio due to the 500 nm gold deposit is drawn for $S_1S_2$, $S_3S_6$, $S_5S_{10}$, $S_7S_{14}$, $S_9S_{18}$, $S_{11}S_{22}$, $S_{13}S_{26}$ ZGV modes, with respect to the ZGV resonance frequencies. The data are extracted from Fig. 4. It can be noticed that the frequency shift is proportional to the mode order, as expected.

Actually, thickness mode frequencies must be proportional to the mass of the deposit. This assumption is used in quartz crystal microbalances to determine the mass of a thin layer. A more precise relation between masses and cut-off frequencies is [12, 13]:

$$\frac{\Delta m}{m} = -\frac{1}{\pi}\frac{Z_{layer}}{Z_{plate}}\frac{f_c}{f'_c}\arctan\left[\frac{Z_{plate}}{Z_{layer}}\tan\left(\pi\frac{f'_c}{f_c}\right)\right] \tag{1}$$

where:

- $\Delta m$: mass of the layer ; $m$: plate mass
- $f_c$: cut-off frequency for the bare plate; $f'_c$: cut-off frequency for the gold-covered plate
- $Z_{layer}$: layer acoustic impedance; $Z_{plate}$: plate acoustic impedance

The formula can be modified. Using $f_c = f_{ZGV}/\beta$ and $\Delta f_c \approx \Delta f_{ZGV}$, it is possible to write:

$$\frac{\Delta m}{m} \approx -\frac{1}{\pi}\frac{Z_{layer}}{Z_{plate}}\frac{f_{ZGV}}{f_{ZGV}+\beta\Delta f_{ZGV}}\arctan\left[\frac{Z_{plate}}{Z_{layer}}\tan\left(\pi\beta\frac{\Delta f_{ZGV}}{f_{ZGV}}\right)\right] \tag{2}$$

Applying this formula to $S_1S_2$ Lamb mode, it was found that sample # 1 layer thickness was 530 nm instead of 570 nm measured with an optical profilometer, and 90 nm instead of 100 nm expected for sample # 2. The error may be caused by the uncertainty on the layer mass-density. As predicted in [12, 13], the deposited mass is too low to allow the determination of $Z_{layer}/Z_{plate}$.

## 6. Conclusion

Using laser-based ultrasonic techniques, we have investigated the effects on zero group velocity resonance frequencies of a thin gold layer deposited on a thick Duralumin plate. Experimental results are in good agreement with theory. The temperature has not been measured in the presented results, but further measurements reveal that in our material, a temperature change of 1 $^{\circ}$C causes a frequency shift of about 430 Hz for $S_1S_2$ resonance, 1.5 kHz for $S_3S_6$ resonance and 2.5 kHz for $S_5S_{10}$ resonance. It was found experimentally that the ZGV resonance spectrum is down-shifted by the deposit, and is quasi proportional to the deposited mass. The higher the order of the mode, the larger the frequency shift. This frequency difference allows to estimate the layer thickness, in agreement with optical measurements. Thicker deposits may enable the measurement of the layer elastic moduli.

## References

[1] Prada C, Clorennec D and Royer D 2008 *J. Acoust. Soc. Am.* **124** 203–212
[2] Tolstoy I and Usdin E 1957 *J. Acoust. Soc. Am.* **29** 37–42
[3] Holland S D and Chimenti D E 2004 *Ultrasonics* **42** 957 – 960 proceedings of Ultrasonics International 2003
[4] Gibson A and Popovics J 2005 *J. Eng. Mech.* **131** 438–443
[5] Clorennec D, Prada C, Royer D and Murray T 2006 *Appl. Phys. Lett.* **89** 024101
[6] Clorennec D, Prada C and Royer D 2007 *J. Appl. Phys.* **101** 034908
[7] Shuvalov A and Poncelet O 2008 *Int. J. Solids Struct.* **45** 3430 – 3448
[8] Balogun O, Murray T and Prada C 2007 *J. App. Phys.* **102** 064914
[9] Royer D and Dieulesaint E Optical detection of sub-angstrom transient mechanical displacements *Proceedings of the 1986 IEEE Ultrasonics Symposium*
[10] Briggs A 1992 *Acoustic Microscopy* (Clarendon Press)
[11] Pavlakovic B and Lowe M 2005 Disperse software, v.2.0.16.
[12] Lu C S and Lewis O 1972 *J. Appl. Phys.* **43** 4385–4390
[13] Benes E 1984 *J. Appl. Phys.* **56** 608–626

# Directivity of guided ultrasonic wave scattering at notches and cracks

**P Fromme[1] and C Rouge[1, 2]**

[1]Department of Mechanical Engineering, University College London, UK
[2]ENSEIRB-MATMECA, Bordeaux, France

e-mail: p.fromme@ucl.ac.uk

**Abstract.** Localized and distributed guided ultrasonic waves array systems offer an efficient way for the long-term monitoring of the structural integrity for large structures. The use of permanently attached sensor arrays has been shown to be applicable to detect simulated corrosion damage. However, the detection sensitivity for fatigue cracks depends on the location and orientation of the crack relative to the transducer elements. Crack-like defects have a directivity pattern of the scattered wave field depending on the angle of the incident wave relative to the defect orientation and on the ratio of the characteristic defect size to wavelength. From FE simulations it has been shown that for cracks and notches almost no energy is scattered in certain directions from the defect, i.e., the data processing algorithm must take into account that for some transducer combinations no change in the signal even for a significant defect will be detected. The directivity pattern of the scattered field for the $A_0$ Lamb wave mode is predicted from 3D Finite Element simulations and verified from experimental measurements at machined part-through and through-thickness notches using a laser interferometer. Good agreement was found and the directivity pattern can be predicted accurately. The amplitude of the scattered wave is quantified for a variation of the angle of the incident wave relative to the defect orientation, the defect depth, and the ratio of the characteristic defect size to wavelength. These results provide the basis for the quantification of the detection sensitivity for defects in plate structures using guided wave array sensors. A hybrid model has been developed, taking the different propagation distances and scattering characteristics into account, in order to predict the relative amplitudes of received pulses for given sensor locations. From a comparison with the signal to noise ratio of the array system, detection capabilities can be predicted for given defect size and orientation.

**Keywords:** Structural Health Monitoring, Guided Ultrasonic Waves, Scattering
**PACS:** 43.35.Zc, 43.35.Cg, 43.35.Yb.

## 1. Introduction

Low frequency guided ultrasonic waves typically have a wavelength that is large compared to the thickness of a structure and can propagate over long distances [1]. This provides an efficient way for the structural health monitoring (SHM) of large technical structures [2]. For structures containing plate-like components, different types of guided ultrasonic wave monitoring systems have been developed. Localized phased array systems typically contain a group of sensor elements at one location of the structure and allow for the detection of defects in a structure based on the analysis of back-scattered waves. Systems have been developed employing the $S_0$ Lamb wave mode excited and measured using EMAT (electro-magnetic acoustical transducers) arrays [3], and the $A_0$ Lamb wave mode using circular [4] and rectangular [5] arrays of piezoelectric elements. Distributed array systems using sensor elements at different locations of the structure have been developed [6-8]. Due to the complexity of the received signals, quite often a baseline subtraction algorithm is used to identify the pulse scattered at a defect. Different concepts for temperature compensation of the signals have been implemented to overcome environmental changes between measurements for the long term monitoring of realistic structures [9].

Both monitoring concepts rely on an additional wave pulse scattered at the defect and were initially tested for circular defects. For the detection of crack-like defects with various, possibly unknown orientations, the scattering pattern must be taken into account to quantify the minimum defect sensitivity for such defects. The scattering of the $A_0$ Lamb wave mode at thickness changes in a plate and at notches has been studied in a 2D configuration through the plate thickness using FE simulations and experiments [10, 11]. Analytical models exist for the scattering of the $A_0$ mode at circular through holes [12], which have been verified from numerical simulations and experiments [13]. Models have been extended to take part-thickness circular defects into account and verified from FE simulations [14]. The change in the scattered field of a guided wave mode due to a fatigue crack emanating from a fastener hole has been reported previously [15]. Analytical models exist for the scattering of mechanical waves at crack-like defects [16], but apply restrictions on the geometry in order to allow for an analytical solution. More recently, FE modeling has been used to study the interaction of shear horizontal waves with a crack in plates [17]. Frequency domain FE models have been developed to investigate the scattering of bulk ultrasonic waves at different types of defects [18]. This allows for the efficient characterization of the defect by calculating the scattered ultrasonic field for all incident and scattering angles [19], called the scattering matrix or scattering amplitude [20]. Preliminary results for the extension of the frequency domain FE approach to 3D models for bulk and guided ultrasonic waves have been shown [21]. It would be beneficial to benchmark the computational efficiency and accuracy against experiments and conventional 3D FE scattering calculations [22, 23].

This contribution investigates the directivity pattern of the $A_0$ Lamb wave mode scattered at a crack-like defect (notch) in a plate, depending on the angle of the incident wave relative to the defect orientation, on the defect length relative to the wavelength, and on the defect depth. The scattered wave fields predicted from Finite Element (FE) simulations and measured experimentally are compared and show good agreement for part-through and through thickness notches. The obtained scattering characteristics for a variation of the geometric parameters are discussed with a view towards the detection of crack-like defects using guided ultrasonic waves.

Based on the FE results a hybrid model is developed to predict the relative amplitudes of guided wave pulses scattered at the defect received at given sensor locations. The model takes the wave propagation and scattering characteristics from the FE simulations for a given defect size into account. Damage maps can be calculated for each sensor pair, specifying the relative amplitude of the defect signal. Using the same data fusion algorithm as the guided wave array sensor evaluation, the defect detection sensitivity for a given defect size and orientation is predicted. This can be compared with the signal to noise ratio (SNR) of an implemented array design and used to optimize the sensor layout.

Journal of Physics: Conference Series **269** (2011) 012018 doi:10.1088/1742-6596/269/1/012018

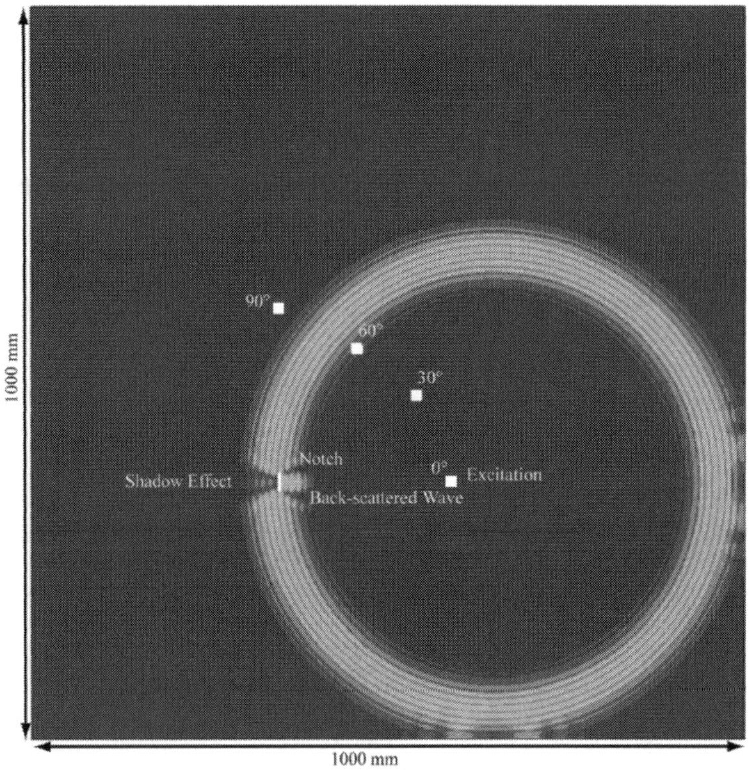


**Figure 1.** Finite Element simulation of $A_0$ Lamb mode wave propagation from 0° excitation location in plate (wave crest propagating radially outwards), scattering at through thickness notch; f = 100 kHz; λ = 19 mm; a = 20 mm; different possible excitation locations shown.

## 2. Finite Element Simulations of Scattered Wave Field

The propagation and scattering of the first anti-symmetric Lamb wave mode $A_0$ in plates was simulated using three-dimensional (3D) Finite Element (FE) models with linear brick elements ($\Delta x = \Delta y = \Delta z = 1.25$ mm) in ABAQUS Explicit. The model of a 5 mm thick aluminum plate (size: 1 m x 1 m) with a notch of varying length and depth was implemented, similar to the model described in [22]. The notch was modeled by removing one row of elements to achieve the desired length and depth. This resulted in a notch with right angle corners and a width of 1.25 mm, matching the experimental configuration described below reasonably well. Explicit time integration was used, and the element size and time step were chosen to adhere to the usual stability criteria. The element size was not changed in the vicinity of the defect; therefore small errors might occur in the near field due to the inaccurate description of the stress concentration.

Point excitation of the $A_0$ Lamb wave mode was introduced at chosen node locations 300 mm from the defect location, allowing for a variation of the incidence angle between 0° (perpendicular) and 90° (parallel), relative to the normal of the notch (see Fig. 1). The excitation pulse was set as a 5 cycle toneburst with a center frequency $f$ of 100 kHz, resulting in a wavelength $\lambda$ for the $A_0$ mode of 19 mm, thus about 4 times the plate thickness. The amplitude of the $A_0$ mode was monitored as the out-of-plane displacement at the center (mid-plane) node on points in a square area of 100 mm by 100 mm around the notch center. The time trace at each monitoring node was time gated to remove reflections from the plate edges. Fast Fourier Transform (FFT) was used to extract the complex magnitude (amplitude and phase information) at the center frequency of 100 kHz for each monitoring node.

Journal of Physics: Conference Series **269** (2011) 012018 doi:10.1088/1742-6596/269/1/012018

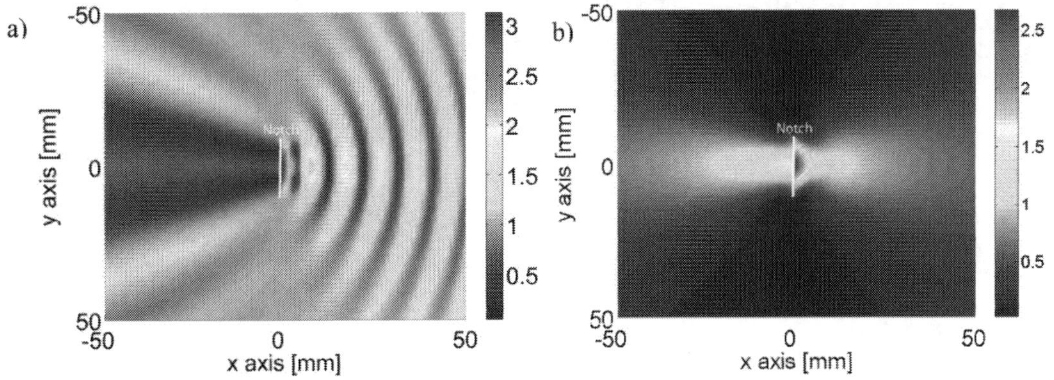


**Figure 2.** Normalized amplitude of scattered ultrasonic wave field ($A_0$ mode) at through thickness notch; f = 100 kHz; λ = 19 mm; a = 20 mm: a) view of combined field; b) scattered wave field only; amplitude of guided wave pulse color-coded (blue = low, red = high amp.).

This captures the combined wave field $U_1$ of the incident wave $U_i$ and scattered wave $U_s$ (Eq. 1), shown in Fig. 2a for a 0° incident wave. In the figure the amplitude of the guided wave pulse at each point is color-coded from low amplitude (dark blue) to high amplitude (red). The typical scattering pattern with large amplitude at the free surface of the defect and shadow area (low amplitude) behind the defect can be seen. Along the incident direction (positive x axis) a periodically alternating high and low amplitude pattern due to the positive and negative interference of the incident wave $U_i$ and back scattered wave $U_s$, according to their phase, can be observed.

$$U_1 = U_i + U_s \qquad (1)$$

$$U_2 = U_i \qquad (2)$$

$$A_s = |U_s| = |U_1 - U_2| \qquad (3)$$

Additional simulations were run to capture the incident wave field $U_i$ only, without a crack present (Eq. 2). The amplitude of the incident wave (not shown) is almost constant in the area of interest (vicinity of the defect). Taking the difference between the complex magnitudes for each point with and without a defect, the amplitude of the wave scattered at the notch (Fig. 2b) can be found [23]. It should be noted that the phase information contained in the complex magnitude is necessary for this subtraction (Eq. 3), as a simple subtraction of the measured amplitudes with and without a defect would not give the amplitude of the scattered wave only.

The scattered wave field (Fig. 2b) shows the expected behavior for a wave incident perpendicular on a long through-thickness defect (notch length a = 20 mm similar to wavelength λ = 19 mm), with a significant shadow area of low amplitude behind the defect (negative x axis) and back scattered wave (positive x axis). Evaluating the scattered field on a radial line (x axis) more than a wavelength (λ = 19 mm) away from the defect (length 20 mm) in Fig. 3a, the amplitude shows a decrease proportional to one over the square root of the distance, as would be expected for a point source of the guided wave mode on a plate. Therefore the amplitude of the scattered wave was extracted on a radius of 30 mm by interpolating between monitoring nodes to obtain an amplitude value every 5 degree. Shown in Fig. 3b is the polar plot of the normalized amplitude of the scattered $A_0$ mode, from which the shadow area behind the defect and the back scattered wave can be clearly identified. The amplitude of the scattered wave is normalized with respect to the amplitude of the incident wave at the defect center. Using the radial square root dependency relative to the monitoring radius of 30 mm, the amplitude of the scattered wave at any location in the plate can be calculated.

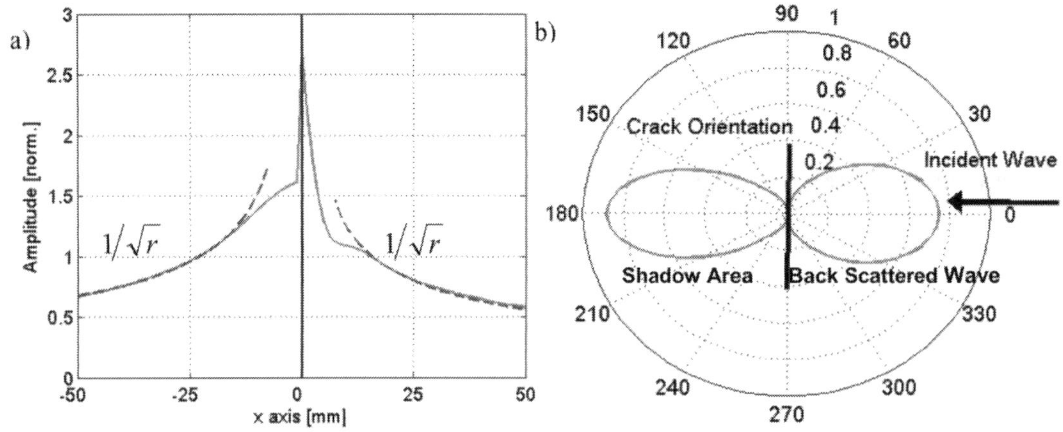


**Figure 3.** Scattered ultrasonic wave field ($A_0$ mode) at through thickness notch; orientation 90°-270°; incident wave from 0°; f = 100 kHz; λ = 19 mm; a = 20 mm: a) normalized amplitude (red, solid) along x axis (perpendicular to defect) and radial dependency (blue, dashed); b) polar plot of normalized amplitude (monitored at 30 mm radius).

The approach to use the local calculation of the scattering characteristics and the knowledge of the wave propagation to predict the ultrasonic wave field in the whole structure has been used in literature for bulk ultrasonic wave scattering [20], but with a different scattered amplitude scaling definition [19, 20].

## 3. Experimental Description

The scattered wave field around a 20 mm long notch was measured on a large, 5mm thick aluminum plate (size: 1.5 m x 1 m). The notch (width: 1 mm) was milled in steps to give a varying depth. Two cases, a part-thickness notch of 2.5 mm depth (1/2 thickness) and a through thickness notch are considered here. The notch had rounded edges due to the milling tool (diameter 1 mm) and a flat bottom for the part-through defect. Three excitation transducers were placed 300 mm away from the center of the notch location to give angles of the incident wave of 0° (perpendicular to notch), 45° and 90° (along notch orientation) relative to the normal of the notch (Fig. 4).

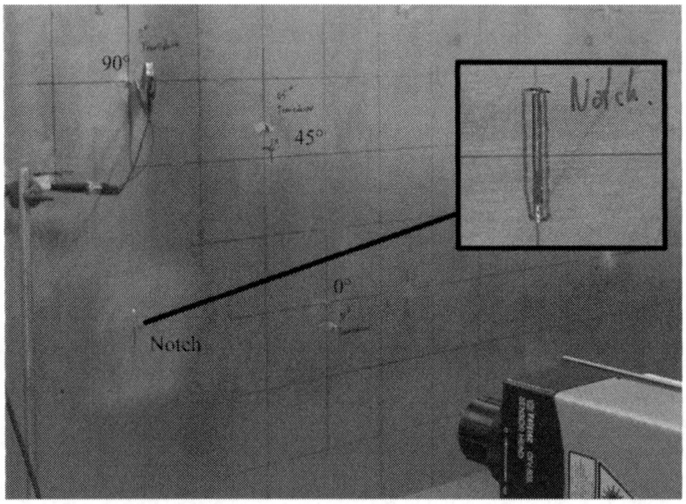


**Figure 4.** Experimental setup: aluminum plate (size: 1.5 m x 1 m) with 20 mm notch (zoom), 3 excitation transducers at 0°, 45°, and 90°, 300 mm from notch, and laser interferometer.

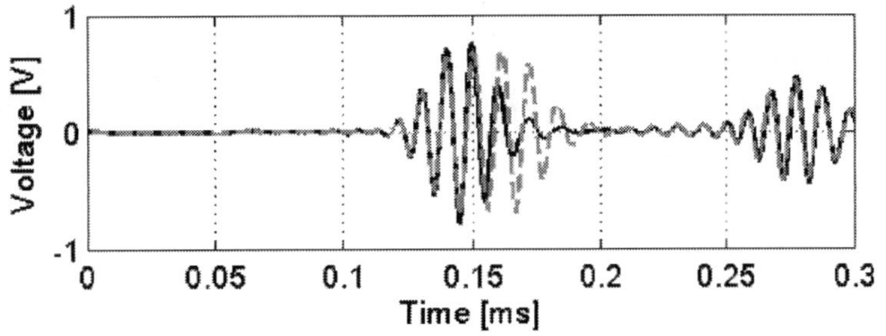


**Figure 5.** Typical experimental time signal, undamaged (black, solid), with notch (red, dashed), f = 100 kHz.

The transducers consist of a PZT disc (Ferroperm Pz27, diameter 5 mm, thickness 2 mm) and a brass backing mass (height 6 mm) and were permanently bonded to the plate using two-component epoxy glue. The transducers act in good approximation as point sources for the excitation of the first antisymmetric Lamb wave mode $A_0$. The excitation signal was a 5 cycle toneburst with a center frequency of 100 kHz modulated by a Hanning window. The signal was generated in a programmable function generator and amplified using a power amplifier. The velocity of the out-of-plane displacement was measured using a laser interferometer every 5 degrees on a circle around the notch with a radius of 30 mm. In the frequency-thickness range of interest the $S_0$ mode has a much smaller out-of-plane component than the $A_0$ mode. Therefore even for the part thickness defect case where mode conversion occurs, the laser measurement will mostly only detect the $A_0$ mode. The full time traces of the measured signals were bandpass filtered (75 kHz – 125 kHz) and averaged (20 averages) in a digital oscilloscope before being transferred to the measurement PC [23, 24].

Similar to the FE simulations, an initial measurement was done for each excitation transducer before the milling of the notch to capture the incident wave field, and then for each notch depth. The measured time traces (Fig. 5) were time gated and the amplitude and phase (complex magnitude) of the wave pulse at 100 kHz determined using FFT. Due to the re-positioning of the plate after each milling step relative to the laser, the repeatability of the phase measurement was somewhat limited. This leads to a noise level of about 5% of the amplitude of the incident wave when taking the difference between complex magnitudes to evaluate the scattered field.

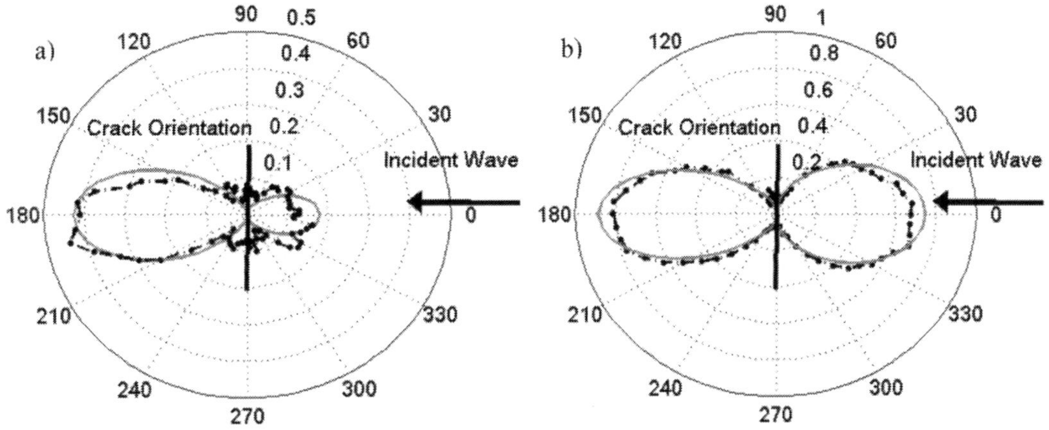


**Figure 6.** Polar plot of normalized amplitude of $A_0$ mode scattered at notch (monitored at 30 mm radius); orientation 90°-270°; incident wave from 0°; f = 100 kHz; $\lambda$ = 19 mm; a = 20 mm; experiment (black, dash-dotted), simulation (red, solid): a) ½ thickness; b) through thickness.

Journal of Physics: Conference Series **269** (2011) 012018 doi:10.1088/1742-6596/269/1/012018

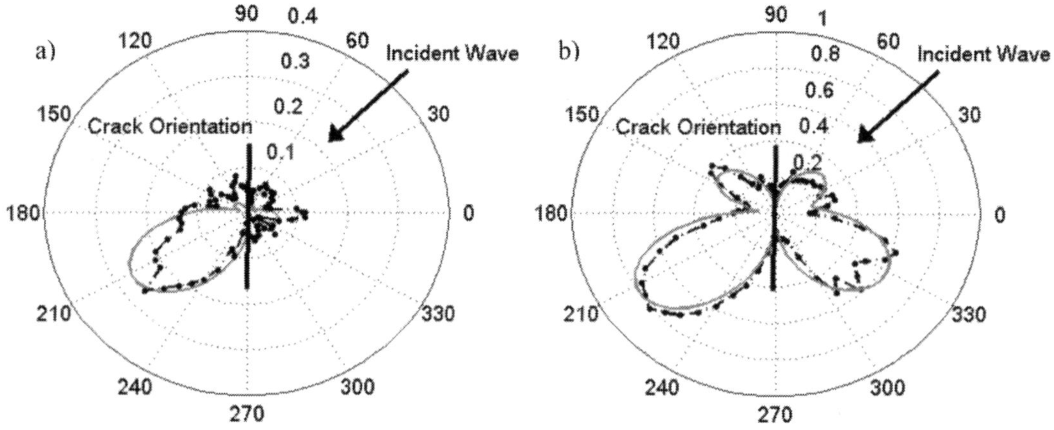


**Figure 7.** Polar plot of normalized amplitude of $A_0$ mode scattered at notch (monitored at 30 mm radius); orientation 90°-270°; incident wave from 45°; f = 100 kHz; $\lambda$ = 19 mm; a = 20 mm; experiment (black, dash-dotted), simulation (red, solid): a) ½ thickness; b) through thickness.

## 4. Experimental Verification of FE Simulation Scattering Directivity Pattern

FE simulation and experimental results were compared for a 20 mm long notch half and fully through the thickness of the 5 mm aluminum plate. For the incident wave perpendicular to the notch (0° incidence angle) good agreement of the comparison between FE predictions and experimental results was found (Fig. 6). The noise level for the complex magnitude difference calculated from experimental results is about 5% as described in section 3, which is evident in Fig. 6a. For the ½ thickness a notch shadow effect behind the defect can be observed, as the path of the incident wave is blocked and mode conversion occurs. Only a small back-scattered amplitude is observed for the ½ thickness defect (about 15%), increasing to about 70% for a through thickness defect (Fig. 6b). This has a significant influence on the ability to detect a crack from pulse-echo type measurements. For a localized array of guided wave transducers with all sensors at one location of the structure, for shallow defects only limited energy is reflected back towards the array location. For the incident wave at a 45° angle relative to the defect orientation as well good agreement was found between experiments and the FE simulations (Fig. 7). For shallow defects (½ plate thickness) mostly the shadow effect behind the notch in the direction of wave propagation was observed (Fig. 7a), with a significant reflected wave only present for a through thickness defect (Fig. 7b). For this case, with the incident wave at a significantly larger angle of 45° relative to the defect orientation, the amplitude of all scattered wave lobes are lower than for the perpendicular incidence case discussed above. For the incident wave propagating along the defect orientation (90° angle), only a very small scattered wave is expected for a part-thickness notch of about 5% and such a defect could not be detected, even from a local laser measurement due to the noise level in the experimental evaluation. For the through thickness notch scattered amplitude of about 10% was predicted and could just about be measured.

## 5. Variation of the Scattered Field due to Defect Geometry and Orientation

The directivity pattern of the scattered wave field depends strongly on the geometric parameters, i.e., angle of the incident wave relative to the defect orientation, defect depth and length relative to the wavelength. As discussed above, a shadow area behind the notch develops and increases with increasing defect depth until most of the energy of the wave is blocked from propagating past the notch for a deep defect. However, for shallow defects only a small back-scattered wave exists. Only for deep defects a significant back scattered wave can be seen which increases in amplitude for a through thickness defect.

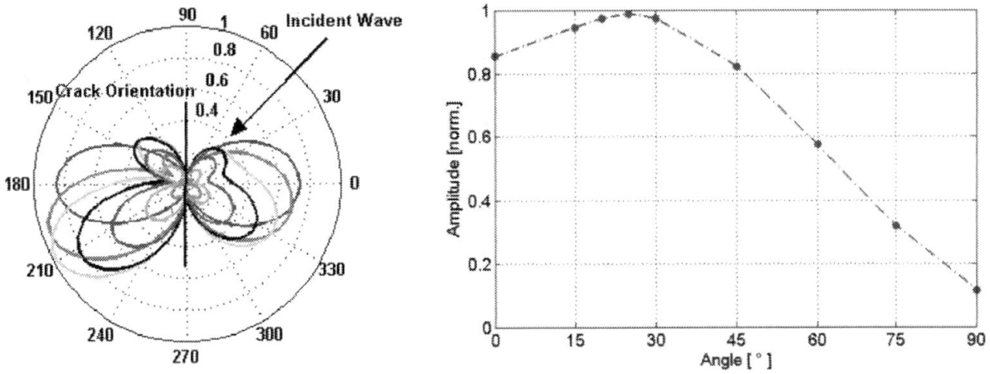


**Figure 8.** Left: polar plot of scattered wave amplitude at 20 mm through thickness notch for variation of incident angle from 0° (perpendicular, blue) in 15° steps (15° red; 30° green; 45° black; 60° brown, 75° yellow) to 90° (along defect, magenta); right: maximum scattered amplitude from polar plot; f = 100 kHz; λ = 19 mm.

Shown in Fig. 8 and Fig. 9 are the polar plots and maximum amplitude of the FE calculated amplitude of the $A_0$ mode scattered at a through thickness defect with lengths a of 20 mm and 5 mm for a variation of the incident angle. It can be seen that the directivity pattern depends on the ratio between the wavelength and the defect size [24]. A pattern with lobes of the scattered field corresponding to the angle between the incident wave and the crack orientation develops (Fig. 8) when the defect length is comparable to the wavelength (λ = 19 mm). Significant maximum amplitude of the scattered wave field is found when the wave is incident on the notch at an angle up to about 45° and then decreases with increase of the incident angle. For defect lengths shorter than the wavelength, the scattered field shows a different directivity pattern with one or two main lobes approximately perpendicular to the crack orientation (Fig. 9). The maximum amplitude for this case is in the shadow area behind the defect and is significantly smaller than for a longer defect. Again, the amplitude of the scattered wave decreases with increasing incident angle and is very small when the incident wave direction is aligned with the notch orientation. Similar patterns can be observed for other variations of the geometric parameters. However, it can be clearly seen that the amplitude of the scattered wave shows a strong angular dependency and that along the orientation of the defect, almost no energy is scattered.

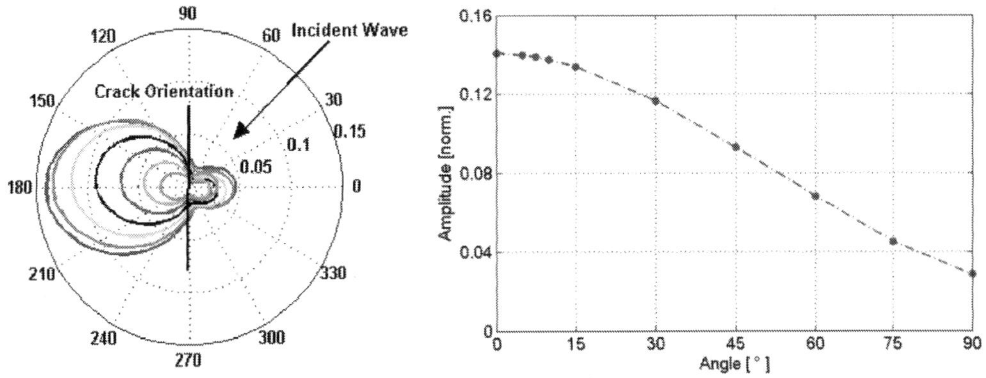


**Figure 9.** Left: polar plot of scattered wave amplitude at 5 mm through thickness notch for variation of incident angle from 0° (perpendicular, blue) in 15° steps (15° red; 30° green; 45° black; 60° brown, 75° yellow) to 90° (along defect, magenta); right: maximum scattered amplitude from polar plot; f = 100 kHz; λ = 19 mm.

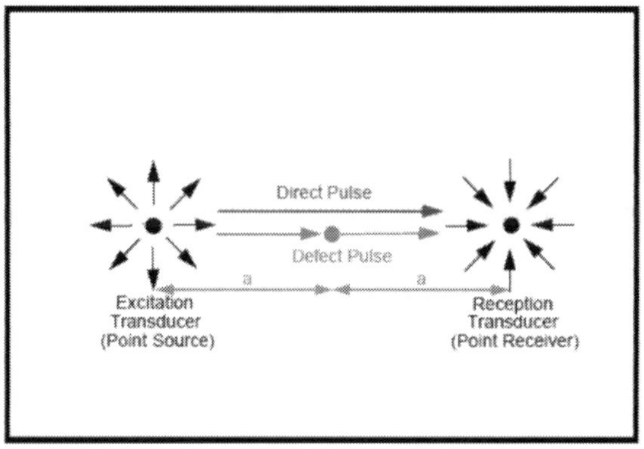


**Figure 10.** Typical geometric configuration for SHM application using distributed array system with 2 sensor locations on plate shown.

## 6. Hybrid Model for SHM Sensitivity Prediction

From the results shown above it can be seen that the amplitude of the scattered wave shows a strong angular dependency and that there are directions in which almost no energy is scattered. This can be problematic for SHM applications as for given transducer locations on a structure, there exist possible defect (crack) positions and orientations, where even for a severe defect no signal from the defect would be detected due to the scattering directivity pattern of the defect. Shown in Fig. 10 is a typical geometrical configuration for a distributed array system with 2 sensor locations.

Based on the findings detailed above, some general recommendations for the placement of sensors on a structure can be given. Shallow defects show a significantly larger shadow effect than back scattered wave and only for deeper defects does a substantial back-scattered wave develop. Furthermore, the scattered wave field is very small when the incident wave direction is approximately aligned with the crack orientation or when measured along that direction for any incident angle. Often the likely crack orientation and location are known due the knowledge of the stress state in the structure, e.g. loading direction and areas of stress concentration. If possible, it is then advantageous to place the sensors in such a way that the incident wave is approximately perpendicular to the likely defect orientation. For a monitoring location placed behind the critical crack location in the shadow area of the incident wave, a significant amplitude change even for shallow defects can be expected. For other monitoring locations, smaller scattered signals are expected. However, this might have the advantage that an additional scattered wave pulse can be detected better than a change in amplitude of the wave transmitted past the defect location. For a distributed array system these conditions are often more easily met than for a localized array system placed at one location of the structure.

Furthermore, assuming all sensors to act in good approximation as point sources and receivers for the guided wave mode, a hybrid model can be developed to predict the amplitude of the wave scattered at a defect at any location on the structure and received at a given transducer location. Taking the decrease in amplitude due to the radial spread as one over the square root of the propagation distance into account, the incident amplitude at any given defect location can be calculated analytically. Using the knowledge of the defect orientation and scattering directivity the amplitude of the wave scattered in the direction of the receiving sensor element is obtained from the FE simulation results. Again correcting for the propagation distance, the amplitude of the received scattered wave is computed. For a distributed array system, the amplitude of the defect pulse relative to the direct pulse propagating from the transmitting to the receiving element is relevant to compare against the signal to noise ratio (SNR) of the system.

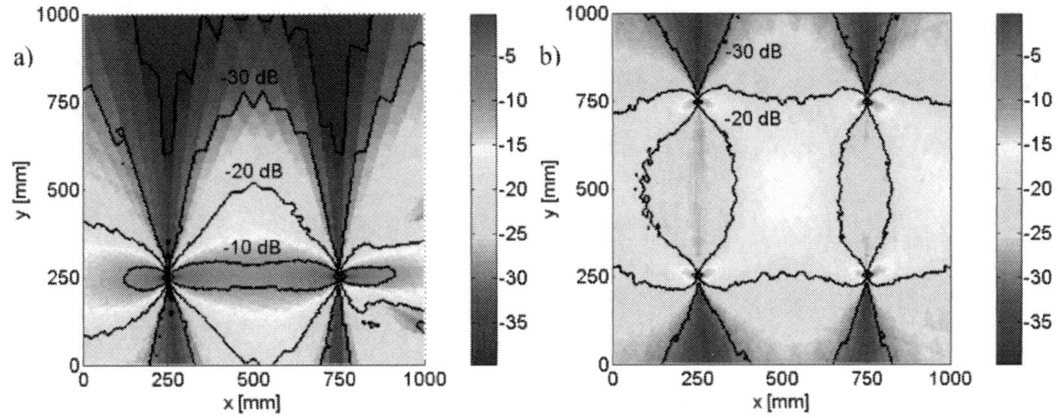


**Figure 11.** Predicted relative amplitude (decibel scale) of defect pulse for 20 mm deep through crack vertically oriented compared to direct pulse between sensors; a) 2 sensors; b) addition of all combinations of 4 sensors.

This calculation is shown in Fig. 11a for excitation and receiving transducers 500 mm apart horizontally on a square 1 m x 1 m plate, using the same parameters as described in the previous sections. For a 20 mm long, through thickness notch, aligned vertically, the predicted relative amplitude of the defect signal is shown color-coded in Fig. 11a. High received scattered amplitude is predicted for the defect located on and close to the line between sensors as the incident wave is approximately perpendicular to the crack orientation. Low received scattered amplitude can be seen for areas further away from the sensor locations and where the incident wave propagates along the crack orientation (x = 250 mm & x = 750 mm). For a distributed array system with multiple sensors, each acting in turn as excitation and receiving element, the same data fusion algorithm as for the measurements can be used to combine the predicted amplitude maps for each sensor pair. Shown in Fig. 11b is the predicted combined detection sensitivity for the same crack size and orientation using 4 sensors placed in a rectangular pattern 250 mm from the plate edges. Good detection sensitivity with relative amplitude above 20 dB is predicted for the center of the area covered by the 4 sensors. However, due to the considered vertical orientation of the crack, lower detection sensitivity can be seen between the vertically aligned transducers at x = 250 mm and x = 750 mm. Here the incident or scattered wave direction for a significant proportion of the considered transducer combinations is aligned along the crack and only very small scattered amplitude is expected.

Based on the knowledge of the guided ultrasonic wave propagation and scattering characteristics, accurate predictions of the sensitivity of a structural health monitoring system employing distributed arrays of sensors can be made. The placement of the sensors can be optimized for expected crack locations and orientations if the stress state in the structure is known. From a comparison to the SNR of the employed array system, the potential to detect crack-like defects can be predicted and the required number of sensors evaluated.

## 7. Conclusions

The scattering of the $A_0$ Lamb wave mode at part-through and through thickness crack-like in aluminum plates has been investigated experimentally and from FE simulations of the wave propagation and scattering. Good agreement was found between the experimentally measured scattered field directivity patterns and amplitudes and FE predictions. This allows for the further investigation of the influence of relative defect size and depth, wavelength, plate thickness and incidence angle from the FE simulations. For all investigated cases of incidence angle and defect size it was found that in certain directions there is only very little scattered wave energy. These findings should be taken into account when developing permanently attached guided ultrasonic wave arrays for SHM, as the sensitivity for defect detection will depend strongly on the chosen geometric configuration in relation to the possible defect locations. Based on the investigation presented in this contribution, a hybrid model has been developed, taking the wave propagation and scattering into account, to predict and quantify the sensitivity for the detection of defects at likely locations and orientations.

**Acknowledgements**
This work was supported by grant EP/D065011/1 from the UK Engineering and Physical Sciences Research Council (EPSRC).

**References**

[1] Chimenti, D.E., *Appl. Mech. Rev.*, **50(5)**, 247-284 (1997).
[2] Rose, J.L., *Mat. Eval.* **60(1)**, 53-59 (2002).
[3] Wilcox, P.D., Lowe, M.J.S., Cawley, P., *IEEE Trans. on Ultrason. Ferroelec. and Freq. Cont.* **52(4)**, 653-665 (2005).
[4] Fromme, P., Wilcox, P.D., Lowe, M.J.S., Cawley, P., *IEEE Trans. on Ultrason. Ferroelec. and Freq. Cont.* **53(4)**, 777-785 (2006).
[5] Yu, L. and Giurgiutiu, V., *Ultrasonics* 48(2), 117-134 (2008).
[6] Croxford, A.J., Wilcox, P.D., Drinkwater, B.W., Konstantinidis, G., *Proc. Roy. Soc. A* **463**(2087), 2961-2981 (2007).
[7] Michaels, J.E. and Michaels, T.E., *Wave Motion* **44(6)**, 482-492 (2007).
[8] Fromme, P., *in Rev. Prog. QNDE, ed. by D.O. Thompson and D.E. Chimenti, AIP Conf. Proc. 975, New York*, **27A**, 78-85 (2008).
[9] Clarke, T. and Cawley, P., *IEEE Trans. on Ultrason. Ferroelec. and Freq. Cont.* **56(12)**, 2666-2678 (2009).
[10] Lowe, M.J.S., Cawley, P., Kao, J.Y., Diligent, O., *J. Acoust. Soc. Am.* **112(6)**, 2612-2622 (2002).
[11] Yeo, F. and Fromme, P., *in Rev. Prog. QNDE, ed. by D.O. Thompson and D.E. Chimenti, AIP Conf. Proc. 820, New York*, **25A**, 202-209 (2006).
[12] Pao, Y.H. and Chao, C.C., *AIAA J.* **2(11)**, 2004-2010 (1964).
[13] Fromme, P. and Sayir, M.B., *J. Acoust. Soc. Am.* **111(3)**, 1165-1170 (2002).
[14] Diligent, O., Grahn, T., Bostrom, A., Cawley, P., Lowe, M.J.S., *J. Acoust. Soc. Am.* **112(6)**, 2589-2601 (2002).
[15] Fromme, P. and Sayir, M.B. *Ultrasonics* **40(1-8)**, 199-203 (2002).
[16] Sih, G.C. and Chen, E.P., *in Plates and shells with cracks: a collection of stress intensity factor solutions for cracks in plates and shells, ed. by G.C. Sih*, Leyden: Noordhoff International (1977).
[17] Rajagopal, P. and Lowe, M.J.S., *J. Acoust. Soc. Am.* **124(5)**, 2895-2904 (2008).
[18] Wilcox, P.D., and Velichko, A., *J. Acoust. Soc. Am.* **127(1)**, 155-165 (2010).
[19] Zhang, J., Drinkwater, B.W., Wilcox, P.D., *IEEE Trans. on Ultrason. Ferroelec. and Freq. Cont.* **55(10)**, 2254-2265 (2008).
[20] Schmerr, L.W., *Fundamentals of ultrasonic nondestructive evaluation: a modeling approach*, Plenum Press, New York (1998).
[21] Velichko, A. and Wilcox, P.D., *in Rev. Prog. QNDE, ed. by D.O. Thompson and D.E. Chimenti, AIP Conf. Proc. 1211, New York*, **29**, 57-64 (2010).
[22] Fromme, P., *in Proc. of SPIE, ed. by T. Kundu*, **6935,** W9350 (2008).
[23] Fromme, P., *in Rev. Prog. QNDE, ed. by D.O. Thompson and D.E. Chimenti, AIP Conf. Proc. 1211, New York*, **29**, 129-136 (2010).
[24] Fromme, P., *in Proc. of SPIE, ed. by T. Kundu*, **7650,** (2010).

Journal of Physics: Conference Series **269** (2011) 012019 doi:10.1088/1742-6596/269/1/012019

# A preliminary study of acoustic propagation in thick foam tissue scaffolds composed of poly(lactic-co-glycolic acid)

**N G Parker [1], M L Mather[2], S P Morgan[2] and M J W Povey[1]**

[1] School of Food Science and Nutrition, University of Leeds, Leeds, LS2 9JT, UK
[2] Electrical Systems and Applied Optics Research Division, University of Nottingham, Nottingham, NG7 2RD, UK
n.g.parker@leeds.ac.uk

**Abstract**. The exclusive ability of acoustic waves to probe the structural, mechanical and fluidic properties of foams may offer novel approaches to characterise the porous scaffolds employed in tissue engineering. Motivated by this we conduct a preliminary investigation into the acoustic properties of a typical biopolymer and the feasibility of acoustic propagation within a foam scaffold thereof. Focussing on poly(lactic-co-glycolic acid), we use a pulse-echo method to determine the longitudinal speed of sound, whose temperature-dependence reveals the glass transition of the polymer. Finally, we demonstrate the first topographic and tomographic acoustic images of polymer foam tissue scaffolds.

## 1. Introduction

Tissue engineering aims at the creation of biological substitutes that replace, restore or improve tissue function [1]. Central to this is the requirement of an artificial extracellular matrix, or scaffold, on which to grow the surrogate tissue, which is either developed *in vitro* with subsequent implantation [2] or transplanted directly for *in vivo* growth [3]. Tissue engineering has demonstrated many successes so far (see [4] for a review). However, significant challenges lie ahead, notably for cell biology, biomaterials, scaffold engineering and imaging/characterization. In recent years acoustics has found diverse applications in assisting tissue engineering. Examples include the use of surface waves to promote cell proliferation in tissue scaffolds [5, 6], monitoring of cell integration [7], sound-induced acceleration of bone tissue growth [8], manipulation of scaffold structure [9], and a route to monitor scaffold formation [10] and polymer degradation [11]. Acoustic waves also offer exclusive opportunities for characterisation of scaffolds due to their ability to probe structural, mechanical and fluidic properties. It is this possibility that motivates the current work.

Of much current focus in tissue engineering are three-dimensional polymer foam scaffolds. These may be composed of a natural polymer, e.g., collagen and chitosan, or a synthetic polymer, e.g., poly(lactic acid) or poly(lactic-co-glycolic acid), all of which satisfy the rudimentary requirement of being biocompatible and biodegradable. The structural and mechanical properties of the scaffold, such as its architecture, transport networks and mechanical integrity, must be precisely controlled to cultivate the required tissue type. As a result, accurate methods to characterize the foams and assure their suitability are crucial to the advancement of this field. Furthermore, approaches to fabricating the foams often feature an inherent randomness in the final structure, e.g., supercritical $CO_2$ and salt leaching/solvent casting, accentuating the need for accurate characterization methods.

Characterisation of polymer foam scaffolds is complicated by the multitude of factors to consider. Typically, the scaffold is characterized by its structural properties, with porosity, pore shape, pore size distribution, pore interconnectivity, total pore surface and tortuosity all cited as key structural parameters for tissue cultivation (see [12] and references therein). Image-based techniques provide the leading route to determine these parameters, with micro *X*-ray CT the most widely used modality [13]. However concerns exist over the variations that arise amongst different imaging modalities, contrast and resolutions [12]. Other work has highlighted the importance of parameterising the scaffold function with dynamic/fluidic information, notably its permeability [14]. These properties require distinct measurement approaches, such as fluid porometry, which are usually invasive. Acoustics has, as yet, been untested as a route to probe the internal characteristics of polymer foam scaffolds, despite its exclusive ability to reveal the structural, mechanical and fluidic properties of foams. These capabilities are beginning to be demonstrated in other areas of tissue engineering. For example, acoustic elasticity imaging has been demonstrated in tissue-mimicking phantoms [15] and hydrogel scaffolds [16], while the mechanical integrity of titanium biomaterials have been analysed acoustically in [17]. Of noteworthy importance for acoustic characterisation techniques is Biot theory, a well-established model of acoustic propagation in fluid-saturated porous media. This model gives the potential to characterise porosity, elastic parameters, permeability [18] and average pore size [19] of foams. Indeed, this theory has proved successful in structural studies of bone [19, 20], which bears many physical similarities to foam scaffolds.

Information on the basic acoustic properties of the synthetic biopolymers is scarce. While monitoring polymer degradation, Wu *et al.* determined the speed of sound at a single temperature [11], while Mather *et al.* monitored acoustic impedance during the scaffold fabrication [10]. Meanwhile, the acoustic propagation through foam tissue scaffolds has, to our knowledge, been untested. It, for instance, remains unknown if the scaffold can be saturated sufficiently with coupling liquid to reliably support acoustic propagation, with capillary resistance, surface chemistry and disconnected regions to contend with. Our motivation here is to address these basic questions. Focussing on foam scaffolds composed of poly(lactic-co-glycolic acid) (PLGA) we first examine the acoustic properties of the bulk biopolymer. A pulse-echo analysis enables us to measure the longitudinal speed of sound over a large temperature range. Notably, the glass transition of the polymer has a marked signature in the speed of sound. We then turn to the foam structures themselves and consider the acoustic imaging of their surface properties and sub-surface structure. The ability to image an object buried within the scaffold is demonstrated, illustrating the feasibility of acoustic propagation through these structures.

## 2. Materials and methods

### 2.1. Tissue scaffolds

Tissue scaffolds were donated by the School of Pharmacy at the University of Nottingham. The scaffolds are composed of PLGA 85:15 (supplied by Lakeshore Biomaterials) with a weight average molecular weight of 53 kDa. The foam structures were generated by the supercritical $CO_2$ method [21, 22]. In brief, PLGA granules are subjected to $CO_2$ under pressure. Entering its supercritical phase the $CO_2$ diffuses through the polymer, lowering its glass transition temperature and causing it to soften. After a suitable time the pressure is vented, the $CO_2$ re-enters its gaseous phase and generates bubbles, to form the final foam. The scaffolds are rigid, opaque and white. They have approximate dimensions of 12 mm in diameter and 6 mm in height, and a mass of around 0.2 g. From reference [22], where similar fabrication parameters were used, we can expect the porosity (the ratio of the volume of the pore space to the overall volume) to be around 80% and the pore sizes to lie in the range 50-250μm. Inherent to the supercritical $CO_2$ foaming method is the formation of a non-porous skin and domed roof, which we remove with a scalpel blade. Although PLGA hydrolyses in water, the degradation can be assumed to be negligible on the immersion timescales involved in this study [23].

The scaffold porosity is a key characteristic and can be derived from the expression,

$$\varepsilon = 1 - \frac{m_{\text{tot}}}{\rho_{\text{frame}} V_{\text{tot}}} . \qquad (1)$$

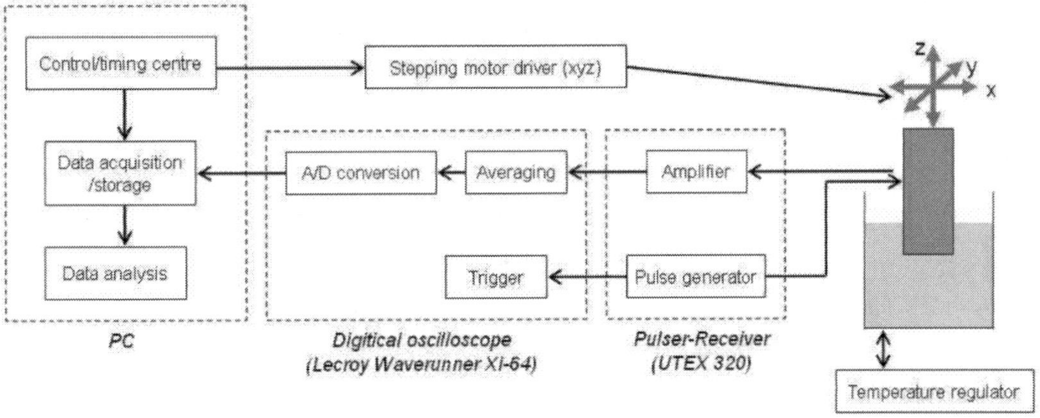


Figure 1: Block diagram of the key components within our acoustic measurement apparatus.

From mass $m_{tot}$ and volume $V_{tot}$ measurements of the scaffold (assuming the scaffold to be cylindrical), and the frame/polymer density to be *as per* the manufacturers specification of $\rho_{frame}$=1250 kg m$^{-3}$, we determine the scaffold porosity to be 81.5% (averaged over four samples).

### 2.2. Scanning acoustic platform

Our acoustic measurements and imaging are performed using a reflection-mode scanning acoustic platform [24]. Two ultrasound transducers are employed in this work, depending on the measurement being performed: a weakly-focussed transducer (Panametrics V311) with a centre frequency of 10 MHz and -6dB bandwidth of 5 MHz, and a tightly-focused transducer (Panametrics V3534) with an operating frequency of 55 MHz and -6dB bandwidth of 20 MHz. The generic operation of the platform is summarized in figure 1. A pulser-receiver (UTEX 320) provides a 300 V rectangular RF pulse with 10 ns duration to excite the piezoelectric transducer. The pulse repetition rate is set to 500 Hz. Acoustic reflections returning to the transducer generate RF pulses which are received and amplified by the pulser-receiver. This signal is then entered into a digital oscilloscope (Lecroy Waverunner Xi-64), where the signal is averaged over 200 sweeps to reduce noise. Following analogue-to-digital conversion the signal is exported to a PC for storage and analysis.

The acoustic platform can perform automated positioning of the transducer in three dimensions. In addition, the platform includes a sample bath which incorporates an outer fluid jacket for temperature control. A water-antifreeze mixture is circulated through this jacket by an external temperature bath/circulator (Haake B5/DC50). This enables us to regulate the temperature of the sample bath over the range 0-80°C. Independent measurements of the bath temperature are performed using a high precision temperature probe (Hart Scientific 5612 probe), accurate to 0.01 °C. From a previous mapping of the temperature profile within our sample unit [24], we anticipate that the variations across the probe-sample region will amount to less than 0.2°C.

### 2.3. Acoustic measurements

#### *2.3.1. Sample preparation*

For measuring the acoustic properties of the polymer we require a homogeneous sample. A foam scaffold is ground into a coarse powder and placed in a cylindrical aluminium mould with diameter 4 mm and height 5 mm. The mould is heated on a hot plate to 150°C for approximately 1 hour. Being above the glass transition temperature (typically 30-60°C at room temperature), the polymer relaxes into the mould. Care is taken to ensure that bubbles are removed from the sample. After cooling, the end result is a transparent disc-shaped solid sample, approximately 3 mm thick and 4 mm in diameter.

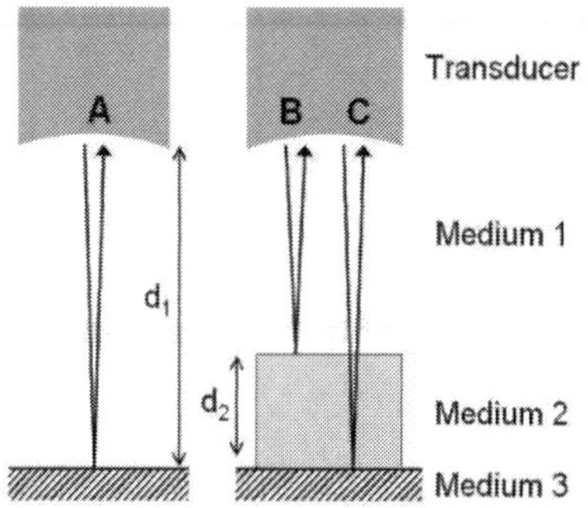


Figure 2: Schematic of the speed of sound measurements. From the time-of-flight of three sound paths (A, B and C) we determine the speed and sound in the sample.

### 2.4. Speed of sound measurements

We employ a time-of-flight technique to measure the speed of sound in this milli-scale polymer sample. Consider a homogeneous sample placed upon a substrate, immersed in a coupling fluid and insonified at normal incidence from above. From the times-of-flight of three sound paths (illustrated in figure 2) we can derive the speed of sound through the sample. Path A corresponds to the reflection from the fluid-substrate interface (in the absence of the sample). For paths B and C, the sample is present. In path B the sound pulse reflects at the first fluid-sample interface. In path C the sound pulses transmits through the sample and reflects at the sample-substrate interface. We denote the speeds of sound in the fluid and sample by $c_1$ and $c_2$, respectively, and the transducer-substrate distance and sample thickness by $d_1$ and $d_2$, respectively. The times-of-flight are then given by,

$$t_A = 2d_1 / c_1 \tag{2}$$

$$t_B = 2(d_1 - d_2) / c_1 \tag{3}$$

$$t_C = 2(d_1 - d_2) / c_1 + 2d_2 / c_2 . \tag{4}$$

Substituting equations (1) and (2) into equation (3) and rearranging gives,

$$c_2 = c_1 \left( \frac{t_A - t_B}{t_C - t_B} \right), \tag{5}$$

from which we can readily determine the sample speed of sound $c_2$, providing the speed of sound of the coupling fluid $c_1$ is well known.

To implement this technique we insert the polymer sample and substrate (a disc of polyether ether ketone) into the sample bath of the acoustic platform, with Millipore water as the coupling fluid. The 10 MHz transducer is employed for this study. This transducer has a focal distance of $F$=54 mm and aperture diameter of $D$=13 mm, giving a semi-aperture angle of $\theta$=$\sin^{-1}(D/2F)$=7°. For rays propagating at this maximal angle the acoustic path length will be larger than those propagating normal to the surface by a factor of $\sec\theta^{-1} \approx 0.007$. Since this is small we are justified in treating the sound beam as paraxial. Assuming an aqueous medium at 20°C our sound beam has a wavelength of $\lambda = c_1/f$=150 μm and a pulse length of 300 μm. In the focal plane the beam diameter will be diffraction-limited to $F_r$=1.22 $F\lambda/D$ = 300 μm. The depth of focus is $F_z$=$8F^2c_1/(D^2f+2Fc_1)$=1.7 cm. Thus, by positioning the focal plane at the substrate surface the full axial extent of the sample lies well within the focal region. The sample thickness (~3 mm) is considerably larger than both the acoustic wavelength and pulse length, ensuring the reflections are well-separated in time and are not coupled.

Each received pulse is demodulated computationally via the Hilbert transform, and its peak used to define its time-of-flight. The speed of sound in water is determined from the temperature-dependent equation of Bilaniuk and Wong [25, 26] (148-point) and the measured temperature of our sample bath.

### 2.5. Acoustic imaging of scaffolds

#### *2.5.1. Scaffold preparation/permeation*

For successful acoustic propagation it is essential to saturate the scaffold with fluid and remove highly-scattering gas pockets. Ideally, this fluid would be water or water-based, due to its biocompatibility and well-characterised properties. However, aqueous permeation of such scaffolds is challenging due to their hydrophobicity, tortuous pore network, capillary resistance and closed pores. To overcome hydrophobicity we use ethanol as a wetting agent, before dilution in water [27]. Scaffolds are immersed in ethanol surrounded by an ice bath. A chilled environment is necessary since ethanol lowers the glass transition temperature of the polymer from around 40°C to 10°C [28]. The scaffold is insonified with high power ultrasound throughout the permeation process using a Hielscher UP100H (which has a maximum power output of 100 W and operating frequency of 30 kHz) on a power setting of 20 W and a duty cycle of 0.1. This is found to dramatically improve the permeation compared to immersion only. Although high power ultrasound has been reported to disrupt the pore structure [9], the repeatability of our measurements and absence of detectable mass loss suggest that such structural modifications are insignificant. Fluid uptake is monitored by measuring the scaffold mass at regular intervals. From this we determine that (94±9)% of the pore space becomes filled with fluid (after 2 hours). The scaffold is then immersed in Millipore water for 6 hours, during which diffusion drives the dilution of ethanol with water throughout the sample.

*2.5.2. Scanning acoustic microscopy*

Acoustic imaging of the scaffolds is performed by employing the acoustic platform as a scanning acoustic microscope. The temperature of the sample is maintained at 25°C±0.2 °C. We employ both the 10 MHz and 55 MHz transducers. For the former the focal region is 300 μm laterally and 17 mm axially. For the latter the focal region is 25 μm laterally and 200 μm axially. The platform performs raster-scanning of the transducer in a plane above the sample and is automated to obtain the received trace at each point. Each trace is demodulated to reveal amplitude information only. From the final data set, various imaging sections can be processed and viewed. Of these we are, firstly, concerned with topographical imaging of the scaffold surface, where the time-of-flight of the returning signal is used to determine the surface height. Secondly, we will obtain C-scan images, where the voltage emanating from a given plane/time-of-flight is viewed.

## 3. Results and discussion

### 3.1. Acoustic measurements

*3.1.1. Speed of sound measurements*

Typical voltage signals in the absence and presence of the solid PLGA sample are shown in figure 3(a). For this example, the ambient temperature and pulse separations are given in table 1. The thickness of the sample can be derived from the times-of-flight via $d_2=c_1(t_A-t_B)/2=2.577$ mm, in firm agreement with a calliper measurement of $d_2=2.58$ mm. From equation (4) we determine the polymer speed of sound to be $c_2=2326$ ms$^{-1}$. Using the quoted density of $\rho_2=1250$ kg m$^{-3}$ we calculate the acoustic impedance of PLGA to be $Z_2=\rho_2 c_2=2.91$ MRayl. The temperature of the sample bath was heated to 60°C and swept down to 5°C, with the pulse times recorded at regular intervals. The variation of the PLGA speed of sound with temperature is presented in figure 3(b). At low temperature (5-25°C) the speed of sound decreases linearly at a rate of 3 m s$^{-1}$ °C$^{-1}$. In the vicinity of 30°C the behaviour becomes nonlinear. Beyond this the speed of sound again decreases linearly but at a greater rate of 15 m s$^{-1}$ °C$^{-1}$. We attribute this change to the glass transition, where the material changes from a solid to a glassy phase. Similar qualitative behaviour has been observed for polystyrene [29]. During the glassy phase pulse path C (Figure 2) becomes strongly attenuated with temperature and beyond 60°C we no longer detect this pulse. We attribute this to the highly viscous nature of this state, which will increase the attenuation [30]. We estimate the glass transition temperature by extrapolating the linear fits in the solid and glassy regimes. From the crossing point we find the glass transition temperature of our PLGA sample to be 31.4°C. This is somewhat lower than that observed elsewhere, e.g. around 50°C for a PLGA 90:10 sample observed in [31], and may

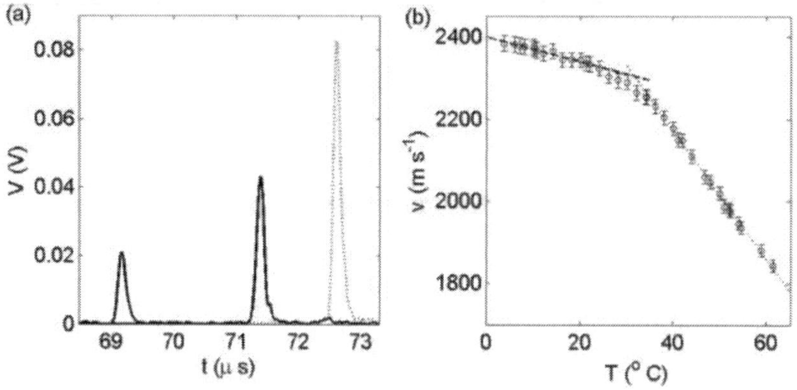


Figure 3: (a) Voltage signal in the absence (dotted line) and presence (solid line) of the PLGA sample. From left to right the pulses corresponds to paths B, C and A. (b) Speed of sound in PLGA as a function of temperature. The dashed and dotted lines are linear fits to the low and high temperature data, respectively.

suggest some polymer scission is occurring during sample formation. Note that we have recently shown this acoustic determination of glass transition temperature to be in very good agreement with the more established approach of differential scanning calorimetry [32].

| Sample | $T$ (°C) | $t_A$-$t_B$ (μs) | $t_C$-$t_B$ (μs) | $c_1$ (ms$^{-1}$) | $c_2$ (ms$^{-1}$) | $Z_2$=$\rho_2 c_2$ (MRayl) |
|---|---|---|---|---|---|---|
| PLGA 85:15 | 24.02 | 3.45 | 2.22 | 1494 | 2326 | 2.91 |

Table 1: Acoustic time-of-flight measurements for the PLGA 85:15 sample.

To our knowledge, only one previous study has measured the speed of sound in PLGA. Wu *et al.* [11] determined a value of 2400 ms$^{-1}$ for PLGA 50:50 at 20°C. This is close to our measured value. Deviations can be expected due to the different polymer ratio and variations in crystallinity.

### 3.2. Acoustic imaging of scaffolds

#### *3.2.1. Scaffold permeation*

#### *3.2.2. Topographic acoustic imaging*

For the topographic imaging we employ the 55 MHz tightly focussed transducer due to its enhanced resolution. The top of the scaffold is aligned with the focal plane of the beam. Figure 4 shows an image of the scaffold surface, over an area of 900×700 μm. The lateral resolution is 25 μm. Vertical variations of the scaffold surface are of approximate size 500 μm. Two large pores are clearly visible at the surface, with a circular profile and diameter of approximately 200 μm. This is consistent with the study of Tai *et al.* [22] who examined similar scaffolds with *x*-ray CT and found the pore sizes to lie dominantly in the range 50-250 μm.

### 3.3. Sub-surface imaging

Attempts to image sub-surface at 55 MHz were unsuccessful and may be due to the beam sensitivity or high attenuation through the polymer. Subsequent imaging was performed with the 10 MHz transducer. A narrow glass slide, of height 0.5 mm and width 2 mm, was inserted transversely through the centre of the scaffold. The object was approximately 3 mm below the surface of the scaffold. The focal plane of the transducer was aligned with the midpoint of the scaffold. Due to the large focal depth the whole axial extent of the scaffold remained in focus.

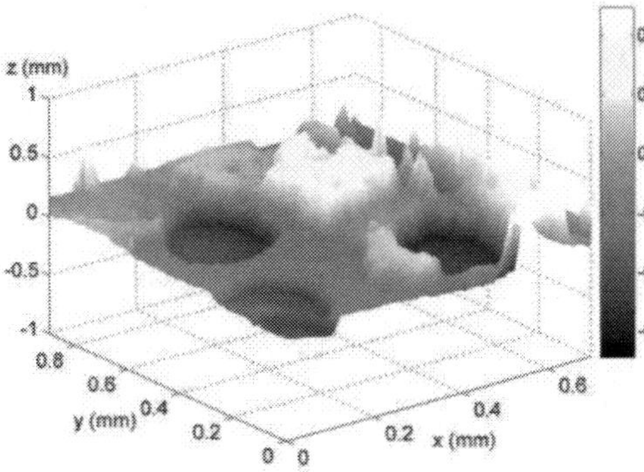


Figure 4: Acoustic image of the scaffold surface, obtained using the 55 MHz transducer. The colour indicates the vertical position, with light/dark regions representing high/low height, respectively.

Various *C*-scans are presented in figure 5(a). The first *C*-scan shows the scaffold surface. While the beam resolution (300 μm) is insufficient to spatially resolve the pores a strongly fluctuating signal indicates the uneven surface. To help identify internal features in the scaffold it is useful to consider the maximum voltage received as a function of depth (figure 5(b)). NB At a given depth this is the maximal voltage arising anywhere in the *x*-*y* plane and so does not correspond to a fixed *x*-*y* position. Beyond the scaffold surface this voltage decays roughly with distance due to the compounded reflections that occur through the foam structure. However, we also see occasional peaks in the voltage. The peak labelled (ii) arises from the presence of a trapped air pocket, as shown in the *C*-scan in figure 5(a)(ii). (A similar bubble is responsible for the peak at $z$~52 mm.) The peak labelled (iii) arises from the inserted glass slide. The rectangular shape of the slide is clearly apparent in the corresponding *C*-scan [figure 5(a)(iii)]. Note that this image is contorted due to the distortion of the beam through the scaffold. Finally, the peak voltage increases at point (iv), due to reflections from the underlying substrate. This is further confirmed from the *C*-scan in figure 5(a)(iv): the substrate surfaces appears as an arc of high voltage due to a slight angle between the surface and the *x*-*y* plane. Additionally, an acoustic shadow is evident in line with the slide position. These results clearly

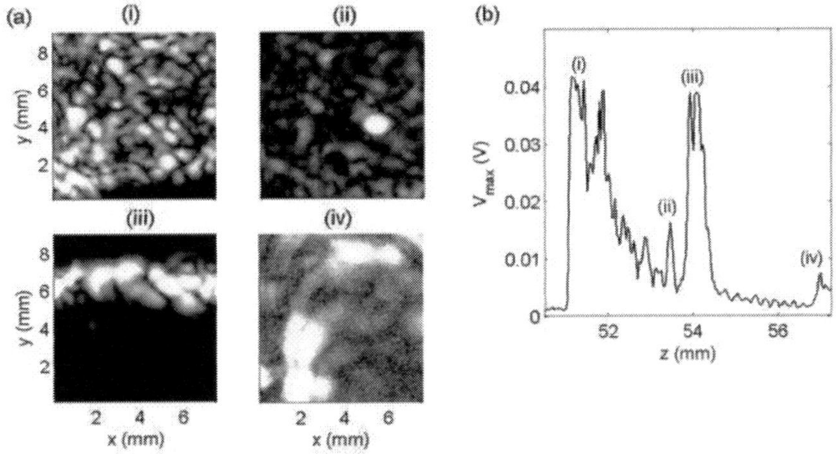


Figure 5: (a) C-scan images, taken using the 10 MHz transducer, at varying depths through the scaffold showing (i) the scaffold surface, (ii) a trapped air pocket, (iii) the embedded glass slide, and (iv) the underlying substrate (which appears as an arc of high intensity due to it being slightly tilted). The focal plane of the beam is aligned with the scaffold midpoint. (b) The maximum voltage in each *x*-*y* plane as a function of depth, with the depths corresponding to images (i-iv) indicated.

illustrate that acoustic propagation, and indeed imaging, is possible through several millimetres of hydrated foam tissue scaffold.

## 4. Conclusions

We have made a preliminary study into the acoustic properties and acoustic propagation in foam tissue scaffolds composed of poly(lactic-co-glycolic acid). Using a time-of-flight analysis we measured the longitudinal speed of sound in the bulk polymer over the temperature range 5-60°C. The marked change in speed of sound at glass transition of the polymer enables us to estimate the glass transition temperature to be 31.4°C. The acoustic sensitivity to the glass-rubber transition [32], as well as the sensitivity to biodegradation observed elsewhere [11], highlight the potential for acoustic characterisation of the bulk polymer properties. Furthermore, knowledge of these parameters is essential for the purposes of acoustically imaging and characterising such scaffolds.

Turning to the foam scaffolds themselves, we demonstrate that the scaffold can indeed be saturated sufficiently with water to support acoustic propagation. We employ an efficient protocol to achieving this involving pre-wetting with ethanol to overcome hydrophobicity and high power ultrasound to promote diffusion. Note that certain fabrication techniques feature a stage in which the scaffold is fluid-saturated, e.g. solvent casting/salt leaching and emulsification techniques. We then performed acoustic imaging of the scaffold surface and sub-surface. Our surface images demonstrate strong imaging contrast of the polymer and the ability to resolve the surface and pore morphology. Meanwhile, our imaging of an object buried several millimetres into the scaffold demonstrates the ability for acoustic waves to propagate through the scaffolds and return spatial information. The possibilities for acoustic scaffold characterization will be pursued in future work. Finally it should be noted that acoustic propagation through foam scaffolds may aid in the functional imaging of tissue development during cultivation. Conventional approaches based on optics fail in thick samples due to the highly scattering nature of both the scaffolds and the growing tissue. The much weaker scattering experienced by acoustic waves may be exploited to enable deep functional imaging through the use of ultrasound contrast agents or the hybrid techniques of ultrasound-modulated optical propagation and photo acoustics.

**Acknowledgements**

We thank Dr Lisa White (Tissue Engineering Group, School of Pharmacy, University of Nottingham) for providing the tissue scaffolds used in this work, and the Biotechnology and Biology Science Research Council for funding (BBSRC ref: BB/F004923/1).

## References

[1] Langer R and Vacanti J P 1993 *Science* **260** 920-6
[2] Teebken O E, et al. 2007 *Zentralbl Chir* **132** 236-46
[3] Zdrahala R J and Zdrahala I J 1999 *J. Biomater. Appl.* **14** 192-209
[4] Khademhosseini A, et al. 2009 *Sci. Am.* **300** 64-71
[5] Li H, et al. 2007 *Biomaterials* **28** 4098-104
[6] Li H, et al. 2009 *Biomicrofluidics* **3** 034102-11
[7] Pavey K D, et al. 2003 *J. Mater. Chem.* **13** 654-6
[8] Moinnes J J, et al. 2006 *Conf. Proc. IEEE Eng. Med. Biol. Soc.* **1** 484-8
[9] Wang X, et al. 2006 *Biomaterials* **27** 1924-9
[10] Mather M L, et al. 2008 *J. Mater. Sci.-Mater. Med.* **19** 3071-80
[11] Wu H-C, et al. 2003 *Biomaterials* **24** 3871-6
[12] Mather M L, et al. 2008 *Biomed. Mater.* **3** 11
[13] van Lenthe G H, et al. 2007 *Biomaterials* **28** 2479-90
[14] Li S H, et al. 2003 Bioceramics 15 (Zurich-Uetikon: Trans Tech Publications Ltd) pp 541-5
[15] Mallidi S, et al. 2006 *Medical imaging 2006: Ultrasonic imaging and signal processing*, ed S Y Emelianov and W F Walker (Bellingham: Spie-Int Soc Optical Engineering) p 1470

[16] Richards M S, et al. 2008 *IEEE Ultrasonics Symposium* (New York: IEEE) pp 536-9
[17] Müllner H W, et al. 2008 *Strain* **44** 153-63
[18] Lin L, et al. 2009 *J. Acoust. Soc. Am.* **125** EL123-EL8
[19] Hughes E R, et al. 2003 *Ultrasonics* **41** 365-8
[20] Haire T J and Langton C M 1999 *Bone* **24** 291-5
[21] Quirk R A, et al. 2004 *Curr. Opin. Solid State Mat. Sci.* **8** 313-21
[22] Tai H Y, et al. 2007 *Eur. Cells Mater.* **14** 64-76
[23] Lu L, et al. 2000 *Biomaterials* **21** 1595-605
[24] Parker N G, et al. *Measur. Sci. Technol.* **21** 045901
[25] Bilaniuk N and Wong G S K 1993 *J. Acoust. Soc. Am.* **93** 1609-12
[26] Bilaniuk N and Wong G S K 1996 *J. Acoust. Soc. Am.* **99** 3257
[27] Mikos A G, et al. 1994 *Biomaterials* **15** 55-8
[28] Ahmed A R, et al. 2008 *Eur. J. Pharm. Biopharm.* **70** 765-9
[29] Bordelius N A, et al. 1973 *Mech. Compos. Mater.* **9** 660-2
[30] Holmes M J, et al. 2010 *J. Phys. Conf. Ser.* (submitted)
[31] Park P I P and Jonnalagadda S 2006 *J. Appl. Polym. Sci.* **100** 1983-7
[32] Parker N G, Mather M L, Morgan S P and Povey M J W, to appear in *Biomed. Mater.*

Journal of Physics: Conference Series **269** (2011) 012020 doi:10.1088/1742-6596/269/1/012020

# Modeling and experimental investigations of Lamb waves focusing in anisotropic plates

**Bastien Chapuis[1], Nicolas Terrien[2], Daniel Royer[3]**

[1] Département Matériaux et Structures Composites, ONERA, 29 avenue de la Division Leclerc, 92322 Châtillon Cedex, France

[2] CETIM, 74 route de la Jonelière, 44326 Nantes Cedex 3, France

[3] Laboratoire Ondes et Acoustique, ESPCI, Université Paris 7, CNRS UMR 7587, 10 rue Vauquelin, 75231 Paris Cedex 05, France

Bastien.Chapuis@onera.fr

**Abstract**. The phenomenon of Lamb waves focusing in anisotropic plates is theoretically and experimentally investigated. An analysis based on a far field approximation of the Green's function shows that Lamb waves focusing is analog to the phonon focusing effect. In highly anisotropic structures like composite plates the focusing of $A_0$ and $S_0$ mode is strong; the energy propagates preferentially in the fibre directions, which are minima of the slowness. This has to be taken into account when developing, for example, a transducer array for structural health monitoring systems based on Lamb waves in order to avoid dead zones.

## 1. Introduction

Over recent years, fiber-reinforced composites have been increasingly used in industrial structures, especially in wind turbines and aircrafts. This has created the need for reliable and economic nondestructive evaluation (NDE) methods for determining the integrity and serviceability of these composite structures. In that field, elastic guided waves are of a great interest since they can propagate over a large distance [1]. Therefore, only a limited number of sensors can monitor wide zones. Structural Health Monitoring (SHM) is a strategy which consists in integrating these sensors permanently in the structure so as to detect damages as frequently as necessary [2]. Thin piezoelectric transducers (PZT) bonded to the surface or embedded between composite plies are generally used as actuators and sensors of Lamb waves for SHM of plate-like structures. A sensor array is built so that the set of acoustic paths between the PZT covers the whole surface to be inspected [3]. However, the propagation of guided waves in these multilayered and anisotropic materials is complex. We will see that the anisotropy induces a focusing of Lamb waves in some directions which can be very strong in composite materials. A precise quantification of this phenomenon is therefore useful in order to avoid any dead zone in the transducer array.

In the first part of this article, a far-field approximation of the 3D Green's function of a multilayered anisotropic plate is introduced. This formulation is based on the analysis developed by Velichko and Wilcox in [4]. As it occurs for bulk waves [5], Rayleigh waves [6], $SH_0$ modes [7] or plate modes [8], the focusing phenomenon is related to the Maris factor. Numerical computations of this factor are provided for a cross-ply composite plate at frequencies typically used in structural health monitoring systems.

In the second part, the field radiated by a PZT disc bonded at the surface of a composite plate is studied. This source of diameter $D$ creates an axisymmetric distribution of radial force $f$. It has been demonstrated [9] that a center of compression is equivalent to a set of three orthogonal dipole of force. In the 2D case studied, the PZT disc is assumed to be equivalent to a set of two orthogonal dipole of strength $fD$. Experiments show that Maris factor provides a good approximation of the far field radiated by the PZT disc for $S_0$ mode and is the dominant term for $A_0$ mode at the frequencies considered.

## 2. Lamb waves focusing in an anisotropic plate

The determination of the far-field approximation of the 3D Green's function of a multilayered anisotropic plate is fully described in [4], in which a strategy for computing this function is also presented. Only the major steps are recalled here.

### 2.1. Far-field approximation of the 3D Green's function

A multilayered anisotropic plate with Cartesian coordinates ($x_1$, $x_2$, $x_3$) is considered with the axis $x_3$ normal to the surface. As shown in Figure 1, the acoustic source is modeled by a force distribution acting on the upper surface of the plate, at $x_3 = 0$. Using the Fourier transform method to solve the wave equation, the displacement field **u** produced by a time-harmonic force $\mathbf{f}(x_1,x_2)e^{-i\omega t}$ is related to the 3D Green's function of the system $\mathbf{g}(x_1, x_2, x_3)$ by:

$$\mathbf{u}(x_1, x_2, x_3) = \iint \mathbf{g}(x_1 - \tilde{x}_1, x_2 - \tilde{x}_2, x_3) \times \mathbf{f}(x_1 - \tilde{x}_1, x_2 - \tilde{x}_2) d\tilde{x}_1 d\tilde{x}_2 \quad (1)$$

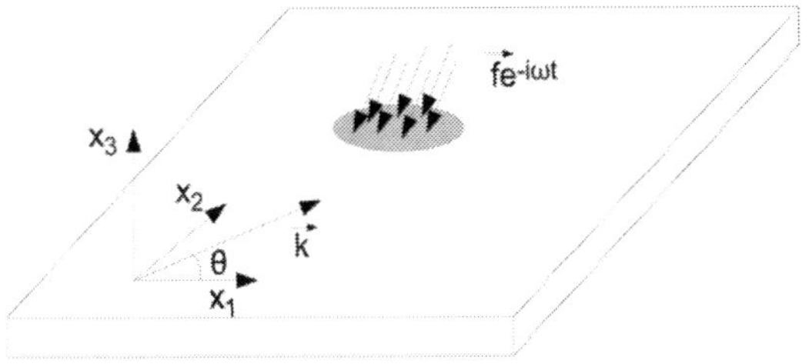


**Figure 1.** System geometry

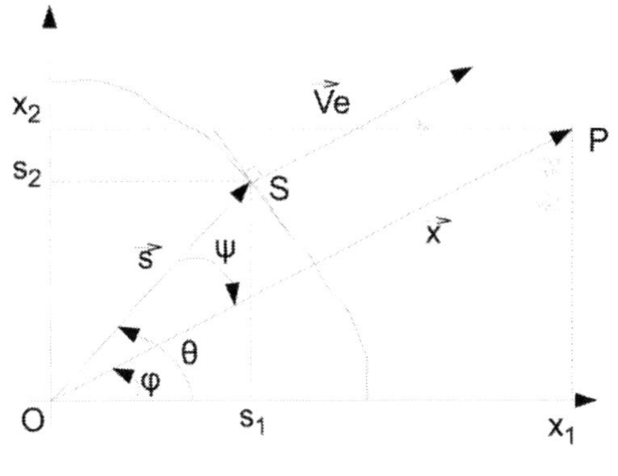


**Figure 2.** Illustration of the stationary phase approximation

A 2D spatial transform of this function gives:

$$\mathbf{g}(x_1, x_2, x_3) = \frac{1}{(2\pi)^2} \iint \mathbf{G}(k_1, k_2, x_3) e^{i(k_1 x_1 + k_2 x_2)} dk_1 dk_2 , \quad (2)$$

where $\mathbf{G}(k_1, k_2, x_3)$ stands for the Green's function for plane waves propagating in the direction given by the wave vector $\mathbf{k}(k_1 = k\cos\theta,\ k_2 = k\sin\theta,\ 0)$, with $\theta$ the angle between the direction of the phase velocity and the $x_1$ axis.

In the following, polar coordinates are used both for displacements and forces. For a given source/receiver configuration defined by the distance $r$ = OP and the angle $\varphi$, $x_1 = r\cos\varphi$, $x_2 = r\sin\varphi$, $x_3 = z$. The Green's function can be written as:

$$\mathbf{g}(r,\varphi,z)=\frac{1}{(2\pi)^2}\int_{\varphi-\frac{\pi}{2}}^{\varphi+\frac{\pi}{2}}\int_{\Gamma\cup-\Gamma}\mathbf{G}(k\cos\theta,k\sin\theta,z)e^{ikr\cos\psi}k\,dk\,d\theta \tag{3}$$

where $\Gamma$ is the real positive half axis and $\psi=\varphi-\theta$.

The residue theorem is then used to evaluate the far field contribution as $r\rightarrow\infty$. The contribution of complex poles in the far field is neglected as these modes are localized near the source and cannot propagate.

The Green's function for propagating modes $M$ is so given by:

$$\mathbf{g}(r,\varphi,z)=\frac{1}{2\pi}\sum_{\mathbf{M}}\int_{\varphi-\frac{\pi}{2}}^{\varphi+\frac{\pi}{2}}\mathbf{G}_{\mathbf{M}}(\theta,z)\left|k_M(\theta)\right|e^{ik_M r\cos\psi_M}d\theta \tag{4}$$

$$\mathbf{G}_{\mathbf{M}}(\theta,z)=i\,\mathrm{res}\left[\mathbf{G}(k\cos\varphi,k\sin\varphi,z)\right]_{k=k_M(\theta)}.$$

However, in the far-field, for each Lamb mode $M$ and for a given direction of observation $\varphi$, only angles $\theta$ for which the phase $\Phi_M(\theta)=k_M r\cos\psi$ of the oscillatory term varies slowly give a significant contribution. This is an application of the stationary phase method. The first derivative of the phase $\Phi_M$ vanishes for values $\theta_M$ such that:

$$\frac{\mathrm{d}k_M}{\mathrm{d}\theta}=-k_M\tan\psi_M,\quad \text{with}\quad \psi=\varphi-\theta. \tag{5}$$

As shown in Figure 2, equation (5) indicates that the propagation direction for which the phase is stationary is such that the observation direction is parallel to the energy velocity $\mathbf{V}^{\mathrm{e}}$, i.e normal to the slowness curve $\mathbf{s}(\theta)=\mathbf{k}/\omega$. In the following, we assume that there will be only one stationary phase point. In this case, the slowness curve is convex for all $\theta$ and equation (5) defines a one-to-one relation between angles $\varphi$ and $\theta$. The application of the stationary phase method gives the following far-field approximation of the 3D Green's function [10]:

$$\mathbf{g}(r,\varphi,z)=\frac{1}{\sqrt{2\pi r}}\sum_M\mathbf{G}_M(\theta_M,z)e^{i[\Phi_M+sign(\Phi''_M)\pi/4]}\sqrt{\left|A_M k_M\cos\psi_M\right|}, \tag{6}$$

where the factor $A_M$ is defined for each Lamb mode $M$ by the relation:

$$A=\frac{\mathrm{d}\theta}{\mathrm{d}\varphi}. \tag{7}$$

The factor A measures the acoustic ray density anisotropy shown in Figure 3. It is called Maris factor since it has been postulated by Maris to explain the phonon focusing effect [11]. This factor is related to the slowness curve of the mode plotted at the angular frequency ω by [8]:

$$A=\left[s^2+(\mathrm{d}s/\mathrm{d}\theta)^2\right]^{-1/2}\left|K_s\right|^{-1} \tag{8}$$

where $K_s$ is the curvature of the slowness curve. Equation 8 gives a geometrical interpretation of the limitation of the approach developed here. When $K_s\rightarrow 0$, i.e. at transition points for which the slowness curve changes from a convex shape to a concave one, Maris factor presents singularities. This is due to the presence of caustics, for which the second derivative of the phase in equation (4) also vanishes.

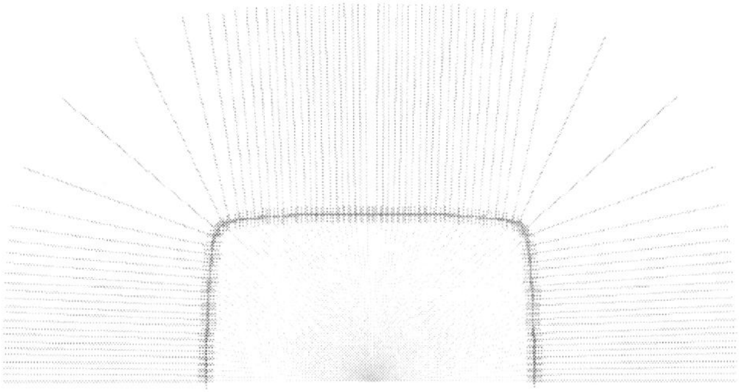

**Figure 3.** Anisotropy of the ray density; green: slowness vectors $s$, blue: slowness curve, red: energy vectors $V_e$

Maris factor has already been used to describe the directivity pattern of bulk [5] and Rayleigh waves [6]. It has also been highlighted for plate modes [7] and $SH_0$ mode [8] at low frequency-thickness product. The formulation given in this paper shows that this factor plays a role for Lamb waves too.

As shown in equation (6) many terms are responsible for the angular dependence of the displacement. However the focusing factor $\sqrt{A}$ (which is the relevant quantity for the displacement) is generally found to dominate the others in highly anisotropic materials [5]. The term $\mathbf{G}_M(\theta_M, x_3)$ is related to the way that Lamb waves are excited and detected. It can be calculated by the procedure given in [4].

The description given here is for a monochromatic wave, the only dispersion involved in the problem is the angular dispersion of Lamb waves. In this case and as noted in [12], it is the direction of the group velocity that is important and not its modulus. However, to fully describe the focusing phenomenon, a study at different frequencies is necessary. The frequency dispersion is expected to soften the phenomenon [13] since the focusing patterns of the spectral contributions in a wave packet are different.

*2.2. Computation of the focusing factor in composite plates*

The computation of the Maris factor (or of the focusing factor) is easily derived from the dispersion curves. In this paper the dispersion curves have been determined using the SAFE method [14]. The plate considered here is a $[0°/90°]_S$-T700GC/M21 cross-ply composite fiber reinforced polymer (CFRP) whose characteristics are given in Table 1. Dispersion curves show that below 500 kHz and in any direction of the plate only three modes can propagate: the two fundamental Lamb modes $A_0$ and $S_0$ and the first shear horizontal mode $SH_0$. However it should be noted that, except for 0° and 90° directions, the polarizations of $S_0$ and $SH_0$ modes are never purely radial nor orthoradial. In the following $S_0$ denotes the quasi-longitudinal mode and $SH_0$ the quasi-shear mode. In the next part we will show that $S_0$ and $SH_0$ modes are in fact coupled in some direction by the axisymmetric source used in the experiments (see also [4] for the case of an out-of-plane source).

**Table 1.** Parameters of the CFRP T700GC/M21 (Hexcel Composites).

| $C_{11}$ (GPa) | $C_{22} = C_{33}$ (GPa) | $C_{12} = C_{13}$ (GPa) | $C_{23}$ (GPa) | $C_{44}$ (GPa) | $C_{55} = C_{66}$ (GPa) | Mass density (kg/m$^3$) | Ply thickness (mm) |
|---|---|---|---|---|---|---|---|
| 123.44 | 11.54 | 5.55 | 6.35 | 2.6 | 4.5 | $1.6\times10^3$ | 0.25 |

Slowness curves and associated focusing factors for $A_0$ mode at different frequencies are plotted in Figure 4. At 40 kHz, the focusing factor is maximal in the direction 0°, but a second local maximum exists in the direction $\varphi = 90°$. The amplitude ratio between the direction $\varphi = 90°$ and $\varphi = 0°$ is 47%, the minimum occurs in the direction $\varphi = 50°$ where the ratio is only 23.4% compared to the direction $\varphi = 0°$. When the frequency increases, the shape of the focusing factor becomes more and more complex.

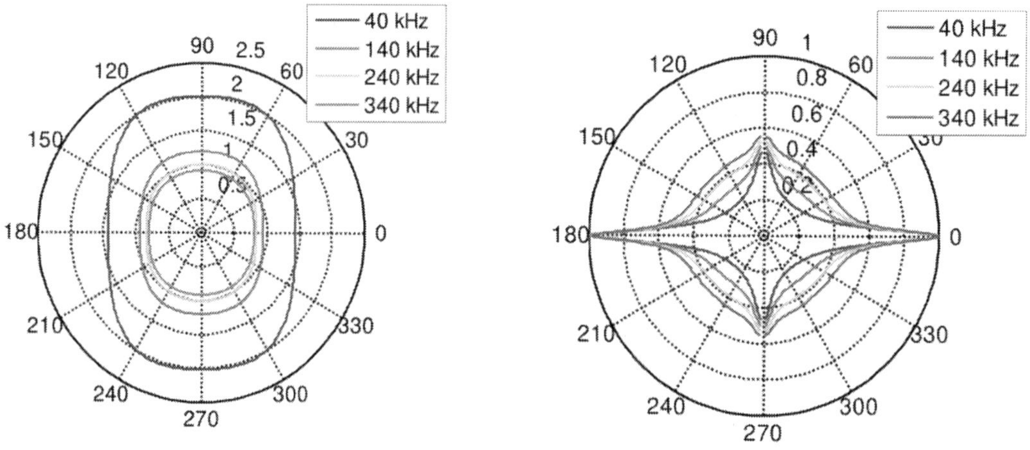


**Figure 4.** (a) Slowness curves in s.km$^{-1}$ in a cross-ply $[0°/90°]_S$ CFRP T700GC/M21 for the $A_0$ mode (b) Corresponding focusing factor $\sqrt{A}$.

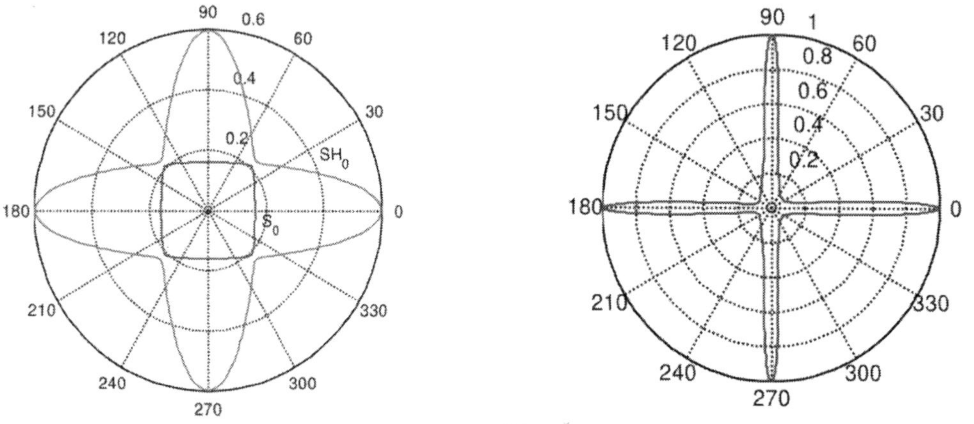


**Figure 5.** (a) Slowness curves in s.km−1 in a cross-ply $[0°/90°]_S$ CFRP T700GC/M21 for $S_0$ and $SH_0$ modes at 300 kHz (b) Focusing factor $\sqrt{A}$ for the $S_0$ mode.

Slowness curve and associated focusing factor for $S_0$ mode at 300 kHz are presented in Figure 5. Since this mode is non dispersive at low frequency-thickness products, the curves obtained for the frequencies below 300 kHz are superimposed. The focusing occurs at $\varphi = 0°$ and 90°, and the minimum is in the direction $\varphi = 45°$. The focusing is very acute in fiber direction since the ratio between the minimum and the maximum of the focusing factor $\sqrt{A}$ is 8%. In the frequency range [40 kHz-300 kHz], $S_0$ mode focusing is much more important than $A_0$ mode focusing.

Journal of Physics: Conference Series **269** (2011) 012020 doi:10.1088/1742-6596/269/1/012020

Slowness curve of $SH_0$ mode is also presented in Figure 5. This curve is an example of situation for which the slowness curve is not always convex. Due to the presence of two caustics at $\varphi = 7°$ and $\varphi = 83°$, the focusing factor becomes singular and the model previously developed is no more valid.

## 3. Acoustic field radiated by a PZT disc

The radiated field from a PZT disc is very different from one mode to the other. The effect of the anisotropy on the propagation of the guided waves has been observed with an air-coupled transducer mounted in a XY-translational stage, which permits to observe the wave front during the propagation in the plate. Figure 6 shows a cartography at a fixed time in the propagation of $S_0$ and $A_0$ modes excited by a thin PZT disc of diameter $D = 10$ mm bonded at the center of a $[0°/90°]_S$-T700GC/M21 cross-ply composite of thickness $d = 1$ mm. The PZT disc is excited with a three cycles tone burst of amplitude 10 $V_{PP}$ and a repetition rate of 10 ms, the detected signal is averaged 200 times at each measurement point. The PZT disc imposes an axisymmetric force at the surface of the plate along its periphery [15]. Lamb waves of wavelength $\lambda = 1.7\,D$, i.e. 17 mm, having a large radial displacement are preferentially launched [16]. Taking into account the dispersion curves, the wave number-thickness product $kd = 0.37$ corresponds to frequencies equal to 40 kHz and 300 kHz for $A_0$ and $S_0$ modes, respectively.

Due to the anisotropy of the material, the $S_0$ mode travels faster in the directions $\varphi = 0°$ and 90°. The $A_0$ mode is less sensitive to the anisotropy. However, for both modes an increase in the amplitude signal is detected at 0° and 90° (less visible for $A_0$ mode in the second case).

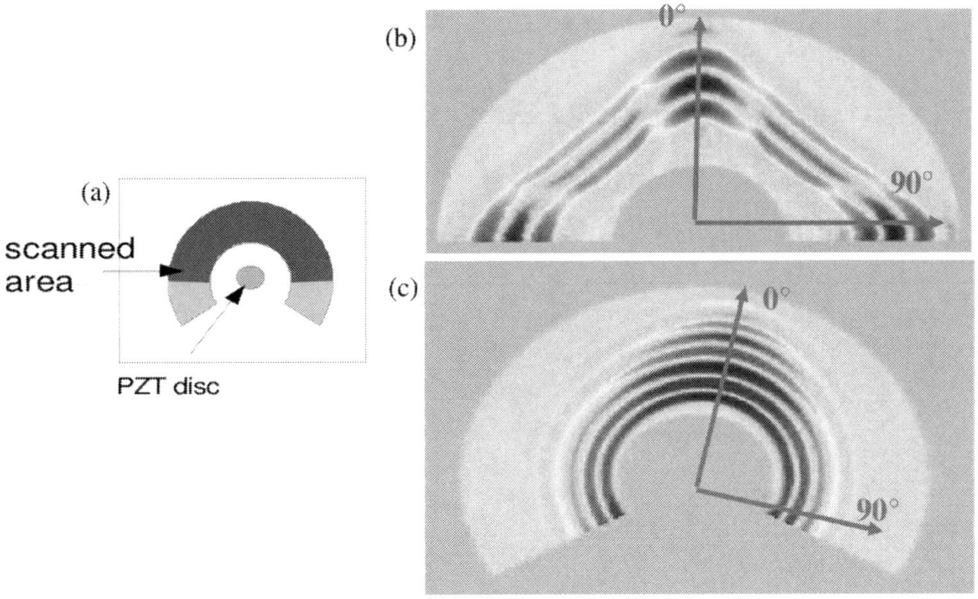


**Figure 6.** Cartography of (b) $S_0$ mode at 300 kHz and (c) $A_0$ mode at 40 kHz propagating in a cross-ply CFRP plate obtained by scanning the plate with an air-coupled transducer. Lamb waves are generated with a thin PZT disc bonded at the center of the plate.

Due to their respective polarization, $S_0$ and $SH_0$ modes are coupled and it is not possible to excite them separately. However, as discussed in [9], the generation by the PZT and the detection techniques used are much less efficient for $SH_0$ mode than for $S_0$ mode. Therefore, due to the shape of the slowness curves presented in Figure 5, the contribution of the shear mode should be significant only in the direction of the two caustics, for $\theta$=35° and $\theta$=55°, i.e. for observation directions $\varphi = 7°$ and $\varphi = 83°$.

Green's function represents the response of the plate to a point-like source. In practice, the lateral dimensions of the source must be smaller than the wavelength. With $D \cong \lambda/2$, this condition is nearly fulfilled and due to the symmetry of the source, the finite dimension of the PZT disc introduces only a frequency filtering. Therefore, the Green's function calculated in the first part is a good approximation of the far-field radiated by the PZT disc. To comply with our model hypothesis, the normal displacement at surface of the plate is measured at $r = 10$ cm from the center of the PZT disc. Since $kr \approx 40$, the far-field assumption is fulfilled. Since the Green's function is calculated at a given frequency, the spectral contribution at the desired frequency is extracted by a Fast Fourier Transform of the incident wave packet measured in each direction.

Results for $S_0$ and $SH_0$ modes at 300 kHz are presented in Figure 7 (a). The measurements have been achieved with an air-coupled transducer and also with a BMI heterodyne interferometer. In this case, amplitude fluctuations can be ascribed to the surface roughness of the plate. The air-coupled transducer is not affected by the roughness, but its spot size tends to enlarge the central lobe. The strong focusing in the direction 0° for $S_0$ mode is however clearly observed and fits well with the prediction. $\sqrt{A}$ is a very good approximation of the field radiated by the PZT disc. The possible contribution of $SH_0$ mode at $\varphi = \pm 7°$ is blurred in the central peak at 0° and consequently cannot be highlighted.

For the $A_0$ mode at 40 kHz, measurements have been performed with a BMI heterodyne interferometer, an air-coupled transducer and a Polytech velocimeter. Results are presented in Figure 7 (b). Small discrepancies exist between the three curves, however they clearly indicates that the focusing occurs mainly in the direction $\varphi = 0°$, and with less intensity in the 90°-direction. The agreement between theoretical and experimental curves is rather good. Conversely to $S_0$ mode case, $\sqrt{A}$ is not sufficient to fully describe the spectral displacement field radiated by the PZT disc for $A_0$ mode at 40 kHz. However, this factor is still the dominant term and indicates the directions of focusing.

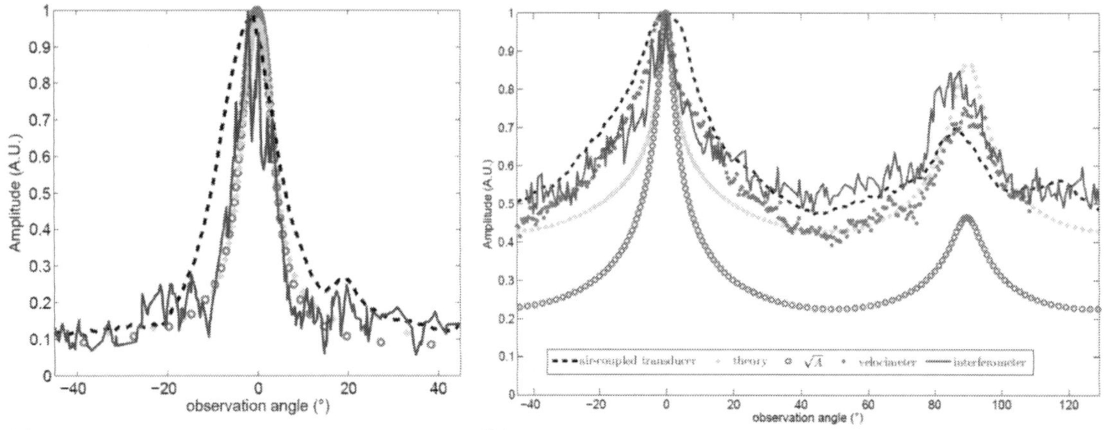


(a) (b)

**Figure 7.** (a) $S_0$ Lamb mode at 300 kHz. Normalized displacement measured at the surface of the cross-ply composite plate with an air-coupled transducer (dashed line) and an heterodyne interferometer (solid line) compared with the theoretical value (cross) and focusing factor $\sqrt{A}$ (circles). (b) $A_0$ Lamb mode at 40 kHz. Normalized displacement measured at the surface of the cross-ply composite plate with a velocimeter (dots), an air-coupled transducer (dashed line) and an heterodyne interferometer (solid line) compared with the theoretical value (cross) and focusing factor $\sqrt{A}$ (circles).

## 4. Conclusion

An approximation of the Green's function giving a general description of the generation of guided waves in a thin anisotropic plate was presented and the role of the Maris factor in the phenomenon of Lamb mode focusing was emphasized. The far-field radiation of PZT discs bonded at the surface of composite plates can be described by this Green's function. It was experimentally and numerically shown that in cross-ply composite plates, focusing occurs mainly in fiber directions and can be very strong, especially for $S_0$ mode. The investigation presented here provides useful information for the development of structural health monitoring systems based on PZT discs array in anisotropic structures.

## References

[1] Alleyne D and Cawley P 1992 *IEEE Trans. Ultrasonics, Ferroelec. and Freq. Control* **39** 381–397
[2] Boller C Structural Health Monitoring - An Introduction and Definitions (*Encyclopedia of Structural Health Monitoring vol 1)* eds C Boller, F-K Chang and Y Fujino, (Chichester: John Wiley & Sons Ltd), chapter 1 p 3–25
[3] Su Z, Ye L and Lu Y 2006 *J. Sound Vib.* **295** 753–780
[4] Velichko A and Wilcox P D 2007 *J. Acoust. Soc. Am.* **121** 60–69
[5] Royer D 2001 *Ultrasonics* **39** 345–354
[6] Maznev A, Lomonosov A, Hess P and Kolomenskii A 2003 *Eur. Phys. J. B* **35** 429–439
[7] Chen H-Y and Man C-S 1994 *Appl. Phys. Lett.* **64** 966–968
[8] Maznev A and Every A 1995 *Acta Acustica* **3** 387–391
[9] Achenbach J D 1980 *Wave Propagation in Elastic Solids* (Amsterdam: North-Holland) p 101
[10] Chapuis B, Terrien N and Royer D 2010 *J. Acoust. Soc. Am.* **127** 198–203
[11] Maris H 1971 *J. Acoust. Soc. Am.* **50** 812–818
[12] Potel C, Baly S, de Belleval J-F, Lowe M and Gatignol P 2005 *IEEE Trans. Ultrasonics, Ferroelec. and Freq. Control* **52** 987–1001
[13] Neau G 2003 PhD dissertation Université Bordeaux 1
[14] Liu G, Tani J, Ohyoshi T and Watanabe K 1991 *J. Vib. Acoust.* **113** 230–234
[15] Giurgiutiu V 2005 *J. Intell. Mater. Syst. Struct.* **16** 291–305
[16] Rose L R F and Wang C H 2004 *J. Acoust. Soc. Am.* **116** 154–171

Journal of Physics: Conference Series **269** (2011) 012021 doi:10.1088/1742-6596/269/1/012021

# Elastic wave diffraction by infinite wedges

**Larissa Fradkin[1], Arthur Gautesen[2], Victor Zernov[1] and Michel Darmon[3]**

[1] Sound Mathematics Ltd.,Cambridge CB4 2AS, UK

[2] Mathematics Department, Iowa State University and Ames Laboratory, USA

[3] CEA-LIST,CEA-Saclay, 91191 Gif-sur-Yvette, France

E-mail: `l.fradkin@soundmathematics.com`

**Abstract.** We compare two recently developed semi-analytical approaches to the classical problem of diffraction by an elastic two dimensional wedge, one based on the reciprocity principle and Fourier Transform and another, on the representations of the elastodynamic potentials in the form of Sommerfeld Integrals. At present, in their common region of validity, the approaches are complementary, one working better than the other at some isolated angles of incidence.

## 1. Introduction

In the nuclear industry one often needs to characterise a smooth planar defect at or near the back-wall of the component. This is easily detected when the defect is normal to the wall because of a specular "cat's-eye" reflection. However, the defect may still be detectable even if it is tilted, through diffracted edge waves from the crack tips. If the defect breaks the back-wall, one of these edge waves comes from the corner or "wedge" formed by the defect and back-wall (Fig. 1). The inspection configuration is normally two dimensional. In order to model it using GTD (Geometrical Theory of Diffraction), it is necessary to know the diffraction coefficients for both the embedded crack tip and the surface-breaking corner. The diffraction coefficients for the embedded crack have been described in [1]. As far as the wedge diffraction coefficients are concerned, for horizontally polarised transversal (TH) waves, this is an acoustic problem for which an analytic solution has been available for a long time [2]. For longitudinal (L) and vertically polarised transverse (TV) waves the problem is elastic and reliable codes have been developed only recently [17] - [18]. It appears that a purely analytical solution is impossible, and instead there have been three major semi-analytical approaches developed so far:

1. In the first approach, the displacement is represented in the integral form form using a form of Green's theorem, otherwise known as the reciprocity principle—the superposition of the

fields radiated by imaginary point sources situated on the faces of the wedge. Their FT (Fourier Transforms) satisfy certain analyticity requirements and the functional equations that are reformulated as Fredholm integral equations of the second kind. The latter are solved numerically (see the above references by Gautesen, Fujii, and Croisille and Lebeau).

2. In the second approach, each elastodynamic potential is represented in the form of a SI (Sommerfeld Integral)—the superposition of plane waves propagating in all (including complex) directions. The amplitudes of the plane waves belong to a certain class of analytical functions and satisfy a system of functional equations that are reformulated as singular integral equations. Budaev and Bogy [7]–[10] have offered a numerical schedule for solving the problem, and for the incident Rayleigh wave they calculated the Rayleigh reflection and transmission coefficients. The schedule has been clarified by Kamotski *et al.* [13]. It has been extended to finding the diffraction coefficients in [18].
3. In the third approach, the functional equations are solved using the Padé approximants, without recourse to any integral equations [15].

The paper is organised as follows: in the next Section we present the mathematical statement of the problem. We then briefly compare and cross-validate the first approach which we call FT with the second approach which we call SI.

## 2. Problem Statement

Let us consider a two dimensional wedge made of an isotropic solid, with the wedge angle $\alpha$ which is less than $\pi$ and introduce the Cartesian wedge coordinate system based on the orthonormal basis vectors $\{\mathrm{e}_1, \mathrm{e}_2\}$, with the origin at the wedge tip and the axis $\mathrm{e}_1$ running along the lower edge. Then an arbitrary point in space $\mathbf{x}$ can be described (see Fig. 1) using the Cartesian coordinates $(x_1, x_2)$ or associated polar coordinates $(r, \theta)$,

$$(x_1, x_2) = r(\cos(\theta + \alpha/2), \sin(\theta + \alpha/2)), \quad -\pi - \alpha/2 < \theta < \pi - \alpha/2. \tag{1}$$

The polar angle is chosen in an unorthodox way to allow for an easy comparison with the results presented in [17]. Let the wedge be irradiated by an incident harmonic plane wave. This means that the incident field and all the resulting scattered fields contain the common factor $\exp(-\mathrm{i}2\pi f t)$, where $f$ denotes the frequency and $t$ time. For simplicity of presentation, we suppress this factor throughout.

If the incident wave has transverse or longitudinal motions, then the incident displacements are respectively,

$$\mathbf{u}^{inc}(\mathbf{x}) = \frac{1}{\mathrm{i}k_0}\nabla_{\perp}(\mathrm{e}^{-\mathrm{i}k_0\mathbf{p}^{inc}\cdot\mathbf{x}}\mathbf{e}_3), \tag{2}$$

$$\mathbf{u}^{inc}(\mathbf{x}) = \frac{1}{\mathrm{i}k_1}\nabla\mathrm{e}^{-\mathrm{i}k_1\mathbf{p}^{inc}\cdot\mathbf{x}}, \tag{3}$$

where $\nabla = (\partial_1, \partial_2)$, $\nabla_\perp = (\partial_2, -\partial_1)$, $\partial_i$ denotes the partial derivative with respect to $x_i$; the wave numbers are

$$k_j = \frac{2\pi f}{c_j} = \frac{2\pi}{\lambda_j}, \; j = 0, 1, \tag{4}$$

with $c_0$, $c_1$, $\lambda_0$ and $\lambda_1$ denoting the speeds and wave lengths of the transverse and longitudinal waves, respectively, and the symbol $i$ when neither a superscript nor a subscript always denotes $\sqrt{-1}$. The direction of incidence $-\mathbf{p}^{inc}$ is shown in Fig. 1 and the unit vector

$$\mathbf{p}^{inc} = (\cos(\theta^{inc} + \alpha/2), \sin(\theta^{inc} + \alpha/2)), \quad -\frac{\alpha}{2} < \theta^{inc} < \frac{\alpha}{2}. \tag{5}$$

Whenever necessary below, the subscript $n$ is used to denote the nature of motions on the incident wave, with $n = 0$ corresponding to transverse motions and $n = 1$ corresponding to longitudinal motions. The geometry of the problem and direction of incidence are shown in Fig. 1 The scattered field consists of the GE (geometrico-elastodynamic) reflected and maybe,

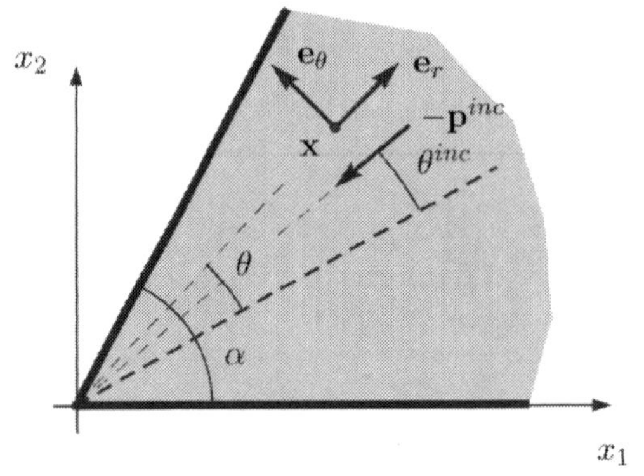


**Figure 1.** Geometry of the traction free elastic wedge.

multiple reflected waves as well as the field diffracted by the wedge tip. It satisfies the standard elastodynamic equations, the zero traction boundary conditions as well as the standard radiation conditions at infinity and the tip conditions of bounded energy. The mathematical description of these equations and conditions can be found in many books and papers, see e.g. [13].

The GE displacement is well known and its description is given in many sources, see e.g. [13] and [18]. According to the Geometrical Diffraction Theory, as $k_j r \to \infty$ the diffracted displacement can represented as:

$$H(\frac{\alpha}{2} - \theta)\mathbf{u}^{diff}(\mathbf{x}) \approx D_n^T(\theta; \theta^{inc}) \frac{e^{ik_0 r}}{\sqrt{r/\lambda_0}} \mathbf{e}_\theta + D_n^L(\theta; \theta^{inc}) \frac{e^{ik_1 r}}{\sqrt{r/\lambda_1}} \mathbf{e}_r, \quad k_j r \to \infty, \tag{6}$$

where $H$ is the Heaviside function and $\mathrm{e}_r$ and $\mathrm{e}_\theta$ are the unit vectors corresponding to the polar coordinate system $(r, \theta)$—see Fig. 1.

### 3. A comparative description of two semi-analytical approaches

In this paper we aim to compare two different semi-analytical approaches to the solution of the above wedge diffraction problem, one based on FT (Fourier Transforms) and reported in [18] and another, based on SI (the Sommerfeld Integrals) reported in [13]. Both methods involve going through eight stages:

1. The first stage of the FT approach involves combining the elastodynamic equation and boundary conditions to produce an integral equation known as a generalised reciprocity principle or else Extinction Theorem for displacements and then applying to it the operations of dilatation and rotation. The first stage of the SI approach involves re-writing the elastodynamic equation and boundary conditions in terms of the elastodynamic potentials $\psi_0(\mathbf{x})$ and $\psi_1(\mathbf{x})$, using the usual two-dimensional decomposition

$$\mathbf{u}(\mathbf{x}) = \nabla_{\perp}[\psi_0(\mathbf{x})\mathrm{e}_3] + \nabla\phi_1(\mathbf{x}). \tag{7}$$

   It is well known that the elastodynamic equation decouples into one Helmholtz equation for $\psi_0$ and another Helmholtz equation for $\psi_1$. In view of that, $\psi_0$ and $\psi_1$ are simply related to the rotation and dilatation of the displacement, respectively.

2. At the second stage, whatever the approach, one employs symmetrisation with respect to the wedge bisectrix. The symmetric and antisymmetric problems decouple and their solutions have to be added up to give the solution of the original problem.

3. As the next stage, the FT method requires application of the Fourier transform in $x_1$ to both the symmetric and anti-symmetric versions of the Extinction Theorem for elastodynamic potentials. The Fourier variable is called $\xi$. Starting with the Extinction Theorem rather than the reciprocity relationship, the FT method allows one to take the full Fourier transform on a line just below the $x_1$ axis. The new unknowns are the one sided Fourier transforms of the displacement components

$$\widehat{u}_i(\xi) = k_1 \int_0^{\infty} u_i(x_1, 0)\mathrm{e}^{ik_1\xi x_1}dx_1,\ i = 1,\ 2, \tag{8}$$

   which are analytic in the upper half of the complex $\xi$-plane. Let us describe their singularities: $\widehat{u}_j(\xi)$ possess branch points $-\kappa_0$ and $-1$ and can be rendered single-valued by a branch cut along the line $\{\xi : \mathrm{Re}\ (\xi) < -1,\ \mathrm{Im}(\xi) = 0^-\}$. Also, on the real axis, each $\widehat{u}_j(\xi)$ has a simple Rayleigh pole at $\xi = -\kappa_R$,where $\kappa_R = c_1/c_R$ and $c_R$ is the Rayleigh speed as well as two other poles, $\kappa_n p_1^{inc}$ and $\kappa_n p_1^{sym}$, where $\mathbf{p}^{sym} = (\cos(\alpha/2 - \theta^{inc}), \sin(\alpha/2 - \theta^{inc}))$. We remind the reader that the subscript $n$ denotes the nature of the motions on the incident wave, with $n = 0$ for shear and 1 for compressional. Given the above two poles, the functional equations can be used to "propagate" them, that is, evaluate successive reflection angles and find the relationships between their residues, which are the corresponding reflection coefficients. All the poles corresponding to the body waves have to be located on the real axis in such a way that when the inverse Fourier transform is taken, they are passed from above, so that their contributions satisfy the radiation condition and represent waves outgoing to infinity. The above branch cut as well as the Rayleigh and incident pole arising in the FT approach are all shown in Fig. 2, all moved down from the real axis by infinitesimal amounts.

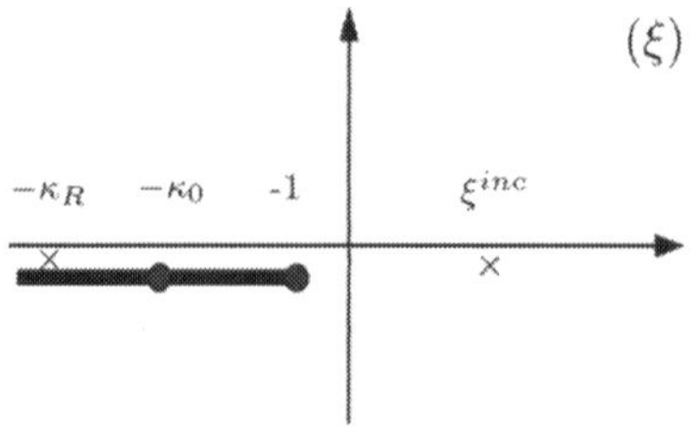


**Figure 2.** The main singularities in the FT method: the incident pole $\xi^{inc} = k_n p_1^{inc}$, the Rayleigh pole, the branch points and a branch cut. Poles corresponding to multiply reflected waves are not indicated.

In the SI the solutions of the Helmholtz equations are represented in the form of the Sommerfeld Integrals

$$\psi_{0,1}(kr,\theta) = \int\limits_{C\bigcup\tilde{C}} \Psi_{0,1}(\omega+\theta)\mathrm{e}^{-ik_{0,1}r\cos\omega}d\omega, \tag{9}$$

the contour C runs from $-\pi/2 + \mathrm{i}\infty$ to $3\pi/2 + \mathrm{i}\infty$; and the contour $\tilde{\mathrm{C}}$ is the reflection of C with respect to the origin (see in Fig. 3). The contour of integration $\mathrm{C}\bigcup\tilde{\mathrm{C}}$ is normally transformed into $C_1\bigcup C_2$, where $C_1$ and $C_2$ are the steepest descent paths indicated in Fig. 3. The singularities lying inside the region bounded by all four contours give rise to reflected and head waves. The integrals along the contours $C_1$ and $C_2$ describe the tip diffracted waves. Only the plane waves with the complex angles of propagation lying in the shaded regions satisfy the radiation conditions. The Sommerfeld Integrals are substituted into the boundary conditions. The new unknown are $\Psi_0(\omega)$ and $\Psi_1(\omega)$, the so-called called Sommerfeld amplitudes.

4. At the fourth stage, assuming that the tip condition and radiation conditions are satisfied, each approach leads to a system of functional equations in the unknown functions of one complex variable, the horizontal component $\xi$ of the wave vector and an angle $\omega$ it forms with the observation direction, respectively. The functional equations can be cast in different forms, with the arguments in different terms related as in the Snell's law. In the FT approach this means rotation of the wave vector through $\alpha$, so that its horizontal

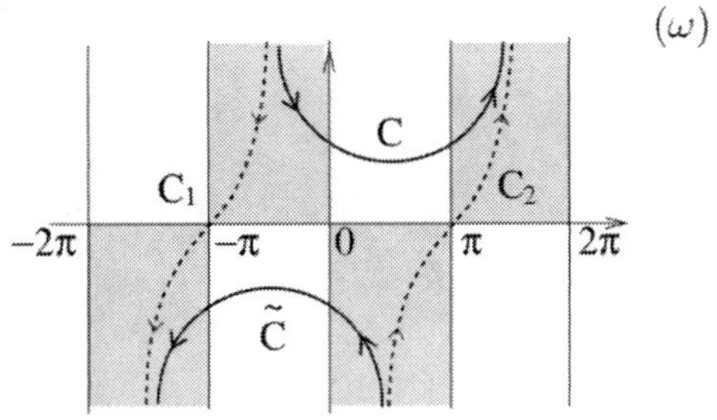


**Figure 3.** The contours used in the SI approach. The shaded areas represent physical regions.

component $\xi$ transforms to $\zeta_1 = \xi\cos\alpha + \sqrt{\kappa_0^2 - \xi^2}\sin\alpha$ for the transverse waves and to $\zeta_2 = \xi\cos\alpha + \sqrt{1 - \xi^2}\sin\alpha$ for the longitudinal waves. In the SI approach, when no mode conversion takes place the underlying transformation is a simple $\alpha$-shift in $\omega$. Otherwise, it is more complicated: prior to the $\alpha$-shift the angles are transformed from the longitudinal to transverse via a multi-valued function $g(\omega) = \cos^{-1}(\kappa_0^{-1}\cos\omega)$ and from transverse to longitudinal via its inverse $g^{-1}(\omega) = \cos^{-1}(\kappa_0\cos\omega)$. The branch cuts of $g(\omega)$ are chosen so that the deformed contour integration $\{\tilde{\omega} = g(\omega) : \omega \in C\bigcup\widetilde{C}\}$ may be transformed back to $C\bigcup\widetilde{C}$ without touching them. They are the segments

$$[-\theta_h + \pi m, \theta_h + \pi m], \quad \theta_h = \cos^{-1}\kappa_0^{-1}, \quad m \ - \ \text{integer}. \tag{10}$$

. The branch of $g(\omega)$ is chosen so that it has the properties

$$\begin{aligned} &g(\frac{\pi}{2}) = \frac{\pi}{2}, \\ &g(-\omega) = -g(\omega), \quad g(\omega + \pi m) = g(\omega) + \pi m, \\ &g(\omega) \simeq \omega - \mathrm{i}\ln\kappa_0^{-1} + O\Big(\mathrm{e}^{-2|\mathrm{Im}\,\omega|}\Big), \quad \text{as Im}\ \omega \to \infty. \end{aligned} \tag{11}$$

The branches of the function $g(\omega)$ lie on the vertical lines Re $\omega = m\pi$.

5. In both approaches, the fifth stage involves recasting the functional equations by inverting a matrix whose determinant is, or is proportional to, the Rayleigh function. In the FT approach, this is matrix $\{A_{ij}\}$ given by

$$A_{11}(\xi) = 1 - 2\xi^2, A_{12}(\xi) = -2\xi\sqrt{\kappa_0^2 - \xi^2}, A_{21}(\xi) = 2\xi\sqrt{1 - \xi^2}, A_{22}(\xi) = 1 - 2\xi^2,$$

where we use the notation $\kappa_0 = c_1/c_0$. In [13], it is the matrix $\{t_{ij}\}$,where

$$t_{11} = \cos 2\omega\ \sin\omega/\sqrt{\kappa_0^{-2} - \cos^2\omega},\ t_{12} = t_{21} = \sin 2\omega,\ t_{22} = -\cos 2\omega.$$

6. At the sixth stage it is assumed that the unknown functions can be decomposed so that the first (known) terms possesses poles corresponding to the incident, reflected and—provided they exist—multiply reflected body waves. The decomposition

$$\widehat{\mathbf{u}}(\xi) = \widehat{\mathbf{v}}^{ge}(\xi) + \widehat{\mathbf{v}}^{ge(E)}(\xi) + D_n^R\widehat{\mathbf{d}}^R(\xi) + D_n^{R(E)}\widehat{\mathbf{d}}^{R(E)}(\xi) + \widehat{\mathbf{v}}(\xi). \tag{12}$$

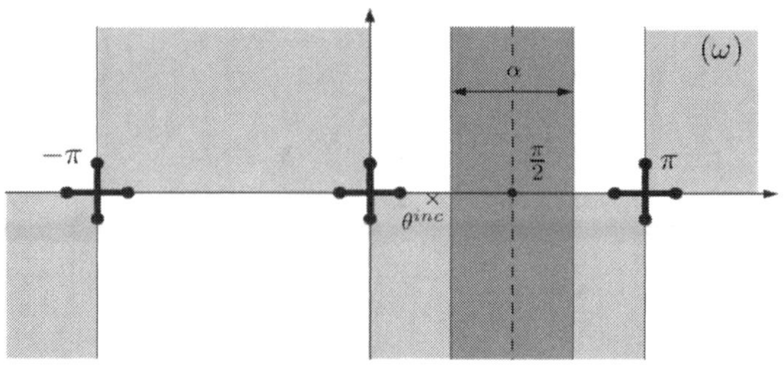


**Figure 4.** The main singularities in the SI method: the incident pole, the branch points, the horizontal branch cuts associated with the function $g(\omega)$ and the vertical branch cuts associated with the function $g^{-1}(\omega)$. The shaded areas indicate which side of a branch cut lies in the physical region. Poles corresponding to multiply reflected waves are not indicated. The dark shaded area is the strip of analyticity described in item 6.

used in the FT approach contains three unknown vector terms, two associated with the Rayleigh pole and one, $\widehat{\mathbf{v}}(\xi)$, which is analytic everywhere outside the branch cuts. Above, the superscript $E$ stands for extraneous poles that appear on the lower non-physical side of the branch cut and $R$ stands for the Rayleigh wave. Note that the unit vectors $\widehat{\mathbf{d}}^R$ and $\widehat{\mathbf{d}}^{R(E)}$ are known. In the SI approach the decomposition contains two types of terms,

$$\Psi_{0,1}^{(\ell)}(\omega) = \widehat{\Psi}_{0,1}^{(\ell)}(\omega) + \widetilde{\Psi}_{0,1}^{(\ell)}(\omega), \quad \ell = 1, 2. \tag{13}$$

Here and everywhere below $\ell = 1$ for the symmetric problem and $\ell = 2$ for the anti-symmetric one; the unknown terms $\widetilde{\Psi}_{0,1}^{(\ell)}(\omega)$ are regular in the strip $\pi/2 - \alpha/2 \leq \text{Re } \omega \leq \pi/2 + \alpha/2$, and the known terms $\widehat{\Psi}_{0,1}^{(\ell)}$ with poles at $\theta_m^{(\ell)}$ are

$$\widehat{\Psi}_{0,1}^{(\ell)}(\omega) = \sum_m \text{Res}(\Psi_{0,1}^{(\ell)}; \theta_m^{(\ell)})\sigma(\omega - \theta_m^{(\ell)}), \quad \frac{\pi}{2} - \frac{\alpha}{2} \leq \text{Re } \theta_m^{(\ell)} \leq \frac{\pi}{2} + \frac{\alpha}{2}. \tag{14}$$

Above, an otherwise arbitrary function $\sigma(\omega)$ should be chosen to be analytic everywhere inside the strip $\pi/2 - \alpha/2 \leq \text{Re } \omega \leq \pi/2 + \alpha/2$, except for a simple pole at zero, where it has the residue one. The position of the analyticity strip is indicated on Fig. 4.

7. At the seventh stage, the above decompositions are substituted into the respective original functional equations to arrive at the functional equations for the new unknowns with prescribed analytic properties.
8. At the eighth stage, the FT approach relies on the Hilbert transform

$$\widehat{\mathbf{v}}(\xi) = -\frac{1}{2\pi i} \int_1^\infty \frac{\mathbf{w}(\zeta)}{\zeta + \xi} d\zeta \tag{15}$$

where the vector unknown, $\mathbf{w}(\xi)$ is the jump of $\widehat{\mathbf{v}}(\xi)$ across the branch cut,

$$\mathbf{w}(\xi) = \widehat{\mathbf{v}}(-\xi + i0) - \widehat{\mathbf{v}}(-\xi - i0). \quad \xi > 1. \tag{16}$$

The SI approach relies on two further Hilbert type transforms $H$ and $\overline{H}$ ,

$$(\mathrm{H}f)(\omega) = \frac{1}{2\alpha\mathrm{i}} V.P. \int\limits_{\pi/2-\mathrm{i}\infty}^{\pi/2+\mathrm{i}\infty} \frac{f(\omega')\mathrm{d}\omega'}{\sin[\frac{\pi}{2\alpha}(\omega' - \omega)]}, \quad \mathrm{Re}\ \omega = \frac{\pi}{2} \tag{17}$$

and

$$\overline{H}f(\omega) = \frac{1}{2\alpha\mathrm{i}} V.P. \int\limits_{\pi/2-\mathrm{i}\infty}^{\pi/2+\mathrm{i}\infty} \frac{(f\omega')g'(\omega')\mathrm{d}\omega'}{\sin\{\frac{\pi}{2\alpha}[g(\omega') - g(\omega)]\}}, \quad \mathrm{Re}\ \omega = \frac{\pi}{2}, \tag{18}$$

(see Eqs. (3.21) and (3.22), [13]. Here $g'(\omega') = dg(\omega')/d\omega'$.

The above transforms allow us to reformulate each system of functional equations that appears in the FT approach as a system involving integral equations with a non-singular kernel and each system of functional equations that appears in the SI approach as an algebraic equation and integral equation with a singular kernel.

Depending on the method and decomposition used, apart from two unknown functions with prescribed analyticity, the equations for the symmetric and antisymmetric problems contain up to two unknown constants each. In the FT approach presented in [18] these are $D_n^{R(\ell)}$ and $D_n^{R(E)(\ell)}$ and in the form of SI approach described in [13] we have $c_1^{(\ell)}$. The $c_1^{(\ell)}$ constants appear as the result of application of the Sommerfeld Integral Nullification Theorem and the fact that the tip asymptotics of the displacement contain a non-zero constant term (see [13]). At the eight stage, in each approach the resulting integral equations are solved by employing solvability conditions that eliminate whatever constants are present. This is achieved by expressing them as integrals of unknown functions. In the FT approach described in [18] the solvability condition takes the form

$$\begin{aligned} D_n^{R(\ell)} &= \int\limits_1^\infty M^{R(\ell)}(z)\mathbf{w}(z)dz + H^{R(\ell)}, \\ D_n^{R(E)(\ell)} &= \int\limits_1^\infty M^{R(E)(\ell)}(z)\mathbf{w}(z)dz + H^{R(E)(\ell)}, \end{aligned} \tag{19}$$

where we use the notation

$$\begin{aligned} M_j^{R(\ell)}(z) &= [\bar{A}^{(\ell)}]_{11}^{-1} M_{1j}^{(\ell)}(-\kappa_R, z) + [\bar{A}^{(\ell)}]_{21}^{-1} M_{1j}^{(\ell)}(-\kappa_R - i0, z), \\ M_j^{R(E)(\ell)}(z) &= [\bar{A}^{(\ell)}]_{12}^{-1} M_{1j}^{(\ell)}(-\kappa_R, z) + [\bar{A}^{(\ell)}]_{22}^{-1} M_{1j}^{(\ell)}(-\kappa_R - i0, z), \\ H^{R(\ell)} &= [\bar{A}^{(\ell)}]_{11}^{-1} H_1^{(\ell)}(-\kappa_R, \widehat{\mathbf{u}}^{ge}(-\kappa_R)) + [\bar{A}^{(\ell)}]_{21}^{-1} H_1^{(\ell)}(-\kappa_R - i0, \widehat{\mathbf{u}}^{ge}(-\kappa_R - i0)), \\ H^{R(E)(\ell)} &= [\bar{A}^{(\ell)}]_{12}^{-1} H_1^{(\ell)}(-\kappa_R, \widehat{\mathbf{u}}^{ge}(-\kappa_R)) + [\bar{A}^{(\ell)}]_{22}^{-1} H_1(-\kappa_R - i0, \widehat{\mathbf{u}}^{ge}(-\kappa_R - i0)). \end{aligned} \tag{20}$$

Above, $[\bar{A}^{(\ell)}]^{-1}_{ij}$ denote the components of the inverse of the matrix $\bar{A^{(\ell)}}$ and matrices $\bar{A}^{(\ell)}$ and $M^{(\ell)}$ are

$$\begin{aligned}
\bar{A}^{(\ell)}_{11} &= \dot{R}(\kappa_R) - H_1^{(\ell)}(-\kappa_R, \widehat{\mathbf{d}}^R(-\kappa_R)),\\
\bar{A}^{(\ell)}_{12} &= -H_1^{(\ell)}(-\kappa_R, \widehat{\mathbf{d}}^{R(E)}(-\kappa_R)),\\
\bar{A}^{(\ell)}_{21} &= -H_1^{(\ell)}(-\kappa_R - i0, \widehat{\mathbf{d}}^R(-\kappa_R - i0)),\\
\bar{A}^{(\ell)}_{22} &= \dot{R}(\kappa_R) - H_1^{(\ell)}(-\kappa_R - i0, \widehat{\mathbf{d}}^{R(E)}(-\kappa_R - i0))\\
M^{(\ell)}_{11}(\xi, z) &= -\frac{1}{2\pi i}(-1)^{(\ell)}\left(-\frac{a(\xi)a(\zeta_1)}{\zeta_1 + z} + \frac{\bar{b}_2(\xi)b_1(\xi)}{\zeta_2 + z}\right),\\
M_{12}(\xi, z) &= -\frac{1}{2\pi i}(-1)^{(\ell)}\left(\frac{a(\xi)\bar{b}_1(\xi)}{\zeta_1 + z} + \frac{b_1(\xi)a(\zeta_2)}{\zeta_2 + z}\right),\\
M^{(\ell)}_{21}(\xi, z) &= -\frac{1}{2\pi i}(-1)^{(\ell)}\left(\frac{b_2(\xi)a(\zeta_1)}{\zeta_1 + z} + \frac{\bar{b}_2(\xi)a(\xi)}{\zeta_2 + z}\right),\\
M^{(\ell)}_{22}(\xi, z) &= -\frac{1}{2\pi i}(-1)^{(\ell)}\left(-\frac{\bar{b}_1(\xi)b_2(\xi)}{\zeta_1 + z} + \frac{a(\xi)a(\zeta_2)}{\zeta_2 + z}\right)
\end{aligned} \tag{21}$$

with $\dot{R}(\xi)$ - the derivative of $R(\xi)$ and

$$\begin{aligned}
&H_1^{(\ell)}(\xi, \mathbf{u}(\xi)) =\\
&(-1)^{(\ell)}\{a(\xi)[-a(\zeta_1)\widehat{u}_1(\zeta_1) + \bar{b}_1\widehat{u}_2(\zeta_1)] + b_1(\xi)[\bar{b}_2\widehat{u}_1(\zeta_2) + a(\zeta_2)\widehat{u}_2(\zeta_2)]\},\\
&\zeta_1 = \xi\cos\alpha + \sqrt{\kappa_0^2 - \xi^2}\sin\alpha,\ \zeta_2 = \xi\cos\alpha + \sqrt{1 - \xi^2}\sin\alpha,\\
&\eta_1 = \xi\sin\alpha - \sqrt{\kappa_0^2 - \xi^2}\cos\alpha,\ \eta_2 = \xi\sin\alpha - \sqrt{1 - \xi^2}\cos\alpha,\\
&\bar{b}_j(\xi) = 2\zeta_j\eta_j,\ j = 1, 2.
\end{aligned} \tag{22}$$

In the SI approach the solvability conditions are

$$\begin{aligned}
c_1^{(1)} &= \int_{-\infty}^{\infty}\left[(Ky^{(1)})(\eta) - q_0^{(1)}(\eta)\right] d\eta\left[\int_{-\infty}^{\infty} q_1^{(1)}(\eta)\, d\eta\right]^{-1},\\
c_1^{(2)} &= \int_{-\infty}^{\infty}\left[(Kx^{(2)})(\eta) - q_0^{(2)}(\eta)\right] d\eta\left[\int_{-\infty}^{\infty} q_1^{(2)}(\eta)\, d\eta\right]^{-1}
\end{aligned} \tag{23}$$

- see Eq. (4.4) in [13]. Above, the new unknown functions are

$$\begin{aligned}
x^{(\ell)}(\frac{\pi}{2} + i\eta) &= \widetilde{\Psi}_0^{(\ell)}[g(\frac{\pi}{2} + i\eta) + \frac{\alpha}{2}] + \widetilde{\Psi}_0^{(\ell)}[g(\frac{\pi}{2} + i\eta) - \frac{\alpha}{2}],\\
y^{(\ell)}(\frac{\pi}{2} + i\eta) &= \widetilde{\Psi}_1^{(\ell)}(\frac{\pi}{2} + \frac{\alpha}{2} + i\eta) + \widetilde{\Psi}_1^{(\ell)}(\frac{\pi}{2} - \frac{\alpha}{2} + i\eta)
\end{aligned} \tag{24}$$

and we use the notation

$$
\begin{aligned}
Kf(\eta) &= \frac{1}{\alpha \mathrm{i}} \int_{-\infty}^{\infty} \left\{ \frac{\tanh^2 2t}{\chi'(t)\sinh\frac{\pi}{\alpha}(t-\eta)} - \frac{\tanh 2t \tanh 2\eta}{\sinh\frac{\pi}{\alpha}[\chi(t)-\chi(\eta)]} \right\} f(t)\,dt,\\
q_0^{(1)}(\eta) &= r_2^{(1)}(\eta) - a'(\eta)\overline{\mathcal{H}} r_1^{(1)}(\eta),\\
q_1^{(1)}(\eta) &= -\tan\frac{\alpha}{2}\frac{\cosh\eta}{\chi'(\eta)\cosh 2\eta} - \frac{\tanh 2\eta}{\alpha\chi'(\eta)}\text{V.P.}\int_{-\infty}^{\infty} \frac{\cosh\tau\, d\tau}{\cosh 2\tau\, \sinh\frac{\pi}{\alpha}(\tau-\eta)},\\
q_0^{(2)}(\eta) &= r_1^{(2)}(\eta) - b'(\eta)\mathcal{H} r_2^{(2)}(\eta),\\
q_1^{(2)}(\eta) &= -\tan\frac{\alpha}{2}\frac{\cosh\eta}{\chi'(\eta)\cosh 2\eta} - \frac{\tanh 2\eta}{\alpha\chi'(\eta)}\text{V.P.}\int_{-\infty}^{\infty} \frac{\cosh\tau\, d\tau}{\cosh 2\tau\, \sinh\frac{\pi}{\alpha}(\tau-\eta)},
\end{aligned}
\tag{25}
$$

in which we have defined

$$
\begin{aligned}
a'(\eta) &= -\mathrm{i}\tanh 2\eta,\\
b'(\eta) &= \frac{2\mathrm{i}\sinh\eta\sqrt{\kappa_0^{-2}+\sinh^2\eta}}{\cosh 2\eta},\\
\mathcal{H}f(\eta) &= \frac{1}{\alpha\mathrm{i}}\text{V.P.}\int_{-\infty}^{\infty}\frac{f(t)\mathrm{d}t}{\sinh[\frac{\pi}{\alpha}(t-\eta)]},\\
\overline{\mathcal{H}}f(\eta) &= \frac{1}{\alpha\mathrm{i}}\text{V.P.}\int_{-\infty}^{\infty}\frac{f(t)\chi'(t)\mathrm{d}t}{\sinh\frac{\pi}{\alpha}[\chi(t)-\chi(\eta)]},\\
r_1^{(\ell)}(\eta) &= -\Big[\widehat{\Psi}_0^{(1)}(g(\tfrac{\pi}{2}+\mathrm{i}\eta)+\tfrac{\alpha}{2}) + (-1)^{\ell+1}\widehat{\Psi}_0^{(1)}(g(\tfrac{\pi}{2}+\mathrm{i}\eta)-\tfrac{\alpha}{2})\Big] -\\
&\quad b'(\eta)\Big[\widehat{\Psi}_1^{(1)}(\tfrac{\pi}{2}+\tfrac{\alpha}{2}+\mathrm{i}\eta) + (-1)^{\ell+1}\widehat{\Psi}_1^{(1)}(\tfrac{\pi}{2}-\tfrac{\alpha}{2}+\mathrm{i}\eta)\Big],\\
r_2^{(\ell)}(\eta) &= -a'(\eta)\Big[\widehat{\Psi}_0^{(1)}(g(\tfrac{\pi}{2}+\mathrm{i}\eta)+\tfrac{\alpha}{2}) + (-1)^{\ell}\widehat{\Psi}_0^{(1)}(g(\tfrac{\pi}{2}+\mathrm{i}\eta)-\tfrac{\alpha}{2})\Big] -\\
&\quad \Big[\widehat{\Psi}_1^{(1)}(\tfrac{\pi}{2}+\tfrac{\alpha}{2}+\mathrm{i}\eta) + (-1)^{\ell}\widehat{\Psi}_1^{(1)}(\tfrac{\pi}{2}-\tfrac{\alpha}{2}+\mathrm{i}\eta)\Big],\\
\chi(\eta) &= \sinh^{-1}(\kappa_0^{-1}\sinh(\eta)),\\
\chi'(\eta) &= \frac{\cosh\eta}{\sqrt{\kappa_0^{-2}+\sinh^2\eta}}.
\end{aligned}
$$

This summary shows that the use of the SI approach involves extra technical complications, such as various integral transforms and their inverses as well as the function $\cos^{-1}(\kappa_0^{-1}\cos\omega)$ with multiple cuts in its domain.

## 4. The testing and cross-validation of codes

As explained in [18], the FT code has been tested in two different ways:

Journal of Physics: Conference Series **269** (2011) 012021 doi:10.1088/1742-6596/269/1/012021

1. The diffraction coefficients defined by Eq. (6) have been computed for $\theta < -\alpha/2$ and $\theta > \alpha/2$. These values of $\theta$ lie outside the wedge. As no energy is diffracted into this region, the diffraction coefficients should vanish. We have found that as a rule, the magnitude of the coefficients for this range of $\theta$ varies from $10^{-8}$ for the larger wedge angles $\alpha$ to around $10^{-3}$ for the smaller wedge angles.
2. When calculating the backscatter diffraction coefficients $D_0^T(\theta;\theta)$ and $D_1^L(\theta;\theta)$, simultaneously the following reciprocity relationship

$$D_0^L(\theta;\theta) = \frac{k_1^2}{k_0^2} D_1^T(\theta;\theta), \tag{26}$$

has been checked. This has been done by calculating the reciprocity discrepancy

$$\Delta(\theta) = |D_0^L(\theta;\theta) - \frac{k_1^2}{k_0^2} D_1^T(\theta;\theta)|, \tag{27}$$

The SI code has been also tested in two ways:

1. Checking that the symmetry properties of the computed Sommerfeld amplitudes are the same as predicted for the corresponding theoretical amplitudes.
2. Using the same reciprocity principle as above.

Note that the above tests involve conditions that are necessary but not sufficient for the diffraction coefficients to be correct.

In addition to results presented in [17] and [18], Fig. 5 shows that in their common region of validity, the outputs of both codes usually coincide. It is argued in [18] that whenever experimental data are available, the computed results are well within the experimental error. Note that due to various recent improvements in the SI code, results in Fig. 5 are in an even better agreement with experimental data than reported in [17]. Note too that the results displayed in Figs. 5 have been obtained using 800 nodes to solve the integral equation in the FT code and 200 nodes to solve the singular integral equation in the SI code.

When the FT code fails the reciprocity test in its region of applicability, the situation can be remedied by changing the incident angle by a fraction of a degree. When the SI code fails the test in its region of applicability, the situation can be remedied by increasing the accuracy - either by allowing for "a safe and slow mode" in which all poles are treated in a careful manner or else by increasing the number of nodes required to solve the underlying singular integral equation or both. At present, the FT code has been optimised to run very fast (with 800 nodes for each particular couple of incident and diffracted angles the run time is under one minute on a modern Pentium), but no such optimisation has been performed in the SI code. The latter takes about a minute to solve the underlying singular integral equations when the number of nodes is 200 and but over 15 minutes when this number is 800.

Several issues with the FT code still remain unresolved:

- sometimes it fails for unknown reasons and the reciprocity test indicates the failure, e.g. when the $T$ wave is incident at 20° on the 80° wedge (and there exists a secondary diffracted $T$ wave that propagates at 180° along a wedge face) - see Fig. 5;
- sometimes it fails for unknown reasons but the reciprocity test does not indicate the failure, e.g. when the $T$ wave is incident at 10° on the 100° wedge (so that the incidence angle is close to critical).

The main issue with the SI code is that when the accuracy is increased it takes a long time to run. Also, when 800 nodes are used for wedge angles smaller than 45° the singular integral equations to be solved by the SI codes become ill-posed.

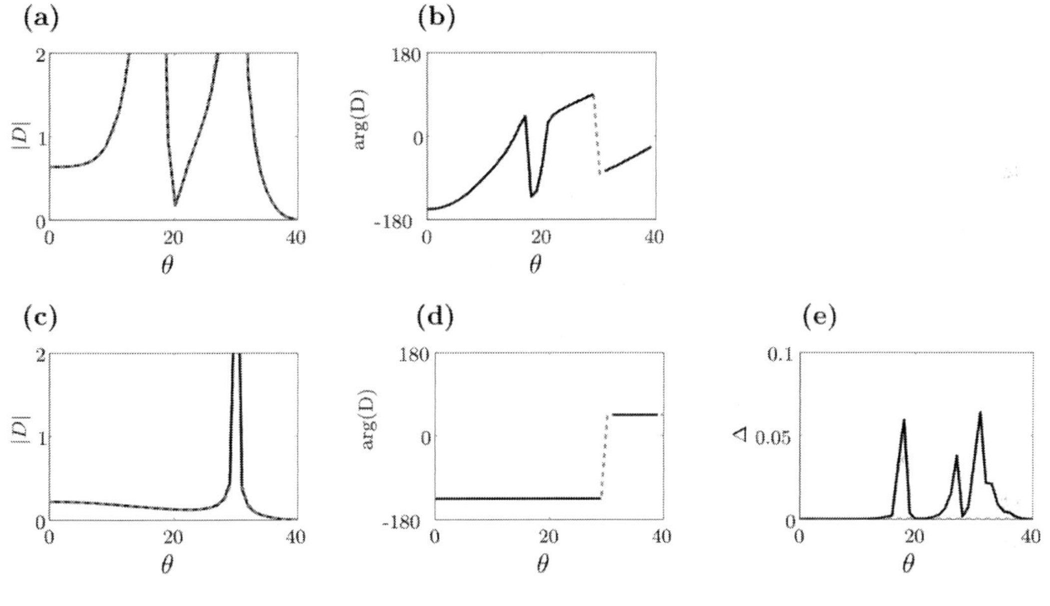


**Figure 5.** Comparison of corrected computations based on the FT approach (dashed line) and corrected computations based on the SI approach (solid line) for the wedge angle 80° and $\kappa_0 = 1.82222222157$. The magnitudes and phases of the backscatter diffraction coefficients $D_1^L$ - (a) and (b) and $D_0^T$ - (c) and (d), respectively. The reciprocity discrepancy for the corrected FT results (dashed line) and corrected SI results (solid line) - (e).

## 5. The uniform GTD

As is to be expected, the magnitudes of the diffraction coefficients blow up at the shadow boundaries of multiply reflected waves, where the standard (non-uniform) GTD is inapplicable. For any given scattered mode, longitudinal or transverse, the non-uniform asymptotics covering the penumbral region are

$$\mathbf{u}^{\text{non\_uniform}} = H\mathbf{C}_{\text{plane}} e^{iks_{\text{plane}}} + \mathbf{C}_{\text{diff}} \frac{e^{iks_{\text{diff}}}}{\sqrt{ks_{\text{diff}}}}, \tag{28}$$

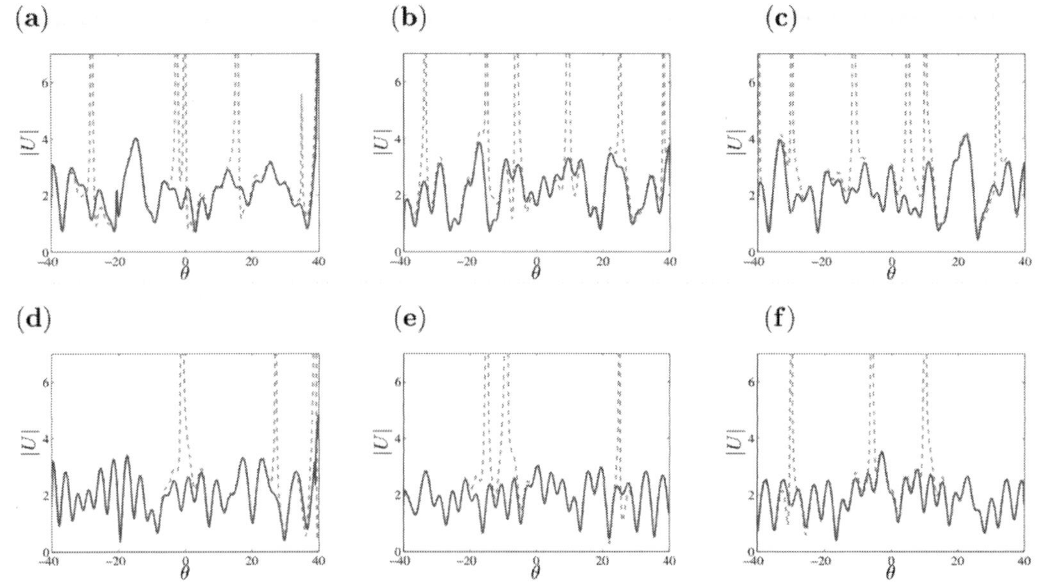


**Figure 6.** The amplitude of displacement (in dimensionless units) versus the observation angle $\theta$ (in degrees) for $80^0$ wedge. Solid line - uniform asymptotics, dashed line - non-uniform asymptotics. (a), (b) and (c) the incident wave - L, with $\theta^{inc} = -5^o$, $10^o$ and $25^o$; (d), (e) and (f) the incident wave - T, with $\theta^{inc} = -5^o$, $10^o$ and $25^o$, respectively.

where $H$ is the Heaviside function, 0 in the shadow and 1 in the illuminated regions and $k$ is the corresponding wave number. This expansion is applicable far from the shadow boundaries, i.e. when $|s_{\text{plane}} - s_{\text{diff}}| \gg 1$. As a shadow boundary is approached, $|s_{\text{plane}} - s_{\text{diff}}| \to 0$, the diffraction coefficient $\mathbf{C}_{diff}$ tends to infinity.

The uniform asymptotics of (28) have in the form

$$\mathbf{u}^{\text{uniform}} = \mathrm{F}\left(\sqrt{k|s_{\text{diff}} - s_{\text{plane}}|}\right)\mathbf{C}_{\text{plane}}\mathrm{e}^{iks_{\text{plane}}} + \widetilde{\mathbf{C}}_{\text{diff}}\mathrm{e}^{iks_{\text{diff}}}, \tag{29}$$

where F is the Fresnel integral

$$\mathrm{F}(x) = \frac{1}{\sqrt{\mathrm{i}\pi}}\int_{-\infty}^{x} \mathrm{e}^{\mathrm{i}\xi^2}d\xi, \tag{30}$$

if $s_{\text{diff}} > s_{\text{plane}}$. The sign of the root $\sqrt{k|s_{\text{diff}} - s_{\text{plane}}|}$ is taken positive in the illuminated region, and negative in the shadow region. The unknown coefficient $\widetilde{\mathbf{C}}_{diff}$ is regular in the vicinity of the shadow boundary.

The simplest way to find $\widetilde{\mathbf{C}}_{\text{diff}}$, is by using the method of matched expansions. As $|s_{\text{diff}} - s_{\text{plane}}| \gg 1$ we expand the Fresnel integral in (29) into the asymptotic series and compare the resulting expression with (28). The higher order terms cancel out and matching the second order terms gives

$$\widetilde{\mathbf{C}}_{\text{diff}} = \mathbf{C}_{\text{diff}}\frac{1}{\sqrt{ks_{\text{diff}}}} + \mathbf{C}_{\text{plane}}\frac{\mathrm{e}^{\mathrm{i}\pi/4}}{2\sqrt{\pi}\sqrt{k(s_{\text{diff}} - s_{\text{plane}})}}. \tag{31}$$

### *5.1. Testing subroutines for calculating the uniform asymptotics*

To test our code we studied the pulse train generated by the interaction of one cycle of sine pulse with an infinite wedge at the observation point which is inside a penumbra. All the arrivals have a bounded amplitude and do not blow up as they would if we used the non-uniform asymptotics (28) - see Fig. 6. This confirms the correct behaviour of the diffraction coefficients in the penumbrae, allowing for the mutual cancelation of the singularities in (31).

## Acknowledgements

Partial support of this work has been provided by CEA LIST within the CIVA2012 framework.

## References

[1] Achenbach J D, Gautesen A K and McMaken H 1982 *Ray Methods for Waves in Elastic Solids; With Applications to Scattering by Cracks* (New York: Pitman)
[2] Jones D S 1964 *The theory of electromagnetism* (Oxford: Pergamon Press) pp 608–612
[3] Gautesen A K 1985 *J. Appl. Mech.* **52** 664
[4] Gautesen A K 1998 *Wave Motion* **10**(5) 393
[5] Gautesen A K 2001 *ASME J. Appl. Mech.* **61** 476
[6] Gautesen A K 2002 *Wave Motion* **35** 91
[7] Budaev B V 1995 *Diffraction by Wedges* (Harlow, UK: Longman Scientific and Technical)
[8] Budaev B V and Bogy D B 1995 *Wave Motion* **22** 239
[9] Budaev B V and Bogy D B 1996 *Wave Motion* **24** 307
[10] Budaev B V and Bogy D B 2002 *Wave Motion* **35** 275
[11] Croisille J-P and Lebeau G 1999 *Diffraction by an Immersed Elastic Wedge* (Berlin: Springer-Verlag)
[12] Fujii K 1994 *Bull. Seismol. Soc. Am.*84 1916
[13] Kamotski V, Fradkin L, Samokish B, Borovikov V and Babich V 2006 *SIAM J. Appl. Maths.* **67**(1) 235
[14] Gautesen A K 2002 *Wave Motion* **36** 417
[15] Abrahams I D 2002 *Wave Motion* **36** 311
[16] Kamotski V V and Lebeau G 2006 *Proc. Roy. Soc. A* **462** 289
[17] Chapman R K, Pearce J, Burch S, Fradkin L and Toft M 2007 *Insight* 93
[18] Gautesen A K and Fradkin L Ju 2010 *SIAM J. Appl. Maths.* **70**(8) 3065

Journal of Physics: Conference Series **269** (2011) 012022 doi:10.1088/1742-6596/269/1/012022

# Simulation of ultrasonic inspection of composite using bulk waves: application to curved components

**S Journiac[1], N Leymarie[1], N Dominguez[2] and C Potel[3]**
[1]CEA, LIST, Département Imagerie et Simulation pour le Contrôle, F-91191 Gif-sur-Yvette, France.
[2]EADS France Innovation Works, ZAC de Saint Martin du Touch, 18 rue de Marius Terce, 31300 Toulouse, France.
[3]Université du Maine, Avenue Olivier Messiaen, 72085 Le Mans, France.

E-mail: sophie.journiac@cea.fr

**Abstract**. The increasing use of composite materials in aircraft, particularly carbon-fiber reinforced samples, presents new challenges for Non-Destructive Testing (NDT) operations to ensure reliable inspections. The aircraft industry is showing a growing interest in the simulation of NDT techniques applied to such materials, which exhibit strong anisotropy. Simulation tools are needed to predict beam deviations that may arise for curved or tilted parts in complex specimen. A main issue is the modelling of the wave propagation and flaw scattering inside a homogeneous medium considering a continuously variable anisotropic orientation. Previous works have presented a way to model these media. This method consists in decomposing them into a set of homogeneous sub-sections with constant orientations. This approximation allows us to use straight paths over homogeneous sections but the main drawback is that it introduces artificial boundaries leading to numerical instabilities. The new model deals with a continuously variable anisotropy. It is based on a ray calculation for an inhomogeneous medium using an iterative time-step method. This method consists in solving for each time step a system of differential equations taking into account the continuous anisotropic orientation variation along the trajectory.

## 1. Introduction

Composites are increasingly being used by the aerospace industry as a substitute for metals. About a quarter of the structure of the Airbus A380 is made up of lightweight carbon-fiber-reinforced plastic (CFRP). The geometry of structures being optimized, they become more and more complex, presenting sometimes tightly curved shapes. Especially concerning stiffeners, the radius dimension can be very small. Simulation tools are of great interest to optimize NDT techniques of such structures.

The CFRP structures are made of several unidirectional layers of carbon fibers embedded in epoxy matrix. These structures exhibit a strong anisotropy between the axis and the plane of the stack. Previous works concerning beam calculation in CFRP were already implemented in CIVA [1], [2], [3]. The approach is based on the search of equivalent energy direction through the stacking sequence of the layered structure. The equivalent mechanical properties of the effective medium are deduced from the equivalent energy direction. In the case of complex geometries, the only way to model beam propagation is to divide the component into a set of homogeneous sub-sections with a local disorientation following the geometry [3]. In the case of curved composites, this approximation can be used but leads to some numerical artefacts due to the presence of fictive interfaces. In this paper, we proposed to extend the ray-based model developed in CIVA [4] to calculate the beam propagation in curved composites without any sample discretization. The idea is to consider a local anisotropic orientation that follows continuously the curvature of the fibers.

First, we present the mathematical formulation used to determine local mechanical properties in CFRP curved composites. Then, the ray based theory adapted to smoothly inhomogeneous anisotropic material is introduced to predict beam field amplitude. Finally, validations and comparisons of beam calculations are performed for cylindrical samples with different radii of curvature.

## 2. Formulation of mechanical properties in a layered CFRP composite with a curved shape

Based on observations of layered curved composites, we considered that each ply follows the global curvature of the sample. Following this assumption, we proposed a general formulation of the mechanical anisotropic properties of a layered CFRP composite in a Cartesian coordinate system. The spatial dependency of the mechanical properties is deduced for curved shaped composites.

Let us denote by $C^{ref}$ the elastic stiffness tensor for a reference flat layered composite structure. In a Cartesian canonical basis $(\vec{e}_1,\vec{e}_2,\vec{e}_3)$, its formulation is:

$$C^{ref} = \sum_{ijkl} c_{ijkl}^{ref} \vec{e}_i \otimes \vec{e}_j \otimes \vec{e}_k \otimes \vec{e}_l \quad (1)$$

where $c_{ijkl}^{ref}$ represent the components of the 4-th order tensor $C^{ref}$ of a reference plate. Considering that the curved sample is obtained as the distortion of a plate into a curved surface, we define the transformation $\Phi$ from the flat to the curved structure (cf. Figure 1).

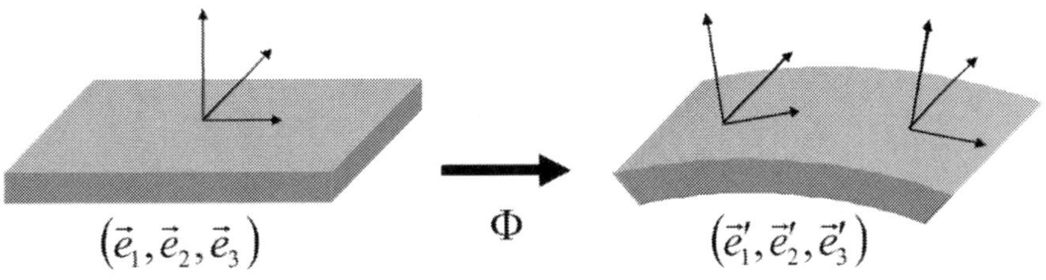


**Figure 1.** Illustration of the $\Phi$ transformation of a flat plate into a curved shell surface

We assume that the basis composed by the normal and tangents of the median surface remains orthonormal after transformation. Applied to a curved composite ply, the physical meaning of this transformation is that the local anisotropy axis is following the fiber curvature.

Considering that the stiffness tensor for the curved plate is conserved in the local orthonormal basis $(\vec{e}'_1,\vec{e}'_2,\vec{e}'_3)$ after transformation, and then we have:

$$C = \sum_{ijkl} c_{ijkl}^{ref} \vec{e}'_i \otimes \vec{e}'_j \otimes \vec{e}'_k \otimes \vec{e}'_l , \quad (2)$$

where the vectors $\vec{e}'_j$ of the local basis verify the relation $\vec{e}'_j \cdot \vec{e}_i = \Phi_{i,j} = J_{ij}$, matrix $J$ being the Jacobian matrix for transformation $\Phi$. Note that this matrix is normalized in order to neglect any dilatation effect due to the curvature. The elastic properties are also invariant with respect to the normal of the median plane before and after transformation. The matrix $J$ is then equivalent to a local transformation matrix.

Combining relations (1) and (2), we obtain a relationship between the elastic constants of the curved sample $c_{ijkl}$ and the elastic constants of the reference plate $c_{ijkl}^{ref}$ :

$$c_{ijkl} = \sum_{mnrs} c_{mnrs}^{ref} J_{im} J_{jn} J_{kr} J_{ls} \tag{3}$$

Finally, we obtain a spatially-dependent relationship of variables $c_{ijkl}$ as a function of the reference elastic constants (spatially invariant) and the Jacobian matrix deduced from the spatial transformation $\Phi$. So the mechanical properties evaluated for a curved shape correspond to a smoothly inhomogeneous anisotropic material since $\Phi$ is a regular function.

## 3. Pencil model in smoothly inhomogeneous anisotropic material

The pencil model is a ray-based model developed for homogeneous anisotropic material [4]. In this case, ray trajectories are straight lines. A pencil is a collection of rays emanating from the point source and diverging slightly. One can define the axial ray, along the geometrical path, and the paraxial rays, belonging to the envelope of the pencil. We proposed here to extend this model to smoothly inhomogeneous anisotropic media.

### 3.1. Ray trajectory evaluation

3.1.1. *Theoretical equation governing ray propagation.* Main theoretical results are presented in depth by V. Cerveny in his book "Seismic Ray Theory" [5] and [6]. V. Cerveny established a general model that describes ray propagation in smoothly inhomogeneous anisotropic media. These ray equations have been developed for seismic applications but they are still valid for ultrasound since wave path is long in comparison with wavelength.

Using the Eikonal equation and the Christoffel equation, a differential expression of the ray position $\vec{x}$ and its associated wave slowness $\vec{S}$ with the travel time of the ray $T$ is obtained:

$$\begin{cases} \dfrac{dx_i}{dT} = c_{ijkl} S_l P_j P_k = V_{e_i} \\ \dfrac{dS_i}{dT} = -\dfrac{1}{2} \dfrac{\partial\left(\rho^{-1} c_{jkln}\right)}{\partial x_i} S_k S_n P_j P_l \end{cases}, \tag{4}$$

where $P_i$ are the components of the normalized polarization associated to the ray mode and $\rho$ is the densitv. Let us notice in the first equation that the expression of the energy velocity vector denoted $\vec{V}_e$ can be identified.

Note that in the homogeneous case the variations of the elastic constants being of zero, the slowness variation with respect to travel time is null too. Therefore the slowness of the ray is a constant and the trajectory is a straight line following the energy direction as shown in Figure 2.

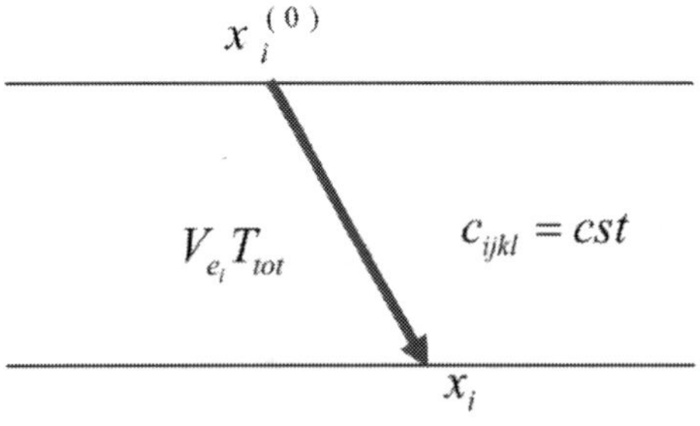


$$\frac{dS_i}{dT} = 0 \Rightarrow S_i = S_i^{(0)}$$

$$\frac{dx_i}{dT} = V_{e_i} \Rightarrow x_i = x_i^{(0)} + V_{e_i} \times T_{tot}$$

**Figure 2.** Trajectory obtained in homogeneous case.

In the general inhomogeneous case, the two equations of the differential system (4) are interdependent. Variations of the position depend on the associated slowness and the variations of the slowness depend on the gradient of stiffnesses with the position. Except for some particular cases, the only way to determine the ray trajectory consists in using an iterative scheme to solve the initial value problem:

$$\frac{dY}{dT} = F(Y). \tag{5}$$

where the variable $Y$ represents the parameters of the ray as a function of travel time such as $Y = (x_i, S_i)$.

3.1.2. *Numerical evaluation of ray trajectory by an iterative scheme.* To solve the differential system (5) with a numerical resolution, a fourth-order Runge-Kutta (RK4) scheme is used [7]. Using this numerical scheme, the error made is of the fourth order with respect to the travel time step $\Delta T$.

The wave mode is specified by initial conditions imposed to the ray tracing system. For an initial source point and an initial phase direction $\vec{n}$, the norm of the slowness vector $\vec{S}$ must be one of the three roots of the Christoffel equation:

$$\left| S^2 \rho^{-1} c_{ijkl} n_j n_k - \delta_{il} \right| = 0. \tag{6}$$

Then, for each step of the calculation, we evaluate simultaneously perturbations of the position of the ray and its slowness. During the iterative process, the slowness vector does not satisfy exactly the Christoffel equation (6). In order to stabilize the iterative scheme, the norm of the slowness is corrected if the determinant of equation (6) is over a certain value. This way, we construct an iterative time-step propagation as represented in Figure 3:

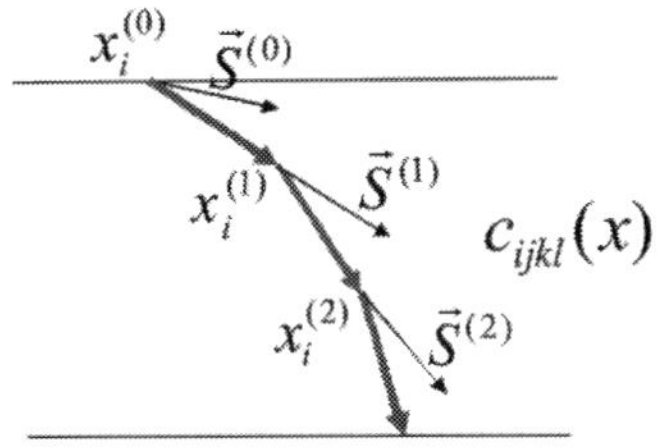


**Figure 3.** Principle of the iterative time-step propagation.

3.2. Extension of the pencil method

For field calculation, after determining ray trajectories, we need to calculate the associated amplitude. To determine the amplitude for each ray we have extended the principle of the pencil method. The pencil method [4] is based on the energy conservation in a ray tube, which is still valid in our case this condition being satisfied by the transport equation [6].

To evaluate the pencil propagation matrix in a smoothly inhomogeneous medium, we consider that the medium is homogeneous between each step of trajectory calculations. This way, we neglect the distortion due to local inhomogeneity between paraxial rays and the axial ray.

So if we introduce the pencil propagation matrix $L_n$ which describes the parameter evolution of the ray tube during its propagation at step *n*, the global propagation matrix is simply the product of matrices $L_n$ for each calculation step. Then after N iterations, the pencil vector $\Psi_N$ at the calculation point and the pencil vector at the source $\Psi_0$ are linked by the formula:

$$\Psi_N = (L_0 L_1 ... L_N)\Psi_0 . \tag{7}$$

## 4. Numerical results and discussion

To validate our continuously variable model, numerical results are compared. We consider this model and a description of the sample as a set of homogeneous sub-sections with a local orientation of anisotropy. In previous work, S. Deydier has shown that the piecewise description remains valid for complex shapes described in the Cartesian basis [3]. But it appears that this generates some numerical artifacts due to fictive interfaces, which can be critical for cases with a strong local elastic gradient.

### 4.1. Application to a cylindrical geometry: the stiffener case

Let us consider the case of a sample with cylindrical geometry, for instance a stiffener. In this case, the fibers follow the component with a constant curvature.

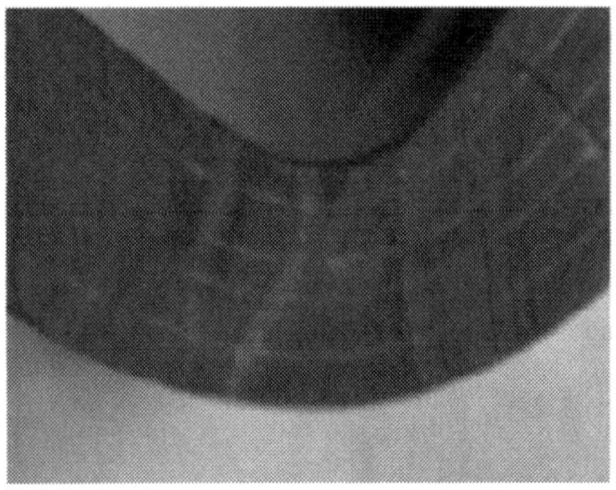

**Figure 4.** Picture of the curved part of a stiffener

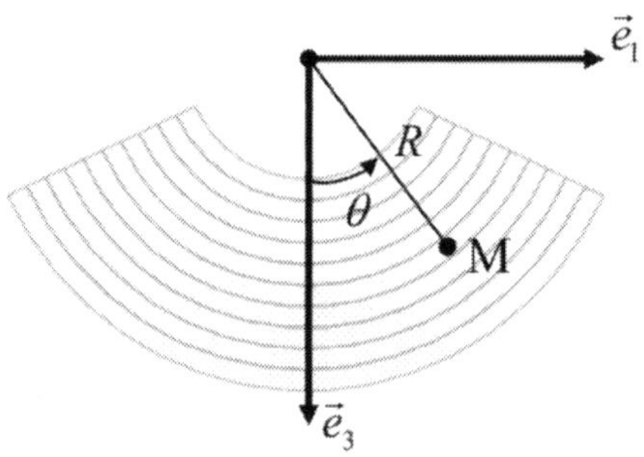


**Figure 5.** Modelling of the sample in Figure 4 by a curved panel with a constant radius

In the case of a stiffener with cylindrical geometry, the transformation $\Phi$ is analytically known. For every point M in the cylindrical sample, the Jacobian matrix $J$ can be defined as a function of angle $\theta$ used in Figure 5:

$$J(M) = \begin{pmatrix} \cos\theta & 0 & -\sin\theta \\ 0 & 1 & 0 \\ \sin\theta & 0 & \cos\theta \end{pmatrix}. \tag{8}$$

Combining this expression of the Jacobian with the general equation (3), we obtain an analytical formulation of the stiffness tensor in a cylindrical sample as a function of point M. To solve the ray tracing system (4), we can also deduce the variations of the elastic constants according to the position with the formula invariant with the variable $R$ such as:

$$\frac{\partial c_{ijkl}}{\partial x_i} = \frac{\partial c_{ijkl}}{\partial \theta} \times \frac{\partial \theta}{\partial x_i}. \tag{9}$$

So all elements to solve the system (4) being known, numerical simulations have been performed in the case of a cylindrical shape. In the following part numerical results are presented.

4.1.1. *Rays trajectories.* In our case, trajectories obtained with piecewise descriptions and calculated with the continuous description are considered. Results for several discretizations are considered: 5, 11, 21 and 39 sectors. In Figure 6, some examples of piecewise description are illustrated showing the variable orientations according to the normal of fibers at the centre of each sector.

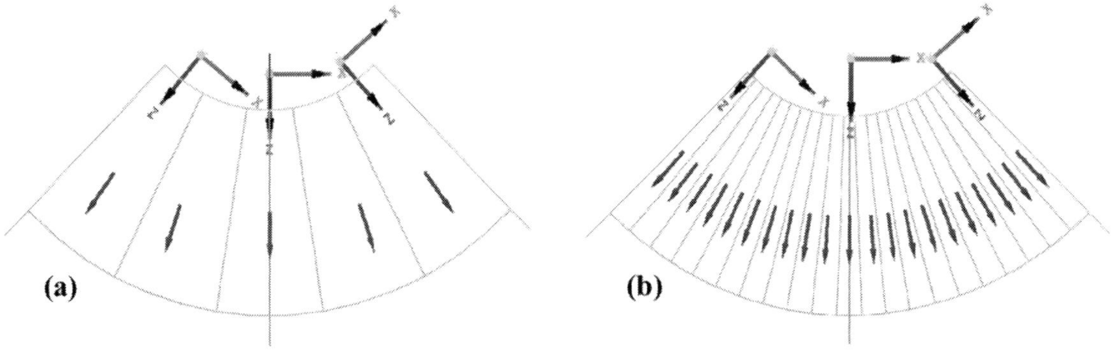


**Figure 6.** Example of piecewise description with (a) 5 and (b) 21 sectors.

Let us consider a stiffener with an inner radius of curvature of 10 mm and a thickness of 13.2 mm, the stiffness tensor is known locally in Cartesian coordinates in an abbreviated Voigt form, numerical values are detailed in table 1.

**Table 1.** Stiffness tensor values used to simulate ray propagation in a CFRP curved composite.

| | $C_{11}^{ref}$ | $C_{12}^{ref}$ | $C_{13}^{ref}$ | $C_{22}^{ref}$ | $C_{23}^{ref}$ | $C_{33}^{ref}$ | $C_{44}^{ref}$ | $C_{55}^{ref}$ | $C_{66}^{ref}$ |
|---|---|---|---|---|---|---|---|---|---|
| **Stiffness (GPa)** | 96.96 | 67.46 | 5.65 | 119.70 | 5.92 | 12.57 | 4.25 | 3.58 | 20.04 |

The material is considered to be orthotropic and shows a strong anisotropy between the x-y plane and the z direction (cf. table 1).

We have also compared the trajectory of rays radiated from a 6.25 mm diameter plane transducer. For piecewise description, the ray propagation corresponds to straight line in each sector and the refraction of the ray is evaluated at each interface. For the continuous description, we use the iterative scheme previously presented in section 3.

Numerical results obtained in the four cases are illustrated in Figure 7. The finer the piecewise description is, the more the plotted trajectories converge to the continuous description trajectories. We observe that for this tiny radius of curvature the localization of the focal point is very sensitive to the piecewise description. It confirms that this curved sample has to be considered with a high degree of precision in order to obtain an accurate beam width.

The trajectories for the description with 39 sectors and for continuous description being quasi identical (cf. Figure 7 (c) and (d)), this study validates our model for trajectories and we look now at the problem of transmitted fields.

Journal of Physics: Conference Series **269** (2011) 012022 doi:10.1088/1742-6596/269/1/012022

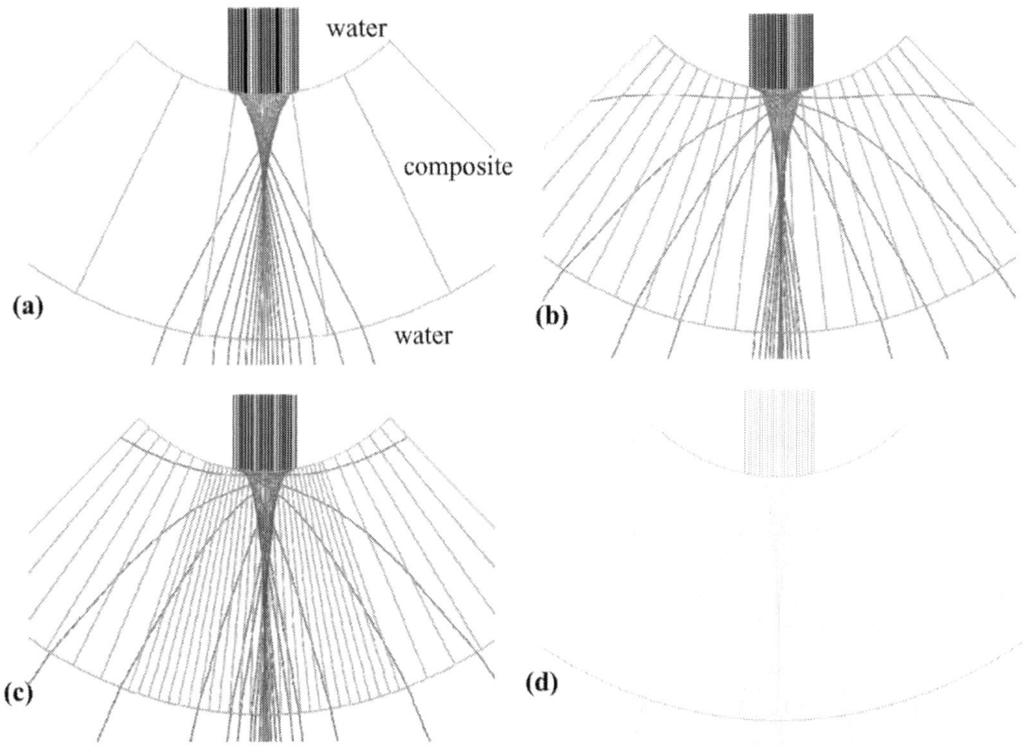


**Figure 7.** Trajectories obtained for the piecewise description with (a) 5, (b) 21, (c) 39 sectors and (d) with the continuous description.

4.1.2. *Transmitted field.* As before, transmitted fields for continuous and piecewise descriptions are compared. The calculation is done on a line at 25 mm form the cylindrical sample in water (cf. Figure 8) for a 6.25 mm diameter plane transducer with a central frequency around 2.25 MHz.

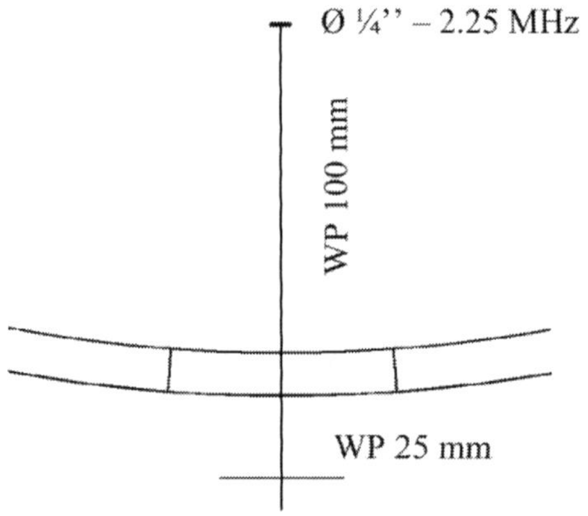


**Figure 8.** Configuration used for validation of the transmitted field calculation

At first, let us consider the case of a sample with a large radius of curvature. For example we choose a radius of curvature of 500 mm. In this situation, effects of the curvature are small and the piecewise description is known as valid [3]. Let us report results for the piecewise description with 5 sectors:

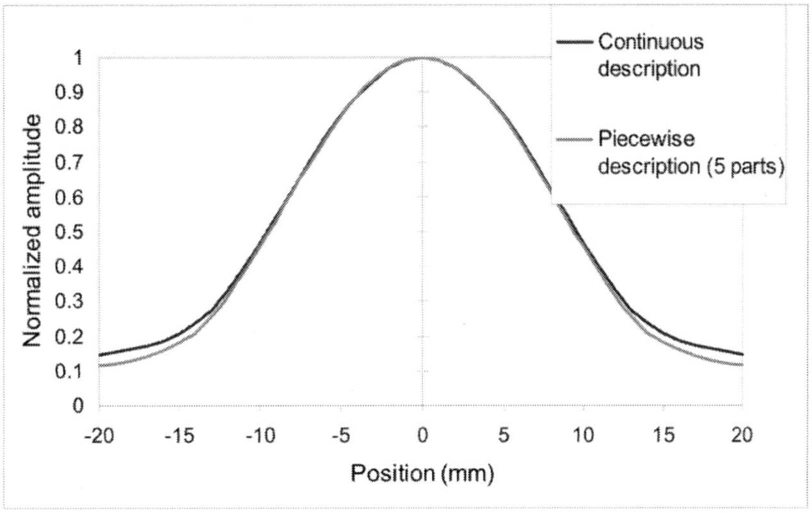


**Figure 9.** Transmitted field for a radius of curvature of 500mm.

There is a very good agreement between the two curves as we can see in Figure 9. In particular beam widths correspond perfectly. With this example we can validate our new model for the case of a large radius of curvature.

Now, let us look the same problem for a smaller radius, of 10 mm. For this case the continuous description and a description with 39 sectors are compared as already done for trajectories (cf. Figure 7 (c) and (d)).

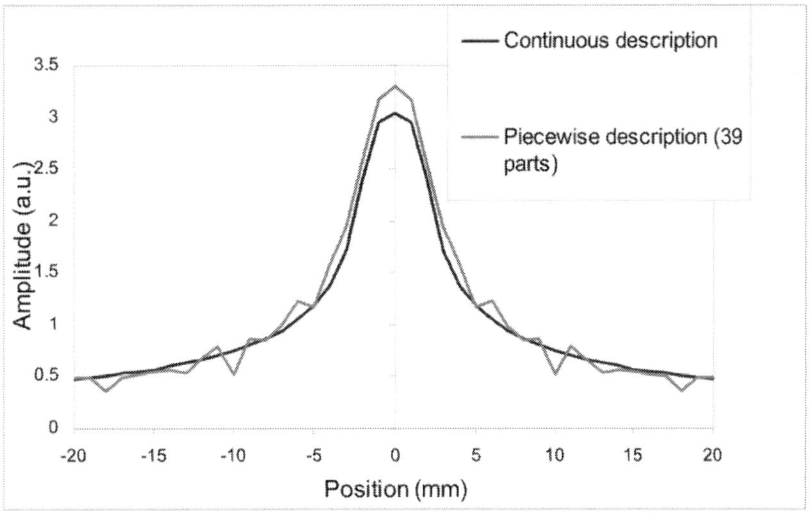


**Figure 10.** Transmitted field for a radius of curvature of 10mm.

Some numerical artifacts appear due to the interfaces created by the discretization of the piecewise description. These fictive interfaces lead to problems for field calculations; rays arriving exactly on the interfaces are not taken into account and numerical artifacts are created. So in the case of small radii of curvature, we need to have a finer description with many sectors to be confident with the result. But this leads to more numerical artifacts due to artificial interfaces.

The Figure 10 shows us that the transmitted fields obtained with the two descriptions are similar in terms of amplitude and beam width.

## 5. Conclusion

In this paper, we propose a ray based model for propagation in curved composites. This continuous model was validated for a quasi-plane specimen by comparing with a piecewise model. In the case of small radii of curvature, results presented here are a first promising step before an experimental validation. There are advantages to this continuous description: numerical artefacts due to fictive interfaces are avoided. But in return the iterative numerical scheme is costly in time. So future works will focus in optimizing the numerical scheme, the time step selected can be adapted with local elastic variations inside the component. Experimental comparisons with composite samples provided by EADS are in progress.

## References

[1] CIVA software platform for simulating NDT techniques (UT, EC, RT) http://www-civa.cea.fr
[2] Lonné S, Lhémery A, Calmon P, Biwa S and Thévenot F 2004 Modeling of ultrasonic attenuation in unidirectional fiber reinforced composites combining multiple-scattering and viscoelastic losses *Review of Progress in QNDE* **23**, ed D O Thompson and D E Chimenti (Melville: AIP Conf. Proc. **700**) 875
[3] Deydier S, Leymarie N, Calmon P and Mengeling V 2006 Modeling of the ultrasonic propagation into carbon-fiber-reinforced epoxy composites, using a ray theory based homogenization method *Review of Progress in QNDE* **25**, ed D O Thompson and D E Chimenti (Melville: AIP Conf. Proc. **820**) 972
[4] Gengembre N and Lhémery A 2000 Pencil method in elastodynamics: application to ultrasonic field computation *Ultrasonics* **38** 495
[5] Cerveny V 2001 *Seismic Ray Theory* chapter 2 (Cambridge: Cambridge University Press)
[6] Cerveny V 2001 *Seismic Ray Theory* chapter 3 (Cambridge: Cambridge University Press)
[7] Numerical Recipes Third Edition http://www.nrbook.com/nr3/

Journal of Physics: Conference Series **269** (2011) 012023 doi:10.1088/1742-6596/269/1/012023

# Prediction of the point of impact in an anisotropic plate

M Koabaz [1], T Hajzargarbashi [2], T Kundu [2] and M Deschamps [1]

[1] Université de Bordeaux, CNRS, UMR 5469, Laboratoire de Mécanique Physique, 351 cours de la libération, Talence, F-33405, France.

[2] Department of Civil Engineering and Engineering Mechanics, University of Arizona, Tucson, AZ 85721, United States.

Co-authors' E-mail addresses:

m.koabaz@lmp.u-bordeaux1.fr
talieh@email.arizona.edu
tkundu@email.arizona.edu
m.deschamps@lmp.u-bordeaux1.fr

## Abstract

Locating the point of impact of a foreign object in a plate is important for continuous health monitoring of structures. A new method based on an optimization scheme has been recently proposed to locate the point of impact in anisotropic plates by analyzing the times of arrival of the ultrasonic signals at the passive sensors attached to the plate. Following this optimization based technique, in this paper the impact point on an anisotropic plate is predicted from the acoustic emission data. Experiments are carried out with a carbon-epoxy plate where the impact point is modeled by an acoustic source. A Parallel Pre-stressed Actuator (PPA) is used as the acoustic source and the acoustic signals at different locations are received by adhesively bonded acoustic sensors. The source point is then predicted and compared with its actual location. Related theory is also presented in the paper.

## 1. Introduction

The point of impact for an isotropic plate can be located after detecting the acoustic emission signal by at least three sensors and applying the triangulation technique. However, if the plate is anisotropic then the triangulation technique does not work, because this technique assumes that the wave velocity is the same in all directions, which is not true for anisotropic plates.

An alternative method based on an optimization scheme was proposed by Kundu et *al.* [1-4] to locate the point of impact in anisotropic plates by analyzing the time of arrival of the ultrasonic signals received by the passive sensors attached to the plate. Recently Hajzargarbashi *et al.* [5] improved the impact point prediction technique. Following their modification, in this paper the impact point on an anisotropic plate is predicted from the acoustic emission data. The method originally proposed in Kundu et *al.* [1] is based on the optimization technique – minimization of a non-linear objective function or error function. However, the proposed objective function in that reference had the inherent problem of multiple singularities which was overcome in Kundu et *al.* [2] by modifying the objective function. The optimization technique was further improved by Hajzargarbashi *et al.* [5]. Using this latest optimization technique the impact point in a Carbon-epoxy plate is predicted and compared with the exact point of impact.

The method proposed in this paper for locating the point of impact works well for any type of anisotropy because it uses the experimentally obtained direction dependent Lamb wave velocity profile.

## 2. Formulation

The formulation presented in Kundu et *al.* [1-2] and Hajzargarbashi *et al.* [5] is repeated here. Let the time of detection of the acoustic signal at the *i*-th station be $t_{id}$. If the time of impact be $t_c$ then the travel time for the signal from the impact point to the station location is:

$$t_i = t_{id} - t_c \quad (1)$$

Note that in (1) both $t_{id}$ and $t_c$ are defined with respect to the same time of reference. If the coordinates of a receiving sensor $S_i$ is $(x_i, y_i)$ with $i = 1, 2, 3 \ldots n$ and the impact point coordinate is $(x_0, y_0)$ then the distance of the sensor from the impact point is given by:

$$d_i = \sqrt{(x_i - x_0)^2 + (y_i - y_0)^2} \quad (2)$$

The time of travel of the wave to the sensor locations is denoted by $t_i$ and the velocity $c(\theta)$ of the wave in the plate is a function of the wave propagation direction $\theta$, therefore, one can write:

$$d_i = c(\theta_i)t_i = \sqrt{(x_i - x_0)^2 + (y_i - y_0)^2} \quad (3)$$

From (3) it is possible to define an error function $E(x_0, y_0)$, also known as the objective function in the optimization literature. Values of $(x_0, y_0)$ can be obtained by minimizing the objective function. The objective function should be expressed in terms of relative times of arrival as defined below, instead of the actual times of arrival that are difficult to obtain:

$$t_{ij} = t_{id} - t_{jd} = (t_i + t_c) - (t_j + t_c) = t_i - t_j \quad (4)$$

It removes the need for explicit knowledge of the time of impact, $t_c$. If the wave velocity is $c(\theta)$ then the relative times of arrival are defined as:

$$t_{12} = t_1 - t_2 = \frac{d_1}{c(\theta_1)} - \frac{d_2}{c(\theta_2)} = \frac{c(\theta_2)d_1 - d_2 c(\theta_1)}{c(\theta_1)c(\theta_2)}$$
$$= \frac{c(\theta_2)\sqrt{(x_1 - x_0)^2 + (y_1 - y_0)^2} - \sqrt{(x_2 - x_0)^2 + (y_2 - y_0)^2}\,c(\theta_1)}{c(\theta_1)c(\theta_2)} \quad (5)$$

The error function $E(x_0, y_0)$ can be defined as:

$$E_1(x_0, y_0) = (t_1 - t_2)c(\theta_1)c(\theta_2) - \sqrt{(x_1 - x_0)^2 + (y_1 - y_0)^2}\,c(\theta_2) + \sqrt{(x_2 - x_0)^2 + (y_2 - y_0)^2}\,c(\theta_1) \quad (6)$$

Ideally, for the correct values of $(x_0, y_0)$ the error function should give a zero value while for wrong values of $(x_0, y_0)$ it should give a positive value. Therefore, we need to minimize the value of this error function or objective function. To give equal importance or bias to three measurements of arrival times at the three sensor locations the error function can be defined in a different manner as shown below. With this definition of the error function, the impact point location prediction should not be strongly influenced by the experimental error in any one time of arrival measurement [5]:

$$E(x_0,y_0)=(t_1-t_2)c(\theta_1)c(\theta_2)-\sqrt{(x_1-x_0)^2+(y_1-y_0)^2}c(\theta_2)+\sqrt{(x_2-x_0)^2+(y_2-y_0)^2}c(\theta_1)$$

$$+(t_1-t_3)c(\theta_1)c(\theta_3)-\sqrt{(x_1-x_0)^2+(y_1-y_0)^2}c(\theta_3)+\sqrt{(x_3-x_0)^2+(y_3-y_0)^2}c(\theta_1) \quad (7)$$

$$+(t_2-t_3)c(\theta_2)c(\theta_3)-\sqrt{(x_2-x_0)^2+(y_2-y_0)^2}c(\theta_3)+\sqrt{(x_3-x_0)^2+(y_3-y_0)^2}c(\theta_2)$$

By some optimization scheme the impact point ($x_0$, $y_0$) can be obtained by minimizing the error function given in (7).

## 3. Experimental investigation

Experiments were carried out on a 910 *mm* x 795 *mm* carbon-epoxy plate where the impact point is modeled as an acoustic source. The plate is 3 *mm* thick with fibers going in the 0° direction. A Parallel Pre-stressed Actuator (PPA) is used as the acoustic source as shown in figure 1, and the acoustic signals at different locations are received by adhesively bonded acoustic sensors. Figure 2 shows the experimental setup.

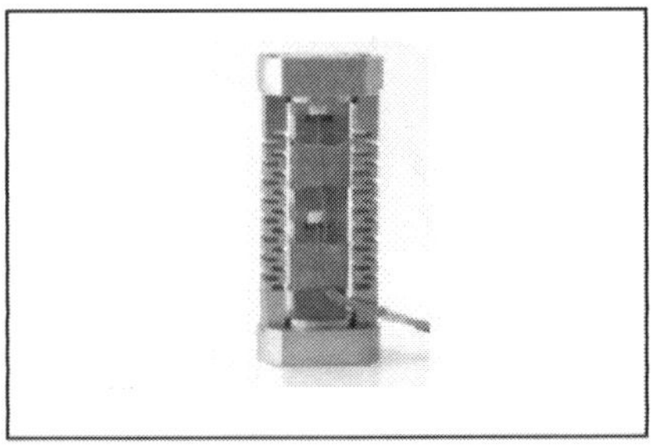

Figure1: view of PPA

Figure 2 : Photograph of the experiment setup

Wave velocity in different directions of the plate are obtained by placing 10 sensors at 10° angular intervals varying from 0° to 90°. Velocity measurements are shown in figure 3 at 150 *KHz.* The signal received from the sensor was visualized on an oscilloscope like the signal shown in figure 4. Signals were then averaged before being recorded on a computer. Angles were measured from the direction of the fibers (0° direction) to its perpendicular direction (90° direction). For every direction, the envelopes were calculated to extract the Energy velocities of the exited modes. Knowing the distance *d*, separating the transmitter and the receiver, and dividing this distance by the arrival times, one can obtain the velocity of the guided wave mode.

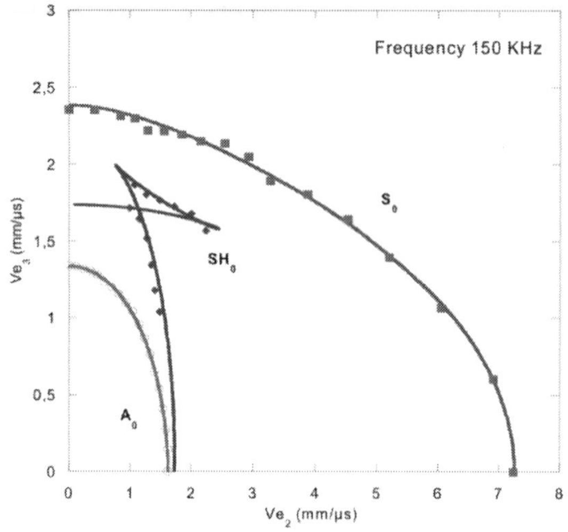


Figure 3: Experimental measurement of the velocity of energy for the fundamental modes. The horizontal axis represents the direction of the fibers.

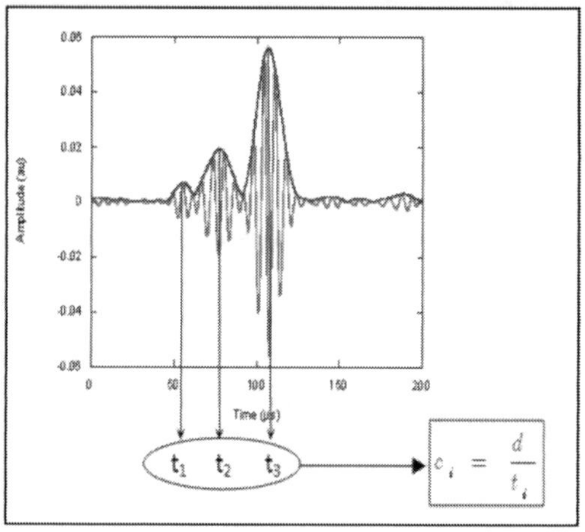


Figure 4: Velocity extraction for experimental data using the envelope.

The experimental signal at 150 kHz is quite simple in the fiber direction, but it becomes much more complex in the cusp range, see figure 5.

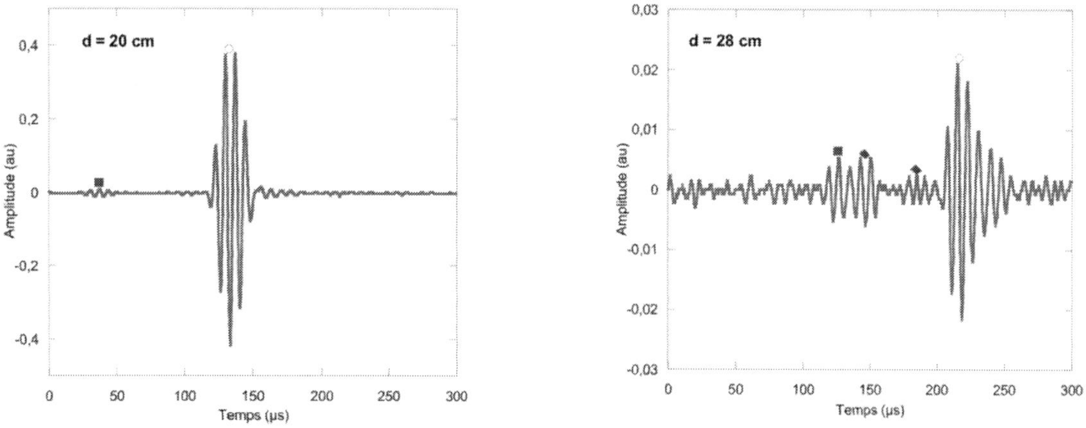


Figure 5: Experimental signals at 0.15 *MHz*. On the left hand side, the signal is quite powerful, in the 0 degree direction, 15 cm away from the source. On the right hand side, it is weaker and more complicated in the 40 degrees direction, 27.5 cm away from the source. Discs, squares and triangles denote the arrival times of $A_0$, $S_0$ and $SH_0$ modes, respectively, to be associated with the polar measurements of Figure 3.

To avoid the complexity that comes from the cusp of the SH mode, the experiment was carried out at 45 kHz frequency that excites only $A_0$ and $S_0$ modes. The experimental measurement of the velocity of $A_0$ mode is shown in figure 6.

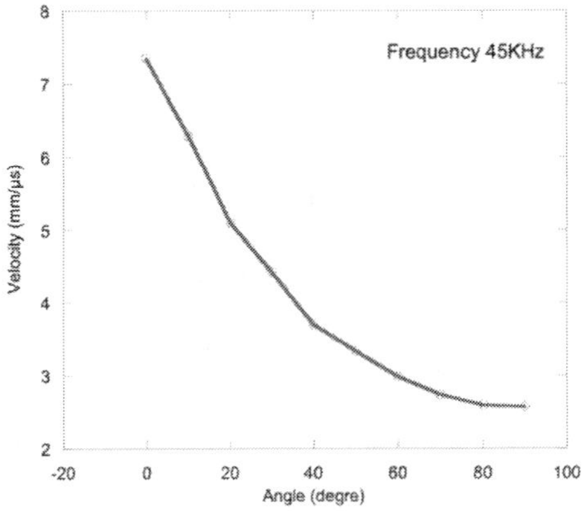


Figure 6: Variation of the guided wave velocity in the Carbone- Epoxy plate as a function of the propagation direction angle.

After obtaining the wave velocity profile in the carbon-epoxy plate a second experiment is carried out to predict the impact point from the acoustic emission data. For this experiment four passive sensors are adhesively bonded to the plate as shown in figure 7. Sensors 1, 2, 3 and 4 (S1, S2, S3 and S4) are placed away from the point of impact. The location of the acoustic source is denoted as S0; it corresponds to the impact point. From the time history of the acoustic source (denoted as S0 in figure 7) the exact time of impact can be obtained.

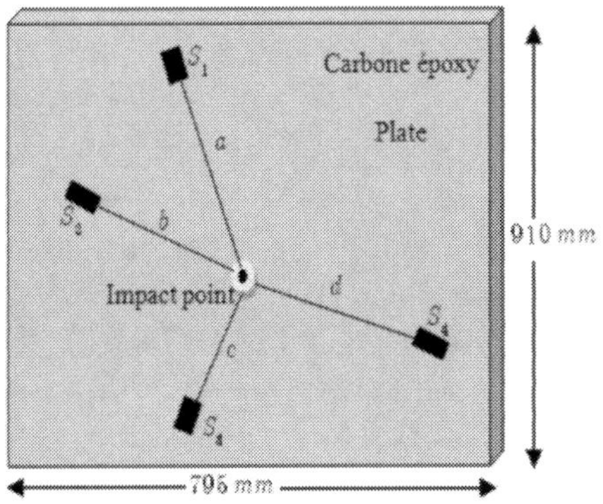


Figure 7: Acoustic emission sensor locations for Experiment # 2, with (a=400 mm, b = 300, c=280 mm, d=320 mm).

It is not easy to measure the exact time of arrival of the signals from the time history plots (envelope of the signals) because there is some ambiguity in these plots about the exact time of arrival or the starting point of the signals. Arrival of the weak extensional mode or the $S_0$ mode is hidden in the low level noise present in the time history plots. The arrival time of the stronger $A_0$ mode is obtained from the plots by the threshold technique. The arrival time of the first noticeable peak that is greater than the noise level is recorded as the arrival time of the ultrasonic energy.

## 4. Optimization of the objective function

Since the objective function (see Eq. (7)) uses the differences in the recorded arrival times the small errors in the arrival time recording, mentioned in the previous section should not affect the final results significantly. If there is redundancy in the number of sensors then the effects of these small errors on the impact location estimate also would tend to be minimized. In saying so it should also be kept in mind that guided waves propagating in a composite plate are dispersive and attenuative. Therefore, if the propagation distances from the impact point to different receivers vary significantly then, due to dispersion and attenuation, the shapes of the signal recorded by different receivers look significantly different and recording the arrival time by the threshold technique may not give accurate results. In that situation time-domain correlation technique [9, 10] applied to two time-domain signals recorded by two sensors can be followed to accurately obtain the difference between the arrival times at the two sensors than what is considered here.

The $x$ and $y$ values corresponding to the minimum of the objective function (see Eq. (7)) should give the impact point or the optimized coordinate values ($x_0$, $y_0$). Because of multiple local minima of the objective function the standard optimization schemes such as the Simplex algorithm [11] do not always converge to the absolute minimum. However, successful convergence is possible by finding the y-coordinate corresponding to the minimum values of the function for different x-values, then among these minimum values the absolute minimum is identified; $x$ and $y$ values corresponding to this absolute minimum are the coordinates of the impact point. Hajzargarbashi *et al.* [5] have shown that if the number of receiving acoustic sensors is increased from 3 to 4 then the accuracy of the impact point prediction is significantly improved. In the objective function it is not necessary to consider all 4 receiving sensors simultaneously, 4 combinations of 3 sensor sets can be used to predict the impact point. To construct a set of 3 sensors from a group of 4, one sensor is left out. In this manner 4 different sets of three sensors can be constructed from a group of 4 sensors. Impact points predicted using 4 different sets of 3 sensors and 1 set of 4 sensors are shown in figures 8 to 11. Different points in these figures are generated by randomly giving a small variation (less than 5 percent) in the recorded arrival time and plotting the point where the optimization scheme converges, figures 8 and 9 correspond to two slightly different objective functions – one given by Kundu *et al.* [2] generates figure 8 and the second objective function proposed by Hajzargarbashi *et al.* [5] produces figure 9. These two objective functions are denoted as "old" [2] and "new" [5] objective functions. Both figures 8 and 9 are obtained from 4 sets of 3 sensors. When the times of arrival of all four sensors are considered simultaneously in the objective function instead of considering 4 sets of 3 sensors separately then figures 10 and 11 are obtained from these two objective functions. The velocity profile shown in figure 6 has been used to predict the impact point. Note that the accuracy of the final prediction is about the same for the two objective functions but the scattering in the predicted data points due to the measurement error in the arrival times decreases with the new objective function.

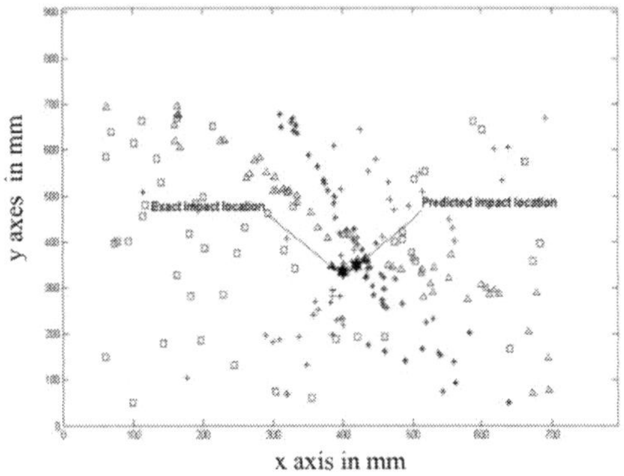


Figure 8: Impact points exact and predicted using the old objective function for 4 sets of 3 sensors.

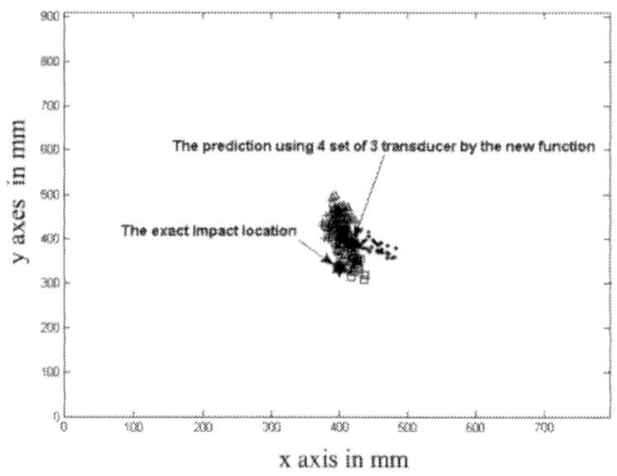


Figure 9: Impact points exact and predicted using the new objective function for 4 sets of 3 sensors.

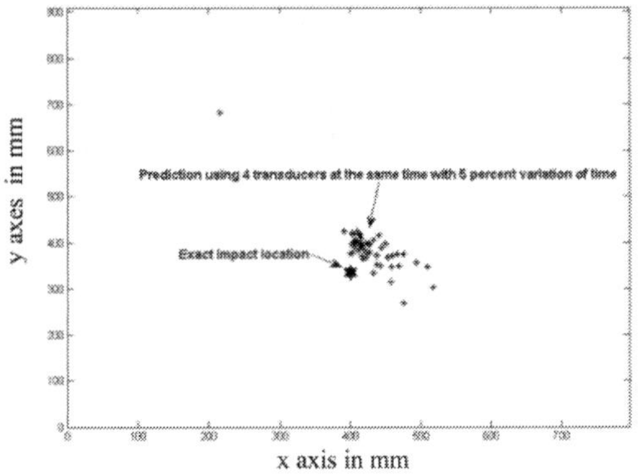


Figure 10: Impact points exact and predicted using the old objective function for 1 set of 4 sensors.

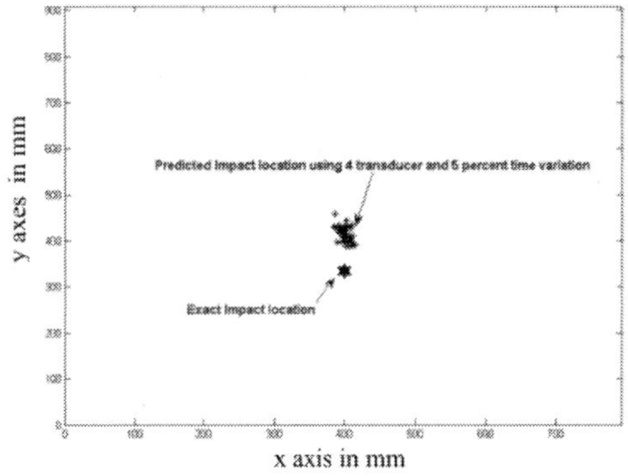


Figure 11: Impact points exact and predicted using the new objective function for 1 set of 4 sensors.

## 5. Conclusion

Optimization schemes proposed by Kundu et *al.* [1-4] and Hajzargarbashi *et al.* [5] to locate the acoustic source in an anisotropic plate by analyzing the times of arrival of the ultrasonic signals are used in this paper. Following this optimization based technique, the acoustic source point on a Carbon-epoxy plate is predicted from the acoustic emission data. Experiments are carried out with an acoustic source (PPA) placed on the plate modeling a point of impact. It is shown that the source point can be correctly predicted by this technique when the anisotropic nature of the velocity profile in the plate is considered.

## Acknowledgement:

Third author T. Kundu would like to acknowledge the financial support he received from the University of Bordeaux where he spent one month as an invited professor during this collaborative research.

## References

[1] Kundu T, Das S and Jata K V 2007 Point of Impact Prediction in Isotropic and Anisotropic Plates from the Acoustic Emission Data *J. Acoust. Soc. Am.* **122 (4)** 2057-2066.

[2] Kundu T, Das S, Martin S A and Jata K V 2008 Locating Point of Impact in Anisotropic Fiber Reinforced Composite Plates *Ultrasonics* **48(3)** 193-201.

[3] Kundu T, Das S and Jata K V 2009 Health Monitoring of a Thermal Protection System using Lamb Waves *Structural Health Monitoring: An International Journal* **8(1)** 29-45.

[4] Kundu T, Das S and Jata K V 2009 Impact Point Detection in Stiffened Plates by Acoustic Emission Technique *Smart Materials & Structures* **18** Article #035006.

[5] HajZargarbashi T, Kundu T and Bland S 2010 A new algorithm for detecting impact points in anisotropic plates by the acoustic emission technique *Health Monitoring of Structural and Biological Systems IV*, Ed. T. Kundu, SPIE's 17th Annual International Symposium on Smart Structures and Materials & Nondestructive Evaluation and Health Monitoring, San Diego, California, March 7-11, 2010.

[6] Kessler S S, Spearing S and Soutis C 2002 Damage detection in composite materials using lamb wave methods *Smt. Mater. Str.* **12** 795–803.

[7] Kundu T, Das S and Jata K V 2007 An improved technique for locating the point of impact from the acoustic emission data *Proc. SPIE* 6532.

[8] Sachse W and Sancar S 1986 Acoustic emission source location on plate-like structures using a small array of transducers *U.S. Patent* No. **4**,592,034.

[9] Kosel T, Grabec I and Muzic P 2000 Location of acoustic emission sources generated by air flow *Ultrasonics* **38** 824–826.

[10] Ziola S M and Gorman M R 1991Source location in thin plates using cross-correlation *J. Acoust. Soc. Am.* **90** 2551–2556.

[11] Nelder J A and Mead R 1965 A simplex method for function minimization *Comp. J.* **7** 308–315.

Journal of Physics: Conference Series **269** (2011) 012024 doi:10.1088/1742-6596/269/1/012024

# Picosecond ultrasonics in single cells: Interface step motion for thin animal cells and Brillouin scattering for thick vegetal cells

**M Ducousso, T Dehoux, B Audoin**

Université de Bordeaux, CNRS, UMR 5469, Talence, F-33405, France

**O Zouani, C Chollet, M C Durrieu**

Université de Bordeaux, INSERM, U 577, Bordeaux, F-33000, France

mathieu.ducousso@u-bordeaux1.fr

**Abstract**. The measurement of the mechanical properties of single biological cells using a picosecond laser–ultrasonic method is proposed. A pump-probe set-up based on ultrafast laser (100 fs pulses) is used to generate and detect acoustic frequencies in the GHz range in a cell on a metallic substrate. The time resolution is about 1 ps and the laser focusing allows a 1 μm lateral resolution. We carry out experiments in both animal and vegetal cells. A semi-analytical simulation model of the physical phenomena involved in experiments is presented. The coupled heat and stress equations are solved including a thermal boundary resistance at the cell/substrate interface and strong acoustic absorption. The optical detection resulting from the interaction between the acoustic wave and the laser light is also modelled. Simulations allow the analysis of experimental signals in both vegetal and animal cells. The results support the potentialities of the non-invasive technique for bioengineering and medical applications.

## 1. Introduction

Cell is the smallest unit of life that is classified as a living organism, and is often called the building block of life. Cell biomechanics seems to be important in cell growth, differentiation, motility and apoptosis, in some diseases proliferation, or in the adhesion mechanics of cell/implant for instance. For this reason, studies of the cell mechanics, or even of subcellular components mechanics, are of growing interest, favored by new technologies, more sensitive instruments, and improvements in bioimaging.

The use of lasers to generate acoustic waves provides a contactless and non-invasive tool to study biological tissues. Ultrasound at MHz frequencies can be generated by taking advantage of the optical absorption of nanosecond laser pulses in the tissue itself. [1] The contrast of the imaging is thus provided by the contrast in tissue absorption. However, many important cell subcomponents have dimensions less than 500 nm and therefore fall below resolution of this technique and of most nondestructive imaging systems. [2]

The picosecond ultrasonics technique [3] uses the absorption of femtosecond laser pulses in a solid transducer to generate acoustic frequencies up to 1 THz. Since 1984, interest in this technique has been growing owing to its wide range of applications in non-destructive control and solid-state physics. By this technique high frequency acoustic waves are both generated and detected with laser light pulses (duration<1ps), allowing the inspection of submicrometric materials. Main applications of picosecond acoustics are thickness measurement [4-5] and bonding control at the nanometer scale, [6] or studies of material microstructures. [7-10] Experiments can be performed in opaque or transparent media. Their specific optical responses permit to determine optical, thermal, or mechanical properties of the sample.

Picosecond acoustics has already been used to study vegetal cells (*allium cepa*) and the potentialities of this technique to improve current cell imaging resolution have been demonstrated. [11-12] The method is applied in this paper to single mouse MC3T3 fixed cells.

In the second section of the paper, a thermo-mechanical model describing a cell of finite thickness, sandwiched between two semi-infinite media is proposed. The third section concerns the modelling of the detection, and a detailed analysis of the influence of the cell thickness on the opto-acoustic interaction is proposed. A comparison between experimental data and simulations is presented in the fourth section.

## 2. Picosecond acoustics in a viscoelastic film sandwiched between two half-spaces.

In picosecond ultrasonics experiments a short light pulse is absorbed in a material. Because of the thermal expansion, a short strain pulse is generated. Picosecond acoustics has already been used to study thin films, [3-6,8] from nanometric to micrometric thicknesses. A detailed analysis of the thermal diffusion and of the acoustic propagation in a viscoelastic thin film is proposed in this section. The model couples the temperature diffusion equation and the wave motion equation. [13] Equations are analytically solved in the frequency domain and time results are calculated using a numerical inverse Fourier transform. Figure 1 shows the considered geometry. The cell is modelled as a thin transparent viscoelastic film of thickness *d* sandwiched between two semi-infinite half spaces, the first one being the opaque substrate and the second one being a transparent physiological serum. Indices $i$=0, 1, 2 refer to the substrate ($z>0$), the cell ($0>z>-d$), and the transparent half space ($z<-d$), respectively.

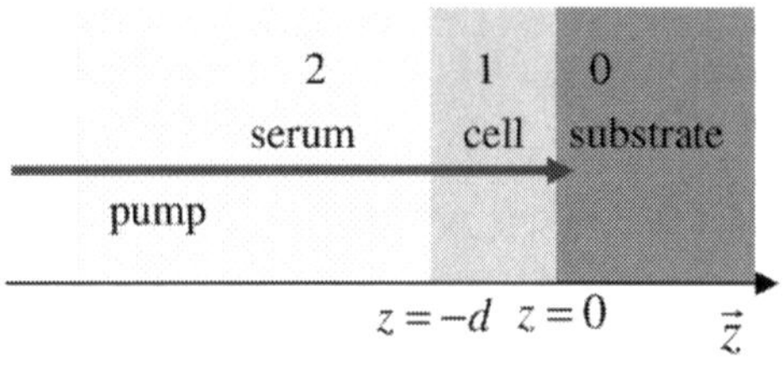


Figure 1: Simulation configuration. The pump beam is absorbed in medium 0.

The temperature fields $T_i$ are determined solving the Fourier heat diffusion equation for each medium $i$:

$$\rho_i C_{pi} \frac{\partial T_i(z,t)}{\partial t} - \kappa_i \frac{\partial^2 T_i(z,t)}{\partial z^2} = \begin{cases} \beta_0 I_0 \delta(t) e^{-\beta_0 z}, i=0 \\ 0, i=\{1,2\} \end{cases}, \quad (1)$$

where $\rho_i, C_{pi}, \kappa_i$ are the density, the specific heat capacity and the thermal conductivity, respectively. $\beta_0$ is the optical absorption coefficient of the substrate and $I_0$ stands for the electromagnetic intensity absorbed in the substrate. Since media 1 and 2 are transparent, the optical generation results only from the absorption of the pump laser beam radiation in the substrate, in the vicinity of the film/substrate interface. No source term is considered in the heat diffusion equation for media 1 and 2. The temperature rise in these media results from the continuity of the temperature and heat flux at the interfaces. A thermal boundary resistance is considered to account for the thermal mismatch at the 0/1 interface.

The thermal parameters of the cell are assumed to be close to that of water. Simulations allow quantifying the maximum temperature rise ~1 K at the 0/1 interface for one laser pulse with the considered experimental fluence, around 16 mJ/m$^2$. Since the thermal diffusivity of the titanium alloy is thirty times higher than that of the cell heat essentially diffuses in the substrate.

The instantaneous temperature rise creates a thermal stress which acts as a source term in the wave equation:

$$C_i \frac{\partial^2 U_i(z,t)}{\partial z^2} - \rho_i \frac{\partial^2 U_i(z,t)}{\partial t^2} = C_i \alpha_i \frac{\partial T_i(z,t)}{\partial z}, \qquad (2)$$

where $C_i$, $U_i$ and $\alpha_i$ are the stiffness coefficients, the displacements and the thermal dilatation, respectively. The product of $\alpha_i$ and $C_i$ defines the thermal rigidity parameter. [14] The boundary conditions are the continuity of the stress and of the displacement at each interface.

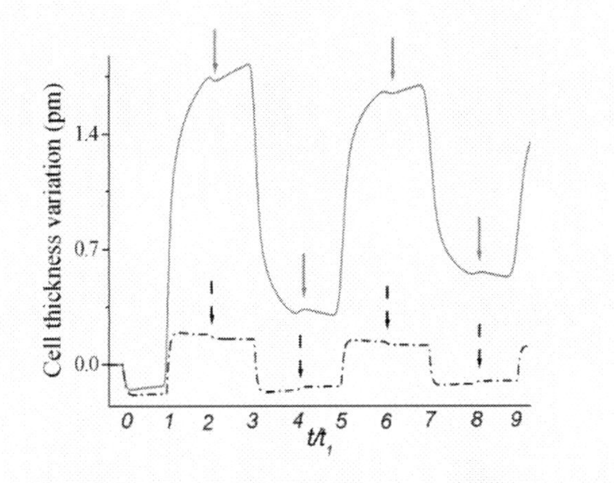


Figure 2: Cell thickness variation as a function of time $t$ with (solid) or without (dash) source term in Eq. (2). The cell thickness changes when $t$ is an integer multiple of the acoustic travel time $t_1 = d/v_1$ in the film. $v_1$ is the velocity in the film.

| | $C_i$ [GPa] | $\rho_i$ [kg/m³] | $\alpha_i$ [1/K] | $\beta_i$ [1/m] |
|---|---|---|---|---|
| TiAl4V | 155 | 4500 | $9\times10^{-6}$ | $5.2\times10^{7}$ |
| MC3T3 | $2.9 + j0.2\omega$ | 1100 | $3.5\times10^{-4}$ | - |

Table 1: Physical parameters used in the simulations shown in Fig. 2.

The difference between the acoustic displacements calculated at the two interfaces 0/1 and 1/2 yields the dynamics of the cell thickness. It is plotted with solid lines in Fig. 2 using data shown in table 1 where medium 0 is TiAl4V and medium 1 is MC3T3 cell. Acoustic attenuation in the cell is modelled by a complex stiffness coefficient. This approach, similar to previous works on ultrasonics attenuation measurements in water, [15] leads to an attenuation factor proportional to the square of the frequency. [16] The signal shows periodic step-like changes with large amplitude caused by the displacement of the 1/2 interface. Periodic step-like changes with smaller amplitude are overlaid, resulting from the smaller displacement (downward arrows) of the 0/1 interface. The periods of both the small and large amplitude contributions is the back and forth propagation time through the cell $2t_1=2d/v_1$, where $v_1$ is the velocity in the cell. In addition, the substrate dilatation creates a negative displacement of the 0/1 interface and a compressive wave in the cell at $t=0$.

The dynamics of the cell thickness has also been calculated when the source term in Eq. (2) is omitted for media 1 and 2. The result is plotted with dashed lines in Fig. 2 using data in table 1. This signal has the same shape with smaller amplitude as the signal plotted with solid lines. This comparison shows that, despite the temperature rise in the cell is small and confined over a distance of a few nanometer from the substrate, the source term can not be neglected since the thermal dilatation of the cell is much higher than that of the substrate.

The rise time $\tau = 1/v_0\beta_0$ of each step depends mainly on the optical absorption length $1/\beta_0$ in the medium 0. The influences of the thermal diffusivity in the substrate and of the viscoelastic properties of the film (medium 1) have negligible influence on $\tau$. The acoustic wavelength in the cell $d_A=v_1\tau$ (about 5 nm in our configuration) at frequency $\tau^{-1}$, will play a key role in the analysis of the reflectivity changes, as discussed in the following section.

## 3. Interface motion and optical scattering contributions to the acousto-optic detection.

The last step in the modelling is to represent the detection, i.e. the change of reflectivity caused by thermo-optic and acousto-optic effects. Several papers deal with the detection in ultrafast acoustics, in either opaque or transparent films, with oblique or normal incidence taking into account 3D acoustic diffraction, for sample thicknesses ranging from a few nanometers to half spaces. [ 6, 8, 17-21] If the optical penetration depth is smaller than the acoustic wavelength, the detection results essentially from the surface displacement. If the sample is transparent, the acousto-optic interaction can no longer be neglected. Part of the probe beam is sensitive to the transient change of the dielectric permittivity induced by the transient perturbation. If the transparent sample is thick enough, the so-called Brillouin oscillation appears. [6-8, 19] This dynamic optical process can be understood in terms of Stokes or Raman scattering. [22] We verified using our simulation that the acousto-optic contribution of the semi-infinite titanium alloy substrate is negligible since the optical penetration depth is around 20 nm. Therefore the acousto-optic detection is considered in the transparent media only. Let us now detail the detection mechanism.

Both thermal and acoustic perturbations create a local modification of the dielectric permittivity $\varepsilon_i^s$ in the medium. Since this perturbation is small, the dependence of $\varepsilon_i^s$ on temperature $T$ and on the acoustic strain $\eta$ is supposed to be linear:

$$\varepsilon_i^s(z)=\frac{\partial n_i}{\partial T}T_i(z)+\frac{\partial n_i}{\partial \eta}\eta_i(z). \tag{3}$$

$\partial n_i/\partial \eta$ and $\partial n_i/\partial T$ are piezo-optic and thermo-optic coefficients, respectively.

In the Fourier domain, the nonlinear Maxwell equation in the perturbed medium is:

$$\frac{\partial^2 E_i(z,\omega)}{\partial z^2}+q^2\varepsilon_i(z,\omega)E_i(z,\omega)=0, \tag{4}$$

where $\varepsilon_i(z,\omega)$, $E_i(z,\omega)$ and $q$ are the dielectric permittivity, the electric field of the medium $i$, and the optical wave vector in vacuum, respectively. Considering a small transient perturbation, the electric field and the dielectric permittivity can be written as the sum of a homogeneous part and of a perturbed part (noted with superscripts $h$ and $s$, respectively). [21] Equation (4) is linearized and two Maxwell equations are obtained: One without source term for the homogeneous part of the electric field and one for the scattered field where the perturbed dielectric permittivity appears as a source term: [21]

$$\frac{\partial^2 E_i^h(z,\omega)}{\partial z^2}+q^2\varepsilon_i^h E_i^h(z,\omega)=0 \tag{5}$$

$$\frac{\partial^2 E_i^s(z,\omega)}{\partial z^2}+q^2\varepsilon_i^h E_i^s(z,\omega)=-q^2\varepsilon_i^s(z,\omega)E_i^h(z,\omega) \tag{6}$$

The homogeneous refractive index $n_i^h$ is obtained using Maxwell formula $\varepsilon_i^h=(n_i^h)^2$ and the homogeneous field can be written as:

$$E_i^h(z,\omega)=M_i^h e^{-jqn_i^h z}+N_i^h e^{jqn_i^h z}, \tag{7}$$

where coefficients $M_i^h$ and $N_i^h$ are determined by expressing optical boundary conditions. The solution of Eq. (6) is: [23]

Journal of Physics: Conference Series **269** (2011) 012024 doi:10.1088/1742-6596/269/1/012024

$$E_i^s(z,\omega)=\frac{-jq}{2n_i^h}\begin{pmatrix}\left(M_i^h\int\varepsilon_i^s(z')dz'+N_i^h\int\varepsilon_i^s(z')e^{2jqn_i^hz'}dz'\right)e^{-jqn_i^hz}\\-\left(M_i^h\int\varepsilon_i^s(z')e^{-2jqn_i^hz'}dz'+N_i^h\int\varepsilon_i^s(z')dz'\right)e^{jqn_i^hz}\end{pmatrix}. \quad (8)$$

Finally, the total electric field is the sum of the homogeneous and the scattered fields. The integral equation of potential scattering is thereby obtained. [24] The relative intensity variation is finally deduced from the electric field. [9]

The homogeneous field is represented by Eq. (7). The continuity of the electric and magnetic fields at each interface is considered. Since the interface displacement due to the transient wave is considered, Eq. (7) describes the optical detection of the interface motion.

The acousto-optic interaction is described by Eq. (8) in which four integral terms are identified. An optical phase reference is defined when there is no perturbation, i.e. when the interface motion is not taken into account. The terms containing the phase shift $\pm 2qn_i^h z$ with respect to the phase reference result from the backscattering induced by the acoustic perturbation. These terms may be understood as Stokes and anti-Stokes scattering modes. [22] If the cell thickness is larger than $d_O = 1/2qn_1^h$, i.e. the half optical wavelength in the cell $\lambda_1/2$, Brillouin oscillations arise and their frequency for a normal probe incidence is: [17]

$$f_1 = \frac{2n_1^h v_1}{\lambda}, \quad (9)$$

where $\lambda$ is the probe wavelength. The other terms, in phase with the reference, do not create interferences but are sensitive to the sign and the amplitude of the perturbation.

Relative reflectivity changes are calculated for several cell thicknesses $d$ and they are plotted in Fig. 3 where medium 0 is TiAl4V and medium 1 is the cell. The influence of the cell thicknesses compared to the critical distances $d_A$ and $d_O$, assuming $d_A < d_O$, is studied. In Fig. 3.(a)-3.(c) the cell has a finite thickness while it is semi-infinite in Fig. 3(d). The signals labelled with stars represent the intensity changes calculated with the solution of the homogeneous Maxwell equation, Eq. (7). The signals labelled with rounds are calculated with the solution accounting for acousto-optic effect in media 1, Eq. (8). Distinguishing these two contributions allows a better understanding of the total signal, plotted with unlabelled lines in Fig. 3. For illustration, medium 2 is assumed to be air.

For films such that $d < d_A < d_O$, Fig 3.(a), the acoustic pulse generated in the substrate acts as a source exciting a resonant layer. Since the film thickness is slightly less than the acoustic wavelength $d_A$, the fundamental resonant frequency of the thin film is excited whereas the harmonics are beyond the acoustic spectrum. Thus the homogeneous solution describes the acoustic ringing of the film at a frequency $f_1 = v_1/4d$. The attenuation is a result of the cell viscoelastic behaviour and of the acoustic losses due to the successive reflections at the cell/substrate interface. In this situation, the film permittivity is quasi-homogeneous within the film and the acousto-optic contribution is modulated at the film resonant frequency.

For larger film thicknesses, the acoustic wavelength $d_A$ in the cell at frequency $\tau^{-1}$ is less than the film thickness and step like waveforms are observed. Successive contributions have opposite signs owing to acoustic impedance ratio between the considered media and the substrate. These features are reproduced in Fig. 3.(b) and Fig. 3.(c) by the waveforms representing the solution of the homogeneous Maxwell equation. For a film thickness $d_A < d < d_O = \lambda_1/2$, the evolution of the reflectivity is modulated by the strain integral along the film thickness. The square-like modulation is caused by the step-like changes in the displacement, Fig. 2. Although the thickness of the cell is too small in Fig. 3.(a) and Fig. 3.(b) to observe Brillouin oscillations, these oscillations arise significantly when the cell thickness is larger than $d_O$, Fig. 3.(c).

Figure 3.(d) illustrates the limit case where the film thickness is semi-infinite. Only the acousto-optic interaction contributes to the change of reflectivity and produces Brillouin oscillations with a larger amplitude.

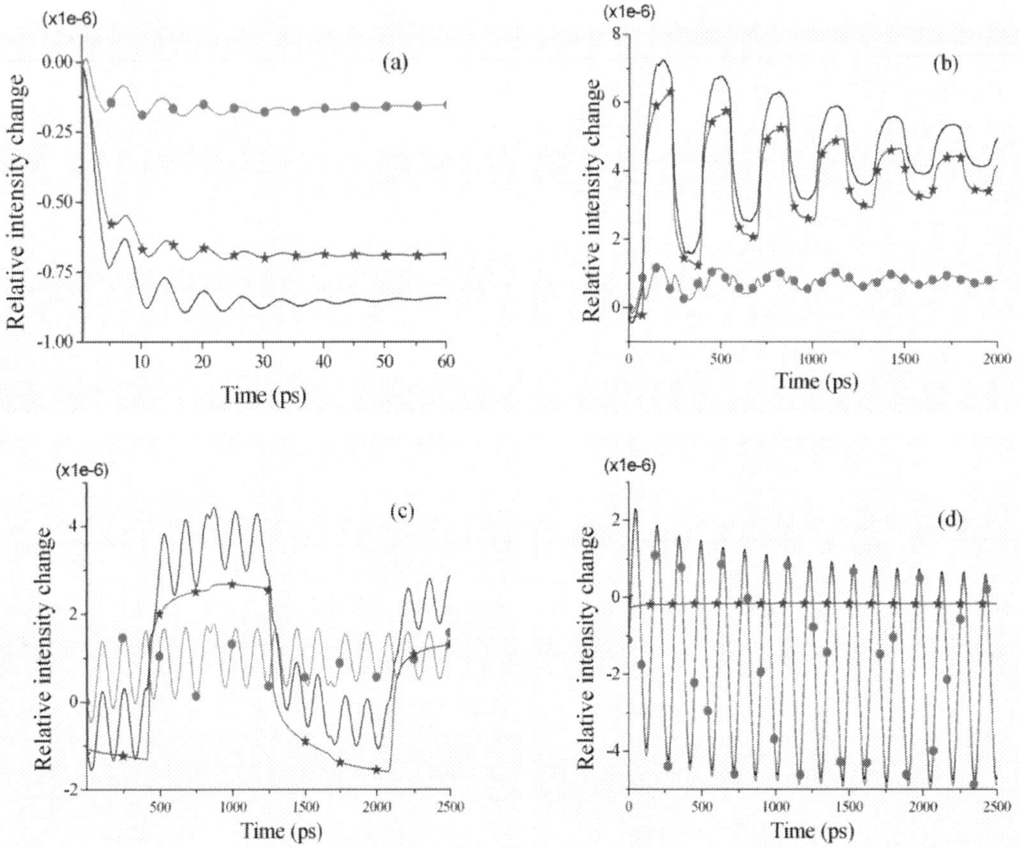


Figure 3: Relative reflectivity change calculated in 4 different cases. (a): $d < d_A < d_O$, (b): $d_A < d < d_O$, (c): $d_A < d_O < d$ and (d): $d \to \infty$. For each plot, signals labelled with round and star markers represent the acousto-optic contribution, Eq. (8), and the interface displacement contribution, Eq. (7), respectively. Unlabelled plot is the sum of the acousto-optic and interface displacement contributions.

## 4. Experimental results for thick vegetal cells and thin animal cells.

### 4.1. Picosecond acoustics experiments

The experimental arrangement is a classical pump-probe setup used in this paper to generate and to detect acoustic waves in a single cell adhering to a TiAl4V substrate. Red pulses (790 nm) of 100 fs duration are generated by a mode-locked Ti:sapphire laser, of repetition rate 82 MHz. A polarizing beam splitter divides the laser beam in pump and probe beams. The pump beam passes through an acousto-optic modulator (330 kHz) to provide the reference signal for lock-in amplification. The pump beam is converted into blue light using a BBO crystal. Its energy is reduced to 0.5 pJ per pulse. A lower energy probe beam passes through a delay line to provide a tunable time delay between the pump and the probe. Both beams propagate through the transparent cell and are focused at a normal incidence with a X50 microscope objective at the TiAl4V surface. The measured sound velocity is an average value over the micrometric probe spot. The full width at half-maximum of the space cross-correlation of the pump and probe beams is approximately 2 µm.

4.2. Thick vegetal cell.

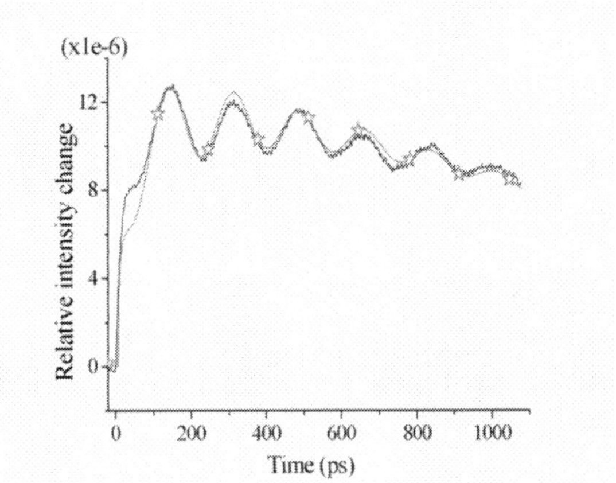


Figure 4: Experimental and calculated (stars) signals in the vacuole of an *allium cepa* cell. The Brillouin attenuation is due to the cell viscoelasticity.

We carry out experiments on *allium cepa* (common onion) cells. The ultrasonic attenuation prevents in-depth inspection above 3µm from the 0/1 interface and the cell thickness is about 5 µm. So this experiment must be compared to the simulated signal shown in Fig.3.(d). The attenuation of the Brillouin oscillations is caused by the viscous properties of the cell only. Figure 4 shows experimental results obtained in the vacuole of a white onion cell and a comparison with the calculated signal. Acoustic data are obtained by fitting the simulation presented in sections 2 and 3 to the experimental curves. The thermal decay is suppressed by subtraction of a polynomial. The simulation result is fitted to the remaining acoustic data (i. e. the Brillouin oscillations) by adjusting the complex stiffness coefficient of the cell to minimize the least mean square residue. [12] Assuming a fixed value for the optical index of the cell $n_1 = 1.4$, the fitting procedure allows obtaining the velocity with an accuracy about 0.1 nm/ps. Piezo-optic and thermo-optic coefficients are adapted from ref. 25. The local inspection in the cell permits to distinguish the vacuole acoustic velocity from the nucleus acoustic velocity and the white onion nucleus velocity from the yellow onion nucleus velocity (See table 2). [12]

| Organelle | Vacuole | Nucleus | |
|---|---|---|---|
| Variety | | White | Yellow |
| $v$ [nm/ps] | $1.6 \pm 0.1$ | $1.8 \pm 0.1$ | $2.0 \pm 0.1$ |

Table 2: Measured velocity in vacuole and nucleus for white and yellow onion cells. [12]

4.3. Thin MC3T3 mouse cell.

A RGD peptide has been added at the surface of the TiAl4V substrate to allow a better cell adhesion on the substrate, resulting in cell spreading and creation of focal adhesion. [26]

Figure 5 presents an optical image of a MC3T3 cell. Optical profilometry along a line across the nucleus of the cell indicates a thickness in other areas than the nucleus less than 200 nm: The experimental condition corresponds to a cell thickness between $d_A$ and $d_O$. This optical scan has been performed in several cells and reproducible results were obtained. An experiment performed on such a cell is compared with a simulation in Fig. 6. The thermal background of the experimental signal has been removed to analyze the acoustic part of the signal only. The interface motion and the acousto-optic interaction arise in the detected signal but the cell is too thin to detect Brillouin oscillations. The main contribution in the signal is the interface motion, which allows determining the acoustic travel time $t_l=d/v_l$=65 ps in the cell. The values of $d$ and $v_l$ cannot be measured independently. However, the couple of values $d$=105nm, consistent with the optical profilometry, and $v_l$ = 1.6nm/ps, agreeing with the values found in litterature, [27] gives a good fit to the data. Compared to measurements performed on onion cells, the signal to noise ratio is degraded in MC3T3 cells. This can be attributed to two main reasons. Firstly, the vacuole of MC3T3 cell is more heterogeneous than that of the *allium cepa* vacuole. This may affect both acoustic and optic wavefronts. Secondly, the MC3T3 upper surface may be rougher than that of the onion cell.

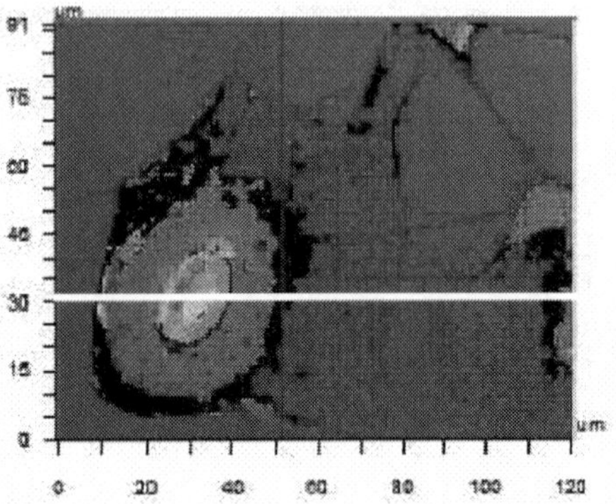


Figure 5: Optical image of a MC3T3 cell. A thickness scan (not shown) along the white line by optical profilometry indicates a thickness in other areas than the nucleus less than 200 nm.

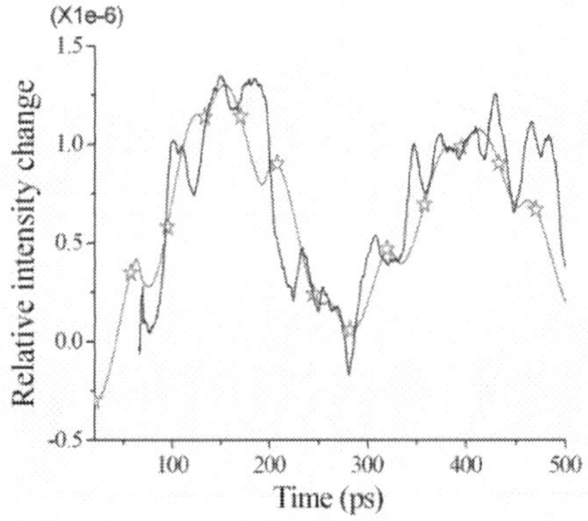


Figure 6: Comparison between the data and the simulation results (stars).

## 5. Conclusion

Picosecond acoustics technique has been presented theoretically in this paper for thin viscoelastic cells. The detection has been described using Maxwell equation and the fundamental processes of the detection have been identified in terms of acousto-optic and interface motion contributions. For the optical detection in the cell, the two detection processes have been studied and acoustic and acousto-optic critical thicknesses have been determined. Experiments have been led in onion and mouse cells with thicknesses respectively larger and smaller than the acousto-optic critical thickness. Good agreement between the simulation and the data is found. The Brillouin oscillations allow the measurement of the acoustic velocity in the onion cell whereas an acoustic travel time is measured in the thin animal cell. Ultrafast acoustics should offer promising perspectives for quantitative cell adhesion measurement on implants or for measurement of cell adhesion during cell cycle and motility.

## References

[1] M. Xu and L. V. Wang, *Rev. Sci. Instrum.* **77**, 041101 (2006)
[2] Tsien R Y 2003 *Nat. Rev. Mol. Cell Biol.* **4** 16
[3] Thomsen C, Strait J, Vardeny Z, Maris H J, Tauc J and Hauser J J 1984, *Phys. Rev. Lett.* **53** 989
[4] Lin H N, Stoner R J and Maris H J 1990, *J. Nondestr. Eval.* **9** 239
[5] Côte R and Devos A 2005, *Rev. Sci. Instrum.* **76** 053906
[6] Rossignol C, Perrin B, Laborde S, Vandenbulck L, De Barros M I and Djemia P 2004, *J. Appl. Phys.* **95** 4157
[7] Shelton L J, Yang F, Ford W K and Maris H J 2005 *Phys. Status Solidi B* **242** 1379
[8] Kashiwada S, Matsuda O, Baumberg J J, Li Voti R and Wright O B 2006 *J. Appl. Phys.* **100** 073506
[9] Wright O B, Perrin B, Matsuda O and Gusev V E 2008 *Phys. Rev. B: Condens. Matter* **78**

024303
[10] Ségur D, Shuvalov A L, Audoin B and Pan Y D 2010 *J. Acoust. Soc. Am* **127** 181
[11] Rossignol C, Chigarev N, Ducousso M, Audoin B, Forget G, Guillemot F and Durrieu M C 2008, *Appl. Phys. Lett.* **93** 123901
[12] Audoin B, Rossignol C, Chigarev N, Ducousso M, Forget M, Guillemot F and Durrieu M C 2010 *Ultrasonics* **50** 202
[13] Antonelli G, Perrin B, Daly C D, Cahill D G, 2006, *Mater. Res. Bull.* **31** 607
[14] Scruby C B and Drain L E 1990 *Laser Ultrasonics Techniques and Applications* (Adam Hilger)
[15] Rouch J, Lai C C, Chen S H 1976 *J. Chem. Phys.* **65** 4016
[16] Auld B A 1990 *Acoustic fields and waves in solids, vol.1* (Krierger Pubishing Compagny)
[17] Thomsen C, Grahn T C, Maris H J and Tauc J 1986 *Phys. Rev. B: Condens. Matter* **34** 4129
[18] Lin H N, Stoner R J, Maris H J and Tauc J 1991 *J. Appl. Phys.* **69** 3816
[19] Wright O B 1992 *J. Appl. Phys.* **71** 1617
[20] Wright O B 1995 *Opt. Lett.* **20** 632
[21] Dehoux T, Chigarev N, Rossignol C and Audoin B 2007, *Phys. Rev. B: Condens. Matter* **76** 024311
[22] Dil J G 1982 *Rep. Prog. Phys.* **45** 285
[23] Jeffrey A and Dai H H 2008 Handbook of mathematical formulas and integrals 4$^{th}$ edition (Elsevier: Academic press)
[24] Born M and Wolf E 2005 *Principles of optics 7$^{th}$ (expended) edition* (Cambridge)
[25] Saubade C 1984 *J. Phys. C: Solid State Phys.* **17** 3507
[26] Hersel U, Dahmen C and Kesser H 2003, *Biomaterials* **24** 4385
[27] Weiss E C, Anastasiadis P, Pilarczyk G, Lemor R M, Zinin P V 2007 *IEEE Trans. Ultrason. Ferroelectr. Freq. Control* **54** 2257

# AUTHOR INDEX


Abrahams, I D ....62
Arca, A ....102
Aristégui, C ....31
Audoin, B ....213
Blarel, F ....1
Callens-Debavelaere, Dorothée ....78
Calmon, P ....110
Campistron, Pierre ....78
Carlier, Julien ....78
Cès, Maximin ....146
Challis, R E ....1, 137
Chapuis, Bastien ....174
Chassignole, B ....88
Chatillon, S ....72, 110
Chollet, C ....213
Clark, M ....102
Clorennec, Dominique ....146
Conoir, Jean-Marc ....130
Corneloup, G ....88
Darmon, M ....110, 182
Dehoux, T ....213
Deschamps, M ....205
Destuynder, Philippe ....51
Dominguez, N ....196
Dubois, J ....31
Duclos, Jean ....41
Ducousso, M ....213
Duflo, Hugues ....41
Dupond, O ....88
Durrieu, M C ....213
El Mouhtadi, Abdelhak ....41
Feuilly, N ....88
Fouquet, T ....88
Fradkin, L J ....110
Fradkin, Larissa ....182
Franceschini, Emilie ....122
Fromme, P ....154
Gautesen, Arthur ....182
Guo, X ....1
Hajzargarbashi, T ....205
Holmes, M J ....95
Inquiété, G ....10
Journiac, S ....196
Koabaz, M ....205
Kundu, T ....205
Laugier, Pascal ....24
Lesselier, D ....72
Leymarie, N ....196
Lhémery, A ....10
Lombard, Bruno ....122
Luppé, Francine ....130
Mahaut, S ....110
Mather, M L ....165
Minonzio, Jean-Gabriel ....24
Morgan, S P ....165
Moysan, J ....88
Ndieguene, Assane ....78
Nongaillard, Bertrand ....78
Parker, N G ....95, 165
Parnell, W J ....62
Pinfield, Valerie J ....137
Piraux, Joël ....122
Poncelet, O ....31
Potel, C ....196
Povey, M J W ....95, 165
Prada, Claire ....146
Puel, B ....72
Rouge, C ....154
Royer, Daniel ....146, 174
Shuvalov, A L ....31
Smith, R ....102
Somekh, M G ....102
Stratoudaki, T ....102
Talmant, Maryline ....24
Taupin, L ....10
Terrien, Nicolas ....174
Unwin, M E ....1
Wang, Shengxiang ....78
Xu, Weijiang ....78
Zernov, V ....110
Zernov, Victor ....182
Zhao, Xingzhong ....78
Zouani, O ....213

**Institute of Physics**
Dirac House, Temple Back
Bristol BS1 6BE UK

CURRAN ASSOCIATES INC
proceedings
.com